THE HISTORY OF ALLELOPATHY

The History of Allelopathy

By

R.J. Willis

*University of Melbourne, Parkville,
Victoria, Australia*

 Springer

A C.I.P. Catalogue record for this book is available from the Library of Congress.

ISBN 978-1-4020-4092-4 (HB)
ISBN 978-1-4020-4093-1 (e-book)

Published by Springer,
P.O. Box 17, 3300 AA Dordrecht, The Netherlands.

www.springer.com

Printed on acid-free paper

All Rights Reserved
© 2007 Springer
No part of this work may be reproduced, stored in a retrieval system, or transmitted
in any form or by any means, electronic, mechanical, photocopying, microfilming,
recording or otherwise, without written permission from the Publisher, with the
exception of any material supplied specifically for the purpose of being entered
and executed on a computer system, for exclusive use by the purchaser of the work.

CONTENTS

Illustrations		vii
Acknowledgements		xi
Foreword		xiii
Chapter 1.	What is Allelopathy?	1
Chapter 2.	Allelopathy in the Classical World – Greece and Rome	15
Chapter 3.	Arabic Works	39
Chapter 4.	Ancient India, China and Japan	53
Chapter 5.	Mediaeval Period and Renaissance	67
Chapter 6.	The Eighteenth Century – Root Excretion	103
Chapter 7.	Augustin Pyramus de Candolle, and His Era	125
Chapter 8.	The Decline of Allelopathy in the Latter Nineteenth Century	159
Chapter 9.	Spencer Pickering, and The Woburn Experimental Fruit Farm, 1894-1921	195
Chapter 10.	The USDA Bureau of Soils and Its Influence	209
Chapter 11.	Approaching the Modern Era	251
Index		301

ILLUSTRATIONS

Figures

Figure 1.1.	Photograph of Prof. Hans Molisch.	2
Figure 1.2.	An allelopathic variety of rice, showing a weed-free zone, being inspected by Dr. Robert Dilday.	4
Figure 1.3.	Grass *Miscanthus floridus* as a dominant in Taiwan.	6
Figure 1.4.	*Eucalyptus tereticornis* plantation in Haryana, India, showing poor understorey growth, attributable, at least partly, to allelopathy.	9
Figure 1.5.	An infestation of spotted knapweed (*Centaurea maculosa*) in California.	10
Figure 1.6.	Cover for *Science*, 31 January 1964 (Volume 143, number 3605), showing an aerial view of the aromatic shrubs *(Salvia leucophylla* and *Artemisia californica*) invading an annual grassland near Santa Barbara, California.	11
Figure 2.1.	Map of the Mediterranean region with some place names associated with the Classical Era.	16
Figure 2.2.	A 16th century version of the tetragram of the elements and principles.	17
Figure 2.3.	A painting by Robert A. Thom portraying Theophrastus, as envisaged teaching his students, *c.* 300 B.C.	19
Figure 2.4.	Cabbage plants from an anonymous 15th century French manuscript, *Livre des Simples Médicines*.	21
Figure 2.5.	Dioskorides receiving a mandrake plant from Euresis, the goddess of discovery.	24
Figure 2.6.	A representation of Caius Plinius Secundus from *Les Vrais et Vies des Hommes Illustrés* by André Thévet (1584).	29
Figure 2.7.	Title page from a rare 1543 edition of the *Geoponika*.	35
Figure 3.1.	Leaf (folio 138a) concerning the sympathy and antipathy of trees from a manuscript copy of *Kitab al-Filaha* by Ibn al-Awwam,	40
Figure 3.2.	Title page from *Le Livre de l'Agriculture d'Ibn al-Awam*, as translated by Clément-Mullet (1864-1867).	45
Figure 3.3.	A leaf depicting the walnut tree from a 12th-13th century Andalusian version of *De Materia Medica* by Dioskorides.	46
Figure 3.4.	A leaf depicting the pomegranate from a 14th century Egyptian or Syrian work by Masâlik-al-absâr al-Umari.	50
Figure 4.1.	Leaf 2 of the manuscript of the *Vrikshayurveda* by Surapala.	54
Figure 4.2.	Examples of a *zhu jian* (bamboo strip book), and a scroll, the latter of a herbal, the *Pen Ts'ao Ching* by Shen Nung.	57
Figure 4.3.	A Han Dynasty tomb tile depicting an agricultural scene.	61

Figure 5.1.	The walnut tree (*Juglans regia*) from *Cosmographia* by Sebastian Münster (1552).	69
Figure 5.2.	Illustration of rue (*Ruta graveolens*) from the 1543 edition of the *Neu Kreuterbuch* by Fuchs.	72
Figure 5.3.	The title page from *Horticultura* by Peter Lauremberg (1631).	75
Figure 5.4.	Illustration of harvesting cabbages, from *Tacuinum Sanitatis*, a 15th century Latin translation of *Taqwim al Sihhah* (Tables of Health) by Ibn Butlan (d. 1066).	76
Figure 5.5.	Illustration from *Phytognomonica* by della Porta (1588), showing the resemblance of plant parts to the human hand.	78
Figure 5.6.	The use of certain sympathetic trees in the culture of grape vines, from *Tacuinum Sanitatis*, a Latin translation of *Taqwim al Sihhah* (Tables of Health) by Ibn Butlan (d. 1066).	80
Figure 5.7.	The Scythian Lamb as figured in the work by Claude Duret (1605), and an engraving of an actual prepared "Scythian Lamb" from the collection of Sir Hans Sloane (Sloane 1698).	93
Figure 5.8.	A woodcut of the Bausor tree (assumedly the upas) from *Hortus Sanitatis* (1491).	95
Figure 5.9.	Engraving of the upas, showing its alleged properties.	95
Figure 6.1.	Section of a rare engraving, after a painting by J.W. van Borselen, depicting Boerhaave as a botanist.	105
Figure 6.2.	Title page of Plappart's thesis on *Juglans nigra*.	114
Figure 6.3.	A lithograph *c.* 1850 of S.J. Brugmans by L. Springer, Leiden.	115
Figure 6.4.	The title page of Coulon's thesis, *De mutata humorum in regno organico indole a vi vitale vasorum derivanda*, 1789.	116
Figure 7.1.	A lithograph of Augustin Pyramus de Candolle by Hébert, after a drawing by Alexandre Calame.	126
Figure 7.2.	The title page of *Principes Élémentaires de Botanique et de Physique Végétale* (de Candolle 1805), and author's inscription.	128
Figure 7.3.	Painting of Jane Haldimand Marcet.	133
Figure 7.4.	A woodcut from the title page to *Lessons on Animals, Vegetables, and Minerals*, by Mrs Marcet (1844).	134
Figure 7.5.	Macaire's signature (1816).	143
Figure 8.1.	A unsigned portrait of Justus Ludewig von Uslar (c. 1805), in Hannover civil service uniform.	162
Figure 8.2.	Detail of the roots of *Poa annua*, showing the excretion of material and a root tip operculum (Gasparrini 1856).	165
Figure 8.3.	A view of the interior of Liebig's laboratory at Giessen, as drawn by Wilhelm von Trautschold, *c.* 1840.	165
Figure 8.4.	Photograph of Prof. Charles Daubeny.	176
Figure 8.5.	Photograph of Dr. Gustav Jaeger *c.* 1880.	181
Figure 8.6.	An apparatus as used by Reveil (1865) to investigate the effects of various toxic substances on plants.	184

Illustrations   ix

Figure 9.1.	A photographic portrait of Percival Spencer Umfreville Pickering taken by Walter Stoneman in 1917.	197
Figure 9.2.	Photograph of the Eleventh Duke of Bedford.	197
Figure 9.3.	Plan of the experimental plots at the Woburn Experimental Fruit Farm as of 1904.	198
Figure 9.4.	Pages from Pickering's copy of the proofs of the *Sixth Report of the Woburn Experimental Fruit Farm* (1906).	199
Figure 9.5.	Advanced seedlings of Bramley variety apple trees grown with and without a grass cover.	200
Figure 9.6.	The effect of various plants on other plants or the same plant.	202
Figure 10.1.	Photograph of Milton Whitney.	211
Figure 10.2	Photograph of Eugene W. Hilgard.	214
Figure 10.3	Photograph of Cyril G. Hopkins.	219
Figure 10.4.	Photograph of Oswald Schreiner as a young man.	223
Figure 10.5.	Structure of picoline carboxylic acid and dihydroxystearic acid.	225
Figure 10.6.	A view of one of the laboratories, USDA Bureau of Soils.	227
Figure 10.7	Glasshouse experiment at the USDA Bureau of Soils in which organic compounds were tested on wheat seedlings.	228
Figure 11.1.	Liebig trading card (*c.* 1937) showing *Azteca* and *Cecropia*	254
Figure 11.2.	Photograph of Gerhard Madaus.	257
Figure 11.3.	Structure of juglone (5-hydroxy-1,4,-naphthoquinone).	267
Figure 11.4.	The effect of black walnut trees on cabbages.	268
Figure 11.5.	Guayule plantings being inspected by Dr Robert Emerson.	272
Figure 11.6.	Photograph of Frits Warmolt Went.	277
Figure 11.7.	Photograph of Gerhard Grümmer *c.* 1957.	282
Figure 11.8.	Grümmer's scheme of allelopathic interactions.	283
Figure 11.9.	Photograph of Sergei Ivanovich Chernobrivenko.	285
Figure 11.10.	Chernobrivenko's scheme of interactions between species.	286
Figure 11.11.	Photograph of Boris P. Tokin	287
Figure 11.12.	Photograph of Hubert Martin.	288

TABLES

Figure 3.1.	Arabic agricultural works of the mediaeval era.	44
Figure 4.1.	Chinese dynasties and their major subdivisions and dates.	58
Figure 10.1.	Principal substances isolated from various soils by the Bureau of Soils.	230
Figure 11.1.	Some species reported as demonstrating "soil sickness".	262

ACKNOWLEDGEMENTS

The following individuals and institutions have all provided invaluable assistance, and I am grateful for their support:

American Academy for the Advancement of Science
Dr Redha Ameur, Arabic Studies, University of Melbourne
Prof. J. Hameen-Anttila, Arabic and Islamic Syudies, Helsinki University, Finland
Bibliothèque Nationale de France, Paris
Prof. Pietro Corsi, Faculty of History, University of Oxford
Ms Gillian Evison, Bodleian Library, University of Oxford
Ms Yasmin Faghihi, Library, University of Cambridge
Dr Vladimir Grakhov, National Botanical Garden of the National Academy of Sciences of Ukraine, Kiev
Dr Harald Grümmer, Germany
Dr Gregory J. Higby, Director, American Institute of the History of Pharmacy, Madison, Wisconsin
Dr Robert Hoeft, Department of Crop Sciences, University of Illinois
Mr Mark Hurd, Aero-Metrics Inc.
Staff of Interlibrary Loans, Baillieu Library, University of Melbourne
Jaeger Company, London
Dr Jinhong Li, Faculty of Biological Sciences, University of Leeds
Prof. S.S. Narwal, Haryana Agricultural University, Hisar, India
The National Portrait Gallery, London
Dr Y.L. Nene, Asian Agri-Historical Foundation, Secunderabad, India
Mr Chris O'Brien, School of Botany, University of Melbourne
Mr Christopher Pasteur, England
Dr. John M. Randall, The Nature Conservancy, University of California, Davis
The Royal Society, London
Dr. Gilles Saindon, Agriculture and Agri-Food Canada
Mr Terry Lee Smith, Pequea, Pennsylvania
Dr Andrew Turner, Centre For Classics And Archaeology, University of Melbourne
United States National Archives at College Park, Maryland
University of California, Berkeley
Ms. Gesine von Uslar, Mainz, Germany
Ms Georgina Vacy-Ash, Melbourne
Ms T. Vlasma, Library, University of Leiden, Netherlands
Prof. Zhaoji Zhou, South China Agricultural University, Guangzhou, China

FOREWORD

This book had its beginnings about thirty-five years ago, when I migrated to Australia from Canada, and began a doctoral study concerning the role of allelopathy in forests of the eucalypt known in Australia as mountain ash (*Eucalyptus regnans*), under the supervision of Dr Kingsley Rowan and the late Dr David Ashton. In first assembling materials for the usual survey of the relevant literature, I came to realise that the relative youth of Australia as a nation and its geographical remoteness were to be barriers in fully dealing with historical concepts. At times, the simplest option was to buy the requisite antiquarian books, if they were not readily available from local libraries. I remember that one of the first such works that I acquired was de Candolle's *Physiologie Végétale*, and it was then that I began to learn that the history of allelopathy had been only superficially investigated.

Allelopathy is a topic which has been very much in the limelight of plant ecology in the past few decades. It is a controversial topic which has a surprisingly large body of literature associated with it, yet the mere existence of allelopathy as an ecological process is still considered doubtful by many.

Most students of allelopathy seem to have assumed that the topic has been commenced in 1937 with the work of Hans Molisch, or to those more historically minded, the theories of A.P. de Candolle in the early nineteenth century are acknowledged as a starting point. It is the aim of this book firstly to show that the concept of allelopathy has been with us for well over two thousand years, and, at least in former times, was relatively well known. It is also an aim of this book to indicate that controversy regarding allelopathy has been with us for almost as long.

It is seldom appreciated that the concept of allelopathy has been addressed not simply in Western culture, but also in the botanical and agricultural literature of ancient China, India and Japan, and the Islamic world. The antecedent of the concept of allelopathy is that of antipathy and sympathy of natural things, although these concepts need not have a chemical basis. Nonetheless, it is seldom realised how pervasive these concepts have been in the past, not simply in natural history, but also in social realms, including literature and religion.

In this book, I have tried to bring to light the majority of the writings that have touched on allelopathy spanning the period from antiquity until about 1957. This book serves both as a simple historical guide and a sourcebook for original relevant material. Much of the material has never been assembled for the student of allelopathy, a considerable amount has not been available before to the English reader, and some has never appeared before in print. I have endeavoured to collect material from sources in languages other than English, including Latin, French, German, Italian, Spanish, Russian, Chinese, Japanese and Arabic. With material quoted from original works in languages other than English, I have taken the liberty of trying to provide a passable translation into English.

For this book, the year 1957 is regarded somewhat arbitrarily as a closing point, simply because the years 1955-1957 represent the dawn of the modern era of allelopathic research with the near simultaneous publication of four important books relating to allelopathy – those by Grümmer (1955), Chernobrivenko (1956), Tokin (1956) and Martin (1957). It is anticipated that a sequel to the present volume will tell the continued story of the history of allelopathy over the past fifty years.

Rick Willis

April 2007
Melbourne, Australia

CHAPTER 1

WHAT IS ALLELOPATHY?

> For the Snark's a peculiar creature, that won't
> Be caught in a commonplace way.
> Do all that you know, and try all that you don't
> Not a chance must be wasted today!
>
> *The Hunting of the Snark*
> Lewis Carroll (1876)

Allelopathy is widely understood as the harmful effect that one plant has on another plant due to chemicals it releases into the environment. However, unfortunately, there has been substantial variation and confusion in defining and using the term over the past fifty years (Willis 1994).

Allelopathy, in concept, dates back well over two thousand years, but the term itself was coined comparatively recently, in 1937. The word "*Allelopathie*" was coined in German by the eminent Austrian plant physiologist Hans Molisch (Figure 1.1), in his last book, *Der Einfluss einer Pflanze auf die andere – Allelopathie*[1], published shortly before his death in 1937. The word originates from the Greek roots, *allelon*, meaning 'mutual' or 'among each other', and *pathos*, meaning 'suffering' or 'feeling'. Many authors have assumed Molisch intended the former meaning for *pathos*, but this is wrong. In coining allelopathy, Molisch wished the term to mean simply the effect of one plant on another. In describing the phenomenon of plant interaction through chemicals, the term allelopathy is far from ideal. Indeed, Molisch originally would have preferred the term 'allopathy', incorporating *allo*, meaning 'other', but this word had already been appropriated by medical science, and thus allelopathy became his second choice. The word 'allelopathy' has consequently caused considerable confusion, as the interactions involved are rarely reciprocal, as *allelon* can suggest, and are not necessarily harmful, as the word ending –pathy usually infers. One could argue vainly that 'allelopathy' etymologically better suits the concept of plant competition. While Molisch is often viewed as the founder of the science of allelopathy, this notion is misguided, as the bulk of Molisch's text is actually

---

[1] An English translation was published in 2000.

Figure 1.1. Photograph of Prof. Hans Molisch (from Grümmer 1955).

concerned with the effects of ethylene, now generally seen more as a plant hormone, than an agent of plant interaction. Molisch did foreshadow the importance of plant substances in ecological interactions, but gave few details.

It is well known that almost any substance that is inhibitory to a plant function at a particular concentration will likely prove stimulatory at some lesser concentration, and vice versa. Molisch, being a plant physiologist, was well aware of this, and stated that he meant the term allelopathy to cover both inhibitory and stimulatory interactions through chemical substances. This duality of substances is sometimes referred to as hormesis, and was recognised in the sixteenth century by Paracelsus, with his phrase "All things are poison and are not poison; only the dose makes a thing not a poison." (Duke *et al.* 2006). The vast majority of allelopathic studies have focused on the inhibitory aspects, but stimulatory effects are probably so routine and likely subtle in nature, that they are generally overlooked. Consequently, allelopathy is commonly viewed as an injurious phenomenon, and most dictionaries and botanical texts have defined allelopathy in this manner. Allelopathy, has sometimes been described as 'chemical warfare among plants', a notion which usually grabs our interest; perhaps the cynic in us likes to think that plants are not always benign.

What is Allelopathy?

In 1968, a subcommittee of the International Biological Program (IBP), the Environmental Physiology Subcommittee of the Division of Biology and Agriculture, attempted to rectify some of the confusion governing allelopathy. They recognised that allelopathy was commonly interpreted as referring to negative interactions, and thus recommended a more global term to include stimulatory effects, "allelochemics". The Subcommittee adopted the view that allelopathy and allelochemics pertained to substances of plant origin that could affect both plants and animals (Environmental Physiology Subcommittee 1971). However, the ground shifted again, as in 1971 an influential paper on allelochemics by Whittaker and Feeny was published in the prestigious American journal *Science*. Whittaker and Feeny (1971) decreed that allelochemics was the domain of all chemical interactions among organisms, a view which has largely persisted. Within this framework, allelopathy then rests as a sub-discipline of allelochemics or what is now known widely as "chemical ecology".

In 1984, E.L. Rice, commonly regarded as the doyen of American allelopathy researchers, reconsidered and redefined 'allelopathy', in light of the dual inhibitory and stimulatory effects of substances, as "any direct or indirect harmful or beneficial effect by one plant (including microorganisms) on another through production of chemical compounds that escape into the environment." While this definition is unashamedly general, it is no more so than any other definition likely found in the realm of ecology, the most interdisciplinary of the sciences. Critics have pointed out that an oil spill from a tanker could thus provide a far-fetched example of allelopathy. However, it is Rice's definition, or something very close to it, that has served until recently the majority of researchers concerned with allelopathy, although oddly, biologists outside the discipline have largely favoured a narrower meaning, as stated at the beginning of this chapter. In the past few years, there has been a trend toward allelopathy researchers adopting once more this earlier, simpler meaning of allelopathy, that is the inhibitory effect of one plant on another due to the release of chemical substances (e.g. Fitter 2003).

Ecologists have become divided in other ways on their understanding of the breadth of the domain of allelopathy. In particular, ecological biochemists such as G. R. Waller, founding President of the International Allelopathy Society, in a thinly veiled attempt to broaden the funding base for allelopathic research, have endeavoured to include plant-animal interactions again under the rubric of allelopathy, and one could argue that physiologically active substances released from plants are indiscriminate in affecting plants, microorganisms or animals. Another confusion has occurred in recent years, and that is due to the word "allelopathy" having been borrowed by zoologists to describe chemically based interactions among sessile animals, especially invertebrates such as sponges and corals[2]. Indeed, the recognition of chemical interactions amongst such animals strengthens the case for allelopathy in plants, in that there are common evolutionary pressures amongst numerous, unrelated types of sessile organisms to defend themselves using secondary metabolites.

---

[2] It has been argued that many sessile marine invertebrates, notably corals, have "plant" affinities, as they host photosynthetic organisms, zooxanthellae (Gross 2003). Primitive sessile animals were once regarded has being plant-like, and were called zoophytes.

Allelopathy is an enigmatic topic. It is well known that plants can be rich both in diversity and quantity of so-called "secondary metabolites". To date, over 100,000 different secondary metabolites have been identified from plants and fungi, and the amount of a common substance may in some cases attain 5% of dry weight. It is only common sense to assume that these substances must exert some effect if they are released from the plant. Farmers, in particular, have long suspected that certain crops "contaminate" the soil for other crops. Conversely, it is indisputable that the growth of legumes is beneficial for following crops because of the eventual release of nitrogen-rich organic compounds. One of the best-documented plants, shown to affect other plants through its metabolites, is the walnut, and many farmers and gardeners can attest to the seemingly poisonous effect that a walnut tree, especially the American black walnut (*Juglans nigra*), can have on certain neighbouring plants. In gardening literature, the concept of allelopathy, in the broadest sense, is embodied in the topic of "companion planting", where paired plantings are seen to be beneficial, e.g. roses and garlic, although most of the evidence is largely anecdotal.

In the past ten years, a great deal of attention has been focused on crops such as "allelopathic rice" (Figure 1.2). Studies with assays of thousands of accessions of different rice varieties, have showed that a small number of accessions demonstrates the native ability to inhibit certain aquatic weeds, and the long-term hope is to use genetic engineering to transfer these herbicidal genes into high-yielding varieties of rice, thereby reducing the need for expensive chemical herbicides (Olofsdotter 2001). Similar studies are underway in diverse systems, such as turf grasses, to evolve plantings, especially in public areas, that can reduce the need for herbicides to manage weed growth (Bertin and Weston 2004).

Other areas of "applied allelopathy" include the study of certain aspects of what is broadly called "soil sickness" or "soil fatigue" or "replant problem". These terms refer to the situation when land, which has been supporting the same crop continuously,

Figure 1.2. An allelopathic variety of rice, showing a weed-free zone, being inspected by Dr. Robert Dilday (Photograph by Joanne Dilday, courtesy of United States Department of Agriculture).

demonstrates declining yield despite adequate physical conditions, such as light, moisture, and nutrients. It has been demonstrated in many fruit crops, notably apple, peach, and citrus, and in herbaceous crops including asparagus, wheat, and legumes such as peas and alfalfa. The causes implicated have been several, including pathogenic microorganisms, nematodes, and allelopathy through the accumulation of harmful organic substances in the soil system. Another area of applied allelopathy is the use of crop residues to control weeds. Crops such as canola (*Brassica napus*) and sunflower (*Helianthus annuus*) are harvested primarily for their seeds, and consequently produce large quantities of vegetative plant material, that is often rich in allelochemicals. Much research, particularly in Europe, have been directed at finding ways of using this green matter in a commercially viable way for the control of pests including weeds. These studies have largely vindicated traditional agricultural practices in which certain "smother crops" such as rye are grown and then turned into the soil to reduce future weed growth in subsequent crop plantings.

In the ecology of natural vegetation, allelopathy has been implicated particularly in situations where vegetation shows unusually strong spatial patterning, as occurs with certain semi-desert shrubs, such as *Salvia leuocophylla*, *Artemisia tridentata* and *Adenostoma fasciculatum* in the United States (see Chapter 11). Allelopathy may also be involved where one plant species displays inordinate dominance or exclusion of other species, as in the case of *Miscanthus floridus* in Taiwan (Figure 1.3) and *Kalmia angustifolia* in Canada. It must also be remembered that allelopathy and competition are always acting simultaneously, and generally the predominance of allelopathy over competitive effects has been found to be rare amongst species. The effects of allelopathy are more likely to be subtle, and allelopathy may affect species and/or their life stages differentially.

As the first organisms to be in contact with allelopathic substances in soil are microorganisms, one should expect that the composition of the soil microbiota will be, in large measure, determined by allelopathic substances, either because the substances are deleterious or because they are viable substrates. During the 1960's and 1970's, E.L. Rice and his students demonstrated that allelopathic substances released from plants could inhibit nitrifying bacteria and thus the balance of ammonium to nitrate in soil. They attempted to develop this into a general theory that because of the high metabolic cost of using nitrates, plants with efficient perennial growth, as found in "climax" vegetation would be more likely to inhibit nitrifying bacteria and thus utilise ammonium as a nitrogen source, particularly as ammonium is less likely to be leached from soil.

Despite all the above, the topic of allelopathy has remained highly controversial among plant ecologists. The first indication of this, perhaps, is suggested by the two foremost dictionaries of the English language. *Webster's Unabridged Dictionary* curiously defines allelopathy as the "*reputed* baneful effect of one plant on another"[3], whereas the word fails to appear at all in the comparatively recent second edition of the *Oxford English Dictionary (OED)*, despite the fact that over 80 monographs and

---

[3] A more recent definition comes from the on-line *Merriam Webster Dictionary*, and is more satisfactory: "The suppression of growth of one plant species by another due to the release of toxic substances."

Figure 1.3. *Grass* Miscanthus floridus *as a dominant in Taiwan. (Photograph by the author).*

about 10000 articles have now been devoted to allelopathy since 1937[4]. Consequent definitions and citations are at best confusing. "Allelopathy" appears only in the 1993 *OED Additions Series*, and is defined there as "The deleterious process by which one organism influences others nearby through the escape or release of toxic or inhibitory substances into the environment: usually restricted to such interaction between higher plants." The same dictionary then provides a illustrative citation using the word "allelopathic", that involves copepods. The best-selling abridgement, the *Shorter Oxford English Dictionary* also omitted the term until the appearance of the Fifth Edition in 2002, and there the definition itself has been abridged to: "the process by which one organism harms or affects others nearby through the release of allelochemicals", which sounds more akin to a description of antibiosis.

To understand some of the problems surrounding the topic of allelopathy, one must look firstly at the processes deemed important in governing plant interactions. Traditionally, competition has been considered to be the foremost factor. Competition is defined as the process in which two or more organisms attempt to utilise the same resource, which is ultimately in limited supply. Thus competition is viewed as a type of negative interaction, or interference, in which the level of some commodity, be it

---

[4] The OED suggests that the first English use of "allelopathy" was in Martin (1957). However, it appeared at least as early as 1949 (Weiss 1949) and was provided in the 5th edition of *A Dictionary of Scientific Terms* (Henderson and Kenneth 1953).

a nutrient, water, light, pollination agents, or simply space, is diminished. Generally, there is little argument that competition is extremely important in plant interrelationships, as it affects primary metabolism. Allelopathy differs conceptually in that it operates through the input of substances, commonly secondary metabolites, into the environment, which then affect other organisms. As the British ecologist John Harper (1975) famously noted, it is impossible to prove that chemicals released by plants do not affect neighbouring plants, and similarly, it is well nigh impossible to prove that any deleterious effect is due to allelopathy rather than to competition for nutrients – this criticism has held great sway with many ecologists. The methodological difficulty of "proving" allelopathy has had the dubious effect of dividing ecologists into "believers" and "non-believers". In some respects, this has had the unfortunate effect of marginalising acceptance of allelopathy, and one could reluctantly draw some parallels with the discipline of exobiology, where the concept of life on other worlds remains intuitively probable, but unproveable at the moment. Biologists, arguably because of the unpredictability often found in their science, often have an inordinate need for order in their science. This is apparent in our discussion, and historically much effort has been spent in attempting to discriminate and partition the effects of competition and allelopathy (e.g. Tinnin 1972). This approach is likely unrealistic, as the effects of physiologically active substances released by a plant are undoubtedly tied with surrounding microorganisms, related symbioses, transport processes involved in nutrient and water uptake, soil chemistry, and so forth (Inderjit and del Moral 1997, Reigosa *et al.* 1999).

Critics of allelopathy, especially in recent times, have made demands on researchers in allelopathic studies for protocols that are unparalleled in other areas of ecology (Harper 1977, Willis 1985, Williamson 1990). I remember once reading the comment of an unashamedly biased critic of allelopathy who, in perhaps unwisely parodying Samuel Johnson, cynically referred to the discipline as "the refuge of scoundrels"[5]. Harper was essentially correct when he wrote that allelopathy was impossible to prove or disprove, but the same can be said of most ecological phenomena. Unfortunately these criticisms and remarks have led to a degree of self-consuming introspection seldom witnessed in other ecological disciplines. Curiously, the study of ecosystems has much in common with an extremely different natural realm, that of subatomic particles. As stated by Werner Heisenberg, in particle physics, the measurement of one attribute changes conditions for the measurements of another attribute, and the same must be said for field studies in ecology. Despite the allowances of the foregoing, there have been many experiments, labeled as allelopathic research, which have been of little merit, and some criticisms has been deserved. The literature of allelopathy is unfortunately replete with studies where crude plant extracts have been administered to germinating seeds or seedlings in Petri dishes and the like, and results have been extrapolated, generally without any basis, as indicating an allelopathic interaction in the field. Recently, one of the editors of the *Journal of Chemical Ecology*, J.T. Romeo (2000), stated it bluntly in calling allelopathy the "poor stepchild"

---

[5] Johnson in 1775 stated that "patriotism is the last refuge of a scoundrel".

within chemical ecology, often lacking legitimacy and respectability, and it needed to move beyond the "grind and find" and the "thrill of the kill" approaches in its research.

In the past few years, some researchers in allelopathy, in the face of virtual ostracism by the community of ecologists, have been rewarded for their perseverance. Important examples come from studies in the genetic variation in allelopathic potential from domestic plants with large gene pools, for example rice and wheat. There is a rapidly growing realisation that allelopathy is likely to have its greatest impact where plants with little coevolutionary history are grown together, as commonly occurs in agriculture, an idea that was first explicitly stated by the Soviet plant ecologist T.A. Rabotnov (1975). The extraordinary success of some plants in new environments has been attributed to what is now dubbed the "novel weapons hypothesis" (Callaway and Ridenour 2004). That is, organisms which possess attributes, such as certain phytotoxins, not seen before in a particular habitat, may experience great short-term success, because of the lack of organisms with the genetic and physiological equipment to deal with such substances. There are many examples of this, of which species of the Australian genus *Eucalyptus* growing overseas are particularly notorious (Figure 1.4).

The above theory has gained substantial currency, in particular through an important study of an invasive weed in the United States, spotted knapweed, *Centaurea maculosa* (Bais *et al*. 2003; Figure 1.5). This heralded study has demonstrated in real time a mechanism for the uptake and action of a plant-produced phytotoxin, racemic catechin, at field concentrations. Firstly, by monitoring the loss of the vital fluorescent stain, fluorescein diacetate from dying cells, Bais *et al.* (2003) were able to demonstrate the cascading toxic effect of (-)-catechin through root tip tissue, exposed to a concentration less than that in the soil environment surrounding spotted knapweed plants. Parallel experiments showed a concomitant series of physiological events, such as a rapid increase in reactive oxygen species, calcium movement, and changes in gene expression. The specificity of the toxicity was remarkable in that (+)-catechin had no allelopathic effect, although it proved to be antimicrobial. The allelopathic effect was highly selective in that roots of *C. maculosa* were unaffected by (−)-catechin, whereas a congener, *C. diffusa*, was strongly inhibited. As stated by Fitter (2003), this study has achieved much in addressing the critics of allelopathy, and has returned a large measure of respectability to the discipline.

Further recent studies from the same research group have provided more remarkable results that highlight the adaptive significance of allelopathy in some species. For example, it has been found that populations of some native species growing in environments subject to long-term invasion by *C. diffusa* are more resistance to the effects of catechin, and that relatively rapid natural selection is involved, both in the case of resistance in native species and invasiveness in *C. diffusa* (Callaway, Bais *et al*. 2005). Also Callaway, Ridenour *et al*. (2005a) have found that some native American species, such as *Gaillardia grandiflora* and *Lupinus argenteus*, when their roots are exposed to catechin released by *C. diffusa*, can block its allelopathic effects through releasing the antioxidant oxalic acid. These are exciting times for students of allelopathy.

What is Allelopathy? 9

Figure 1.4. Eucalyptus tereticornis *plantation in Haryana, India, showing poor understorey growth, attributable, at least partly, to allelopathy. (Photograph by the author).*

A remarkable feature of the concept of allelopathy is its persistence in the literature through the ages, in spite of often scathing criticism. From the earliest times, the notion of one plant harming another through toxins evidently has had an innate attractiveness to the human mind and has fired the imagination in many sectors of society. Allelopathy, perhaps to its detriment, was once intertwined with the occult sciences, and has been used as a metaphor for the sinister in both religion and literature. It is also interesting that there have been recent attempts to utilise the term 'allelopathy' within the domain of psychology (Gibeau 1997), where it has been used to describe the situation where a person's behaviour has become negatevely affected due to the 'contaminating' influence of their immediate environment[6].

---

[6] This metaphorical transposition is not new for allelopathic and allied phenomena. The best known example is *upas*, the name of a poisonous tree (see Chapter 5), which during the late 19th and early 20th centuries, came to represent any activity or thought which had a pernicious effect on society. Curiously "Allelopathy" has also been adopted as the name of a Japanese compact disc (CD) label for jazz music

*Figure 1.5. An infestation of spotted knapweed (*Centaurea maculosa*) in California (photograph by John M. Randall, courtesy of The Nature Conservancy).*

At first, allelopathy was considered to be a factor in plant interaction likely because of comparisons between plants and animals, and some plants were well known to be capable of poisoning animals. This comparative approach reached its epitome in the nineteenth century when the notion of allelopathy was coupled to the idea that plants excreted waste substances inimical to themselves, as witnessed in animals. In more recent times, allelopathy has achieved popularity on the coat-tails of the success of antiherbivory chemistry in plants. It is now universally accepted that the diverse array of toxic secondary metabolites in plants is largely in response to the pressures of herbivores, notably insects, and pathogenic organisms. Allelopathy has been rationalised as a largely accidental outcome of plants maintaining an arsenal of biologically active substances. Nonetheless, the notion of the immobile plant, in the face of both animal and plant enemies, engaging in "chemical warfare" seems to have great popular appeal. It is noteworthy that interest in allelopathy gained remarkable acceleration after 1964, when a photograph of zones of inhibition around *Salvia leucophylla* appeared on the cover of the journal *Science* (Figure 1.6) in conjunction with an article about allelopathy (Muller *et al*. 1964). As the author C.H. Muller observed, it was not the data that attracted attention and subsequent research funding - it was the image (Muller 1982).

We are learning that plants are far more complex entities than ever thought. The selective toxicity of allelopathic substances and the seasonal and developmental diversity of these substances are areas that have received comparatively little attention. Studies, such as those by Bais *et al*. (2003) now illustrate that allelopathy may be

---

by Evan Parker, John Zorn, Yoshiaki Kinno and others, perhaps because of the significance of "mutual effect" in jazz. Jane Holtz Kay, author of *Asphalt Nation*, has recently used the term in the following curious context: "the automobile also has been incredibly allelopathic to other forms of transportation."

much more adaptive than previously credited. Whereas it has been realised for some time that herbivory can lead to changes in the secondary metabolites of a plant, it has also been demonstrated recently that the presence of another plant, for example barnyardgrass, or its metabolites, can cause an increase in the concentration of the allelopathic substances in rice (Kong *et al*. 2004). It has been suggested in the past, although with little data to support it, that allelopathy may play a role in maintaining species diversity; recent studies with phytoplankton species indicate that toxic species can help to the prevent competitive exclusion of species, and thus maintain a species mix (Roy *et al*. 2006).

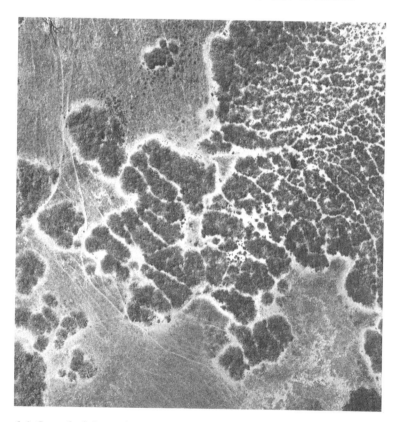

Figure 1.6. *Cover for* Science, *31 January 1964 (Volume 143, number 3605), showing an aerial view of the aromatic shrubs (*Salvia leucophylla *and* Artemisia californica*) invading an annual grassland near Santa Barbara, California. The cover was captioned unusually as "Chemical Plant Competition". (Reproduced with permission of the American Association for the Advancement of Science, and Mark Hurd, Aero-Metrics Inc.).*

The history of allelopathy begins with the records emanating from societies with early written records of agriculture, including the ancient civilisations of China, India, Greece, and Rome. The phenomenon of allelopathy was expressed historically within the framework of the antipathy and sympathy of plants, and these topics were controversial in their own eras, as they are today. The various emanations, exudations and excrescences of plants were generally viewed with distrust, and events such as blight and pestilence were often mistakenly ascribed to the excretions of plants, as late as in the nineteenth century. The popularity of allelopathy has waxed and waned repeatedly. Notable peaks of interest occurred in the Classical era, the late eighteenth and early nineteenth centuries, and the early twentieth century, and we are undoubtedly amid a peak of activity at the present time. The scepticism accorded to allelopathy in some quarters in recent decades is very much a legacy of the well publicised failure of botanists in the 1830's and 1840's to find evidence in support of allelopathy, particularly the notion of root excretion, and similarly the disfavour that devolved concerning relevant experimental results found in England, the United States and elsewhere during the first two decades of the twentieth century.

It is the aim of this book to explore in detail the vicissitudes of the concept of allelopathy from the earliest recorded times until the arrival of allelopathy as a recognised component of plant ecology – this occurred arguably in the mid-1950's with the almost synchronous publication of three books on allelopathy: a monograph on allelopathy by Grümmer (1955)[6] in German, a book on the effects of allelopathic substances in agriculture by Chernobrivenko (1956)[7] in Russian, and a little known but valuable monograph in English by Hubert Martin (1957) entitled *Chemical Ecology in Relation to Agriculture*.

REFERENCES

Bais, H.P., R. Vepachedu, S. Gilroy, R. Callaway and J.M. Vivanco. 2003. Allelopathy and exotic plant invasion: from molecules and genes to species interaction. *Science* **301**: 1377-1380.

Bertin, C. and L.A. Weston. 2004. Alternative weed management strategies for landscape and turf settings. In: Inderjit (ed.) *Weed Biology and Management*, pp. 403-422. Kluwer Academic Publishers, Dordrecht.

Callaway, R.M., H.P. Bais., T.L. Weir, L. Perry, W.M. Ridenour and J.M. Vivanco. 2005. Allelopathy and exotic plant invasion: from genes to communities: synopsis, updates, and implications. In: J.D.I. Harper, M. An, H. Wu, and J.H. Kent (eds), *"Establishing the Scientific Base": Proceedings of the Fourth World Congress on Allelopathy, 21-26 August 2005*, pp. 33-38. Charles Sturt University, Wagga Wagga.

---

[6] This has been translated into Russian and was published in 1957 as *Zvaimnoye Vliyaniye Vysshikh Rastenii – Allelopatii*, by Inostrannoi Literatury, Moscow.

[7] It is little known that much of this book has been translated in Chinese, and was published in 1961 with the title (in Pinyin): *Zhi wu fen mi wu di sheng wu xue zuo yong he jian zuo zhong di zhong jian xiang hu guan xi*.

Callaway, R.M. and W.M. Ridenour. 2004. Novel weapons: a biochemically based hypothesis for invasive success and the evolution of increased competitive ability. *Frontiers in Ecology and the Environment* **2**: 436-443.
Callaway, R.M., W.M. Ridenour, T. Laboski, T. Weir and J.M. Vivanco,. 2005. Natural selection for resistance to the allelopathic effects of invasive plants. *Journal of Ecology* **93**: 576-583..
Chernobrivenko, S.I. 1956. *Biologicheskya Rol' Rastitel'nykh Vydellennii i Mezhvidovye Vzaimnootnosheniya Smeshannykh Posevakh*. Sovetsakya Nauka, Moscow.
Duke, S.O., N. Cedergreen, E.D. Vellini, and R.G. Belz. 2006. Hormesis: is it an important factor in herbicides use and allelopathy? *Outlooks on Pests Management* **19**: 29-33.
Environmental Physiology Subcommittee. 1971. *Biochemical Interactions Among Plants*. National Academy of Sciences, Washington, D.C.
Fitter, A. 2003. Making allelopathy respectable. *Science* **301**: 1337-1338.
Gibeau, E. 1997. Allelopathy and the depressive object. Paper presented to the Michigan Psychoanalytical Council.
Gross, E.M. 2003. Allelopathy of aquatic autotrophs. *Critical Reviews in Plant Sciences* **22**: 313-339.
Grümmer, G. 1955. *Die gegenseitige Beeinflussung höherer Pflanzen – Allelopathie*. Gustav Fischer, Jena.
Harper, J.L. 1975. New Biological Books: *Allelopathy* by E.L. Rice. *Quarterly Review of Biology* **50**: 493-495.
Harper, J.L. 1977. *Population Biology of Plants*. Academic Press, London.
Henderson, I.F. and J.H. Kenneth,. 1953. *A Dictionary of Scientific Terms.* Fifth Edition. Oliver and Boyd, Edinburgh.
Inderjit and R. del Moral. 1997. Is separating resource competition from allelopathy realistic? *Botanical Review* **63**: 221-230.
Kong, C., X. Xu, B. Zhou, F. Hu and C. Zhang 2004. Two compounds from allelopathic rice accession and their inhibitory activity on weeds and fungal pathogens. *Phytochemistry* **65**: 1123-1128.
Martin, H. 1957. *Chemical Ecology in Relation to Agriculture*. Research Monograph 1, Science Service. Canada Department of Agriculture, Ottawa.
Molisch, H. 1937. *Der Einfluss einer Pflanze auf die andere – Allelopathie*. Gustav Fischer, Jena
Muller, C.H., W.H. Muller and B.L. Haines. 1964. Volatile growth inhibitors produced by aromatic shrubs. *Science (Washington, D.C.)* **143**: 471-473.
Muller, C.H. 1982. This Week's Citation Classic. *Current Contents: Agriculture, Biology & Environmental Science* **13**(45): 20.
Olofsdotter, M. 2001. Getting closer to breeding for competitive ability and the role of allelopathy – an example from rice. *Weed Technology* **15**: 798-806.
Rabotnov, T.A. 1982. 'Importance of the evolutionary approach to the study of allelopathy' [in Russian]. *Ekologiya* **1982**(3): 5-8. Translated into English in *Soviet Journal of Ecology* **12**: 127-130.
Reigosa, M.J., A. Sánchez-Moreiras and L. Gonzalez. 1999. Ecophysiological approach in allelopathy. *Critical Reviews in Plant Sciences* **18**: 577-608.
Rice, E.L. 1984. *Allelopathy*. Second Edition. Academic Press, Orlando.
Romeo, J.T. 2000. Raising the beam: moving beyond phytotoxicity. *Journal of Chemical Ecology* **26**: 2011-2014.
Roy, S., S. Alam and J. Chattopadhyay. 2006. Competing effects of toxin-producing phytoplankton on the overall plankton populations in the Bay of Bengal. *Bulletin of Mathematical Biology* **68**: 2303-2320.
Tinnin, R.O. 1972. Interference or competition. *American Midland Naturalist* **106**: 672-675.
Weiss, F. 1949. Weeds, fungi, and the education of botanists. *Scientific Monthly* **68**: 257-261.
Whittaker, R. H. and P.P. Feeny. 1971. Allelochemics: chemical interactions between species. *Science (Washington, D.C.)* **171**: 757-770.
Williamson, G.B. 1990. Allelopathy, Koch's postulates and the neck riddle. In: J.B. Grace and D. Tilman (eds), *Perspectives on Plant Competition*, pp. 143-163. Academic Press, San Diego.
Willis, R.J. 1985. The historical bases of the concept of allelopathy. *Journal of the History of Biology* **18**: 71-102.
Willis, R.J. 1994. Terminology and trends in allelopathy. *Allelopathy Journal* **1**: 6-28.

CHAPTER 2

ALLELOPATHY IN THE CLASSICAL WORLD – GREECE AND ROME

> Roots are the branches down in the earth.
> Branches are roots in the air.
>
> *Stray Birds*
> Rabindranath Tagore (1916)

INTRODUCTION

The observation that certain animals were venomous to others, and that some plants were poisonous to livestock and even humans, likely led some individuals to wonder whether some plants were actually toxic to other plants. This concept is the core of allelopathy, that is, the chemical interaction of plants, although today we also acknowledge that many plants may benefit others through the chemicals they release. In any case, the concept that one plant could poison another plant was well known to the classical authors of Greece and Rome (see Figure 2.1 for a map of the Classical world). Furthermore, the idea that one plant was inimical to another fitted comfortably within the ancient concepts of antipathy and sympathy. The literature from ancient Greece and Rome, as it concerns antipathy (Pease 1927), or more specifically allelopathy, has been broached on a few occasions (Rice 1983, Willis 1985, Aliotta and Mallik 2004, Petriccione and Aliotta 2006); however, it is the intent of this chapter to investigate this matter more fully. What emerges is that the concept of allelopathy was well known to a wide range of classical authors, and not simply those remembered for their works on natural history.

GREECE

The history of allelopathy centers on two issues. Firstly, do plants release excreta? Of course, this phenomenon was known well through animal examples, and it was common knowledge that animal excreta were typically noxious to the same or related species. Ironically, it was well established in ancient times that animal excreta were beneficial to crops as fertiliser. Secondly, do plants produce something harmful to other plants (the notion that some plant substances may be beneficial to other plants appears not well developed in the classical literature)? These two precepts, one essentially based more on theory, and one based primarily on practice, were later often at odds with one another, but, in any case, are two fundamental roots of the concept of allelopathy. As we shall see, the Greeks were more concerned with theory, and the Romans more with practice.

It is important to retain the development of ideas concerning plant nutrition, as with these are tied concepts concerning plant excretion, which played an extremely important role in thoughts concerning allelopathy in the eighteenth century, and laid the theoretical basis of allelopathy from the eighteenth to the early twentieth centuries.

While a cogent atomic theory was expressed by Demokritos of Abdera (460-360 B.C.), it was the four element theory of Empedocles (500-430 B.C.), elaborated by Plato (*c.* 428-347 B.C.) and Aristotle (384-322 B.C.), which achieved dominance for the next two thousand years. Empedocles held that all matter was built from four basic elements: earth, water, air and fire, and that the association of these elements was based upon particular circumstances of attraction or repulsion, wherein lies the earliest concepts of sympathy and antipathy, themes which achieved great development much later and also influenced the concept of allelopathy. Aristotle and others held that all terrestrial matter was formed from earth, water, air and fire, and furthermore these were conjoined through four basic principles: cold, moistness, heat and dryness respectively (Figure 2.2). A fifth element, ether, was invoked to explain the workings of the heavens above.

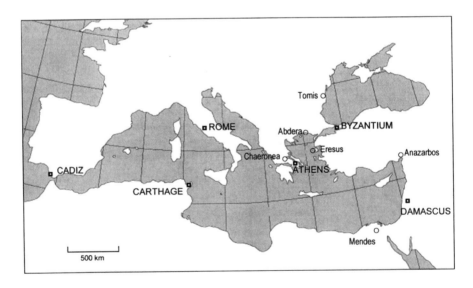

Figure 2.1. Map of the Mediterranean region with some place names associated with the Classical Era.

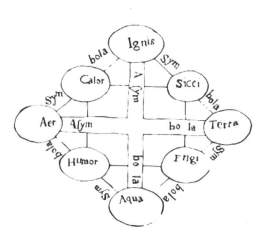

Figure 2.2. *A 16th century version of the tetragram of the elements (clockwise from top: fire, earth, water air) and principles (clockwise from top-right: dryness,cold, moistness, heat).* Symbola *has the meaning of "in common", and* asymbola *has the meaning of "not in common".*

Apart from noting the origins of the concept of sympathy and antipathy of things, what is important here to note is the Aristotelian view of plant nutrition. The only surviving botanical work from this school is *De Plantis* (On Plants), a short tract once included amongst the works of Aristotle (384-322 B.C.), but now generally believed to have authored considerably later by someone such as Nicholas of Damascus (64 - *c*.4 B.C.). Nonetheless, this work provides a view of early Greek thinking on plants. It was regarded that plants are living organisms comparable to animals, showing growth and decay, and possessing a soul, but separable principally by their inability to move[1]. Indeed, the structure of plants was compared directly to that of animals, as much as was possible, and plants were viewed as sorts of upside-down animals with their mouth (roots) in the soil (Robin 1928). The origin of this idea is to be found in Aristotle's *On the Parts of Animals*:

> ... plants take their food, already processed, by their roots from the earth (which is why plants have no excrement, since they use the earth and the heat in it in lieu of a stomach)... (Book II, Section 3)

This comparative approach persisted through the ages, and botanists are still encumbered with zoological terms for certain plant parts: e.g. vein, ovary, ventral and dorsal surface, etc. The early Greeks believed that plants, like animals, simply ingested their required, preformed foodstuffs, albeit feeding from the soil with roots, and converted these substances into plant tissue within the plant, using the properties of water and fire (sun) in a manner akin to creating earthenware. The simplistic notion, of plants feeding in the soil, became refined and better known much later as

---

[1] According to this work, Anaxagoras, Demokritos and Empedocles believed that plants possessed intelligence.

the "humus theory". It remained substantially unchanged for two thousand years, and was a hindrance to advancement in plant physiology.

Of the vital processes in plants, excretion was not a large consideration, as it was argued that because the food of the plant was preformed in the soil, there was little waste produced[2]. This concept appears to have been incompletely developed, and one could interpret a very early allusion to excretion by plant roots in Arisotle's *On the Parts of Animals*:

> Just as in the bathroom the heat attracts the moisture, and transforms it into steam, and this, being light, when it is in excess condenses into drops of water, so also in animals and plants the waste product rises from the lower to the upper parts, and descends again from the upper to the lower. (Book II, Chapter 1)

A related notion which has its recorded origins in early Greek writings is that of the release of substances from organisms. This is variously translated as exhalations, effluences or effluvia, terms which remained in various common usage until the nineteenth century. The writings of Empedocles are known only from fragments (Wright 1981), and one statement reads: "there are effluences from all things in existence." While this could be interpreted as a statement supporting, amongst other things, the idea of plant excretion, it should be borne in mind that Empedocles believed that all natural objects, whether living or not, released tiny particles, and that all objects correspondingly possessed tiny pores. He believed that the interaction of the particles with the pores was responsible for a range of phenomena, from magnetism to sensory perception; in the case of the sense of smell, he was close to the truth. The release of effluvia in the soil was commonly associated with evil, and strong smells from the soil were to be feared. They were taken to indicate either death had occurred or was about to occur. It is not surprising then that root effluvia were associated with disease or pestilence.

Despite the commentary above, it is possible that the oldest record, that does concern allelopathy, originated from 594 B.C., the date attributed to the formulation of the Laws of Solon. Solon (*c.* 638 – *c.* 559 c. B.C.) was an Athenian who is chiefly remembered for introducing a code of laws that replaced the harsh and unfair legal system of Drakon[3]. The original records, which were likely engraved on wooden tablets or *axones*, are long lost, but some details are recalled in the biography of Solon by the historian Plutarch (46- *c.* 120 A.D.) in his *Lives*. Plutarch, who travelled widely, was born and lived most of his life in Chaeronea in Boeotia, Greece, which during his lifetime was part of the Roman Empire. While Plutarch wrote in Greek, it is likely that he has embellished the report with information of Roman origin:

> He [Solon] showed skill in his orders about planting, for any one that would plant another tree was not to set it within five feet of his neighbour's field; but if a fig or an olive not within nine; for their roots spread farther, nor can they be planted near all sorts of trees without damage, for they draw away the nourishment, and in some cases are noxious by their effluvia.

---

[2] This view is stated more clearly in Theophrastus' *De Causis Plantarum*, Book 6, Chapters 10 and 11.

[3] Hence the term "draconian" or "draconic", meaning severe or harsh.

Statements, undoubtedly concerning the concept of allelopathy, that are early, were written by Theophrastus of Eresus (*c*. 370 – *c*. 286 B.C.). Theophrastus was a student of Plato and Aristotle, and eventually established his own school in Athens (Figure 2.3). Theophrastus was apparently the author of numerous works, of which only a few have survived and are known today. The two principal surviving works, which in themselves are incomplete, are *Historia Plantarum* (known in the English translation as *Enquiry into Plants*) and the recently translated *De Causis Plantarum*, which have earned Theophrastus recognition as the "Father of Botany". Theophrastus adopted most of Aristotle's teachings on plants; however, the question of plant excretion was not completely resolved with him. He concluded in *De Causis Plantarum*:

> Furthermore, since a plant has no excrement, it is not likely to attract to itself and draw in what is non-nutritive, since this would then have to be somehow excreted. (Book VI, Section 10.3)

This latter concept was to prove central two thousand years later to A.P. de Candolle (see Chapter 7). Theophrastus on the other hand recognised the contradiction that some plants are distinctly salty, and that in many cases the salt originates from the soil, and is deposited on the outer layers of the plant (*De Causis Plantarum*, Book VI, Section 10.5).

Figure 2.3. A painting by Robert A. Thom portraying Theophrastus, as envisaged teaching his students, c. 300 B.C. This painting comprises part of a series entitled "A History of Pharmacy in Pictures" commissioned by Parke, Davis and Co., and was reproduced in 1951.

There is considerable overlap in the content of *Historia Plantarum* and *De Causis Plantarum*, and with the exception of the texts relating to allelopathy below, I have given the text from the larger *De Causis Plantarum* and indicated the corresponding reference in *Historia Plantarum*.

In *De Causis Plantarum*, Theophrastus wrote:

> It would doubtless not be difficult to set down injuries in many other encounters, since injuries are far more numerous than benefits, as in animals. Indeed a few plants are even injured by odours, as the vine by the odours of bay and cabbage, and it shows that this is so from the moment it sends out shoots. For when the vine is near cabbage or bay, its shoot curves its tip and (as it were) turns back because of the pungency of the odour. For the vine is sensitive to smell, just as wine too is apt to attract the odours of objects placed near it, wine drawn off in jars doing this more and faster because of its small quantity and of its exposure. But (as we said) effects of this sort are easily seen in many instances. (Book II, Section 18.4)

A very similar passage appeared in *Historia Plantarum*:

> Again some things, though they do not cause death, enfeeble the tree as to the production of flavours and scents; thus cabbage and sweet bay have this effect on the vine. For they sat that the vine scents the cabbage and is infected by it. Wherefore the vine-shoot, whenever it comes near this plant, turns back and looks away, as though the smell were hostile to it. Indeed Androkydes used this fact as an example to demonstrate the use of the cabbage against wine, to expel the fumes of drunkenness[4] for, said he, even when it is alive, the vine avoids the smell. (Book IV, Section 16.6)

In considering these passages by Theophrastus, one should bear in mind the difficulty of a translating works from ancient Greece. One is likely to conjure a mental image of a robust and odiferous head of cabbage somehow causing the demise of a grapevine. In fact the cabbage we find in today's supermarkets is very different from what the Greeks knew as "cabbage" in 300 B.C. The cabbage of ancient Greece was likely more similar to kale or colewort, a leafy, non-heading form of cabbage, and was possibly richer in allelochemicals (Figure 2.4). Indeed the Greek word ραφανοσ (raphanos), as used by Theophrastus, refers to something close to the wild form of *Brassica oleracea*, sometimes known as *Brassica cretica*, an edible, but bitter, herb of coastal regions. In any case, it is this reference to the interaction between the cabbage and the vine that is the source of very many related statements that appear in natural history works over the next two thousand years[5].

---

[4] This belief has persisted to modern times. Edmund Spencer (1834) wryly noted that the popularity of sauerkraut in Germany was linked to the potential to drink more wine. In the United States, sauerkraut juice was alleged to relieve inebriation (Olybrius 1934).

[5] In Theophrastus' *Enquiry into Plants*, at least five plants, all bitter tasting, have very similar Greek names: ραφανοσ = *Brassica cretica* ('cabbage'); ραφανοσ η αγρια = *Raphanus raphanistrum* ('charlock'); ραφανοσ η ερεια = *Euphorbia apios* (spurge); ραφανισ = *Raphanus sativus* (raddish); and ραφανισ η αμωρεα = *Armoracia lapathifolia* (horseradish?). In works which have evidently copied Theophrastus, and been translated into another language, it is not surprising that the plant names vary. In Pliny, we find that "the vine abhors all coleworts", and "the radish and the laurel are harmful to the vine." Gerard (1597) wrote: "Divers think that this Horse Radish is an enemie to Vines."

Classical Greece and Rome 21

Figure 2.4. *Cabbage plants from an anonymous 15th century French manuscript,* Livre des Simples Médicines *(MS Francais 9137, fol. 111v, Bibliothèque Nationale de France).*

However, Theophrastus was also well aware that competition was a major factor in plant interactions, as for example, shown in *De Causis Plantarum*, (see also Book III, Section 10 where both competition for light and soil factors are considered, and *Historia Plantarum* Book IV, Section 16.5):

> Destruction coming from neighbours that are planted or grow up spontaneously near by is due to their removing the tree's food; and the destruction is more rapid if the neighbours are stronger and more numerous, as is the case when they are wild, or when they have many roots and take much food, or branch out and entwine about the tree, choking it, or grow into it, like ivy. Indeed mistletoe too, and in general all plants that sprout in the tree, are held to kill it. Tree-medick and tree-purslane kill by their great consumption of food and by their salinity; tree-purslane is the stronger because it has more.[6] (Book V, Section 15.4)

Thus, in the first known books devoted to botany, we encounter the dichotomy between injury caused through allelopathic interaction and through competition, an issue which still stirs debate among plant ecologists. A statement often regarded as

---

[6] C.-H. Chou (1999), citing D.B. Jelinek, stated that, according to Theophrastus, pigweed inhibited alfalfa; Kohli (1998) has given the reciprocal interaction. There is little mention of either pigweed (either *Chenopdodium* spp., *Amaranthus* spp. or *Portulaca* spp.) or alfalfa (*Medicago sativa*) in Theophrastus. The error arose likely from a wrong translation of plant names here, and in the parallel text from *Historia Plantarum*: "Again an overgrowth of ivy is dangerous, and so is tree-medick, for this destroys almost anything. But *halimon* is more potent even than this, for it destroys tree-medick." (Book IV, Chapter 16.5) Tree-medick (*Medicago arborea*) is not the crop alfalfa, and halimon or tree-purslane is not a true purslane (*Portulaca* spp.), sometimes known as pigweed, but is usually regarded as the salt-tolerant shrub, *Atriplex halimus*.

the starting point for allelopathic literature originated in *Enquiry into Plants*, and concerns the chickpea:

> There is the fact that in general it does not reinvigorate the ground, since it exhausts it, but it destroys weeds, and above all and soonest caltrop[7]. (Book VIII, Chapter 7)

In close affinity to allelopathy is "companion planting"[8], which today is commonly practiced by plant growers, especially those who wish to eliminate the use of pesticides, in so-called "organic" agriculture. Companion planting recognises that certain plants grow unusually well in association with others, in acting as "nurse" plants, or perhaps offering allelochemicals which help to minimise damage by herbivores or pathogens in a neighbouring species. These concepts were also known to Theophrastus, who wrote in *De Causis Plantarum*, (see also Book III, Section 10, and *Historia Plantarum* Book VII, Section 5.4):

> That among plants too some collaborate to preserve and propagate others can also be seen from the following: among the wild the deciduous help the evergreen, since it happens that the earth is manured (as it were) by the decomposing leaves, and this is useful for good feeding and making the seeds sprout; among the cultivated there are the plants sown among the young vines when the growers wish to reduce their excess of fluid, and the plants sown among vegetables either to do this or to keep them free of the pests that arise, as bitter vetch is sown among radish to help against the flea-spider, and any similar case where a plant of this kind is sown with others. (Book II, Section 18.1)

Similarly, he recognised the efficacy of plant constituents against herbivores, and the hazards of domestication; for example, in *De Causis Plantarum*, (see *Historia Plantarum* Book IV, Sections 14.2, 14.4), he wrote:

> Pungent trees are the least liable to get grubby, not only because they resist decomposition, but also because their pungency prevents the breeding of animals when decomposition occurs. Proof of this is the case of the bay; it is quick to decompose but not quick to the same degree to get grubby. Indeed this is why the wild fig suffers less from them than the cultivated fig; its juice is more pungent. For in general the sweet fruit trees decompose faster, since the savour, being weaker, is more subject to change. This is why sweet apple and pomegranate decompose faster than acid, and the sweet spring apple and pomegranate more than the rest by reason of their juice and their whole nature as well. When planted in a pine-thistle[9] all trees are less liable to grubs because of its heat and its odour. (Book V, Section 9.4-5)

A later Greek writer, Athenaeus, who authored a work entitled *Deipnosophistae*[10] in about 200 A.D. during the time of the Roman Empire, alluded to Theophrastus' remarks:

---

[7] Caltrop or caltrops can refer to plants that entangle the feet, such as brambles, or to those having entangling fruits or seeds, such as the water chestnut (*Trapa natans*). Here, the plant may be *Tribulus terrestris*, a weed with spiny seeds.

[8] Theophrastus was actually the first to use this term; "Another exception are the companionable plants, as olive and myrtle are held to be among trees" (*De Causis Plantarum*, Book III, Section 10.4).

[9] Possibly *Atractylis gummifera*, a toxic thistle, also known as gum thistle.

[10] A deipnosophist is a master of the art of dining.

> Theophrastus also has written; he alleges that the growing vine loathes the smell of cabbage. (Book I, Section 34)

However, more importantly, he cited from the work of Theophrastus an example of companion planting:

> Theophrastus also says, that the fig tree if planted among squills grows up faster, and is not liable to be destroyed by worms; and in fact, that everyhting which is planted among squills both grows faster and is more sure to be vigorous. (Book III, Section 13)

Another Greek author, of whom very little is known, is Bolos Demokritos[11] of Mendes in Egypt, who was active about 200 B.C., and is believed to be the author of at least one important agricultural work, commonly referred to as his *Georgics*. Columella mentions that Demokritos was the author of a work entitled *On Antipathies*, and assumedly this was a book within the former. However, these writings are known only through citation in later works, including those by Pliny, Columella, Varro, and Ibn al-Awwam, and the *Geoponika*. The various fragments have been collected by Wellmann (1921). The reputed *Georgics* of Demokritos are of interest for their original content concerning early ideas of applied allelopathy and biological control. For example, Demokritos suggested that forest may be cleared by soaking lupin-flower in hemlock[12] juice and sprinkling the solution on the tree roots (Pliny, Book XVIII, Chapter 8), and that planting branches of laurel could keep vines free of rust (see Ibn al-Awwam, Book 1, p. 589). Bolos Demokritos apparently provided some of the earliest sources concerning the concept of sympathy and antipathy among living organisms, including plants (e.g. sympathy of pomegranate and myrtle, as quoted much later in the *Geoponika*, Book X, Section 29, and by Ibn al-Awwam, Book 1, p. 254).

There are several Greek authors who lived during the period of the Roman Empire. The Jewish scholar, Philo of Alexandria (20 B.C. – 50 A.D.) was the author of *De Animalibus*, a work which rejected the idea of the reasoning capacity of animals, and which cited plant antipathies in support of his argument (Terian 1981):

> Be not misled. That these things are altogether doubtful may be illustrated by the trees as well as the bushes. Even though such have not partaken of soul, they manifest no less intimacy or indifference; they move and grow, they kiss and embrace each other lie lovers, such as the olive tree and the vine. And there are certain things which they reject and turn away from. They not only raise themselves against other plants, openly and face to face, but also turn way, as if they had feet, and never come closer. Furthermore they do not even put forth buds. If they happen to be in bloom, some might bear, but the rest drop out of sight or wither away gradually.
> Likewise the vine shuns the cabbage and the laurel too. But I do not think that anyone, however foolish, would dare to sat that any one of them behave in a friendly or hostile manner. (Sections 94-95)

---

[11] This is not the Greek philosopher Demokritos of Abdera (*c.* 460 to 370 B.C.) previously mentioned.

[12] Hemlock is the common name for two very different plants. It is the name of a conifer, *Tsuga* spp., and as here, it is the name of a poisonous herb of the family Apiaceae, *Conium maculatum*. All parts of the plant are poisonous, due largely to the alkaloid coniine, and the fruits are especially deadly, and were used in classical times by the Greeks as a means of effecting execution, as in the famous instance of Socrates.

One of the most revered of the classical botanic authors was Pedanius Dioskorides (c. 40 – c. 90 AD; Figure 2.5), a Greek born in Anazarbos (now Nazarba near Tarsus in present-day Turkey), which was ruled by the Roman Empire at the time. Dioskorides wrote a five volume work known as *De Materia Medica*, originally written in Greek, which described the properties of plants and animals useful in medicine. He appears to have originated one example of antipathy that has been passed to subsequent writers. The juice of the fern (*Dryopteris filix-mas*[13]) was supposed to be useful in curing wounds caused by reeds (and thus arrows), and Dioskorides (1934)[14] wrote that this antipathy was of such strength that:

Figure 2.5. Dioskorides receiving a mandrake plant from Euresis the goddess of discovery. This engraving is based on a miniature painting which appeared in the Codex Vindobonensis, *a famous illustrated manuscript copy of Dioskorides'* De Materia Medica, *which dates from the Byzantine period, c. 512 A.D. The engraving appeared firstly in* Commentarii de Augustissima Bibliotheca Caesarea Vindibonensi, Band II *by Peter Lambeck (1669). It was reputed that the mandrake was either poisonous, or its shriek could kill when the plant was removed from the soil. Thus a dog was tethered to the mandrake and sacrificed in collecting the mandrake. The engraving depicts the dead dog, but has omitted the cord.*

---

[13] Formerly known as *Aspidium felix-mas* or *A. filix mas*.

[14] The *De Materia Medica* of Dioskorides, also known as Dioscorides, was not published in English until 1934. The work had been translated into English in 1655 by John Goodyer, but was never published. The manuscript languished unnoticed in the collection of Magdalen College, Oxford, until published in 1934.

And ye root being drank with Axungia[15] & laid on is good for such as are hurt with a Reed. The proof is this. Where there is much *reed*[16], & much fern encompassing there ye fern vanisheth.. (Book IV, § 186)

ROME

The earliest of the Roman writers to consider the concept of allelopathy was Varro (Marcus Terrentius Varro). Although Cato's *De Agri Cultura* (*c*.149 B.C.) is regarded as the earliest agricultural work, and indeed the earliest prose work in Latin, there is only mention therein of plants that are harmful through competition. Varro (116-27 B.C.) was regarded as one of the greatest of the Latin scholars, but only a fraction of his work survives including much of *De Rerum Rusticarum* (commenced *c*. 36 B.C.). In this we find:

> Again, the products of the farm are influenced by the way in which your neighbour's land is planted. If, for instance, he has an oak-grove on the common boundary, you would be wrong to plant olive trees on the edge of such a wood, for these have a natural antipathy to it so great that, not only do they bear worse, but even, in their efforts to escape, bend away inwards toward the farm precisely as does the vine planted near cabbages. Like oak trees, walnut trees near your farm, if of large size and standing a little distance from one another, make its margins totally unproductive. (Book I, Chapter 16)

Many unexpected Roman authors have incorporated natural history lore into their works. For example, the statesman and philosopher Marcus Tullius Cicero (106-43 B.C.), in discussing the characteristics of living organisms wrote:

> Indeed it is even said that if cabbages have been planted near them, the vines shrink from them as from something deadly and injurious, and come nowhere into contact with them. (Book II, Chapter XLVII)

A Roman writer who is commonly overlooked in the history of natural science is Publius Vergilius Maro, known commonly as Virgil (70-19 B.C.), as his surviving works, notably *The Georgics*, are in verse. It is believed that *The Georgics* were completed in about 29 B.C. While they are rich in social and historical commentary, they also provide sound observations and practical advice to the farmer. A detailed analysis of Virgil's Georgics has been provided by Billiard (1928), who considers the concepts of soil toxicity and detoxification (as put forward in the twentieth century by the USDA Bureau of Soils) vis-à-vis the Georgics and other contemporary Roman writings. A passage, as translated by C. Day Lewis (Virgil 1940), indicates Virgil's awareness of the value of crop rotation and legumes, the harmful effects of certain plants, and early concepts of soil exhaustion or soil sickness and its cure:

> See, too, that your arable field lies fallow in due rotation,
> And leave the idle field alone to recoup its strength:

---

[15] Axungia is a type of grease, likely animal in origin.

[16] According to Gunther's edition of Dioskorides (1934), Goodyer's manuscript provided the word "seed", but this was likely a misreading; other editions have provided "reed".

> Or else, changing the seasons, put down to yellow spelt
> A field where before you raised the bean with its rattling pods
> Or the small-seeded vetch
> Or the brittle stalk and rustling haulm of the bitter lupin.
> For a crop of flax burns up a field, and so does an oat-crop,
> And poppies drenched in oblivion burn up its energy.
> Still, by rotation of crops you lighten your labour, only
> Scruple not to enrich the dried-up soil with dung
> And scatter filthy ashes on fields that are exhausted.
> So too are the fields rested by a rotation of crops,
> And unploughed land in the meanwhile promises to repay you.
> Often again it profits to burn the barren fields,
> Firing their light stubble with crackling flame: uncertain
> It is whether the earth conceives a mysterious strength
> And sustenance thereby, or whether the fire burns out
> Her bad humours and sweats away the unwanted moisture,
> Or whether that heat opens more of the ducts and hidden
> Pores by which her juices are conveyed to the fresh vegetation,
> Or rather hardens and binds her gaping veins against
> Fine rain and the consuming sun's fierce potency
> And the piercing cold of the north wind. (Book I)

A statement in *The Eclogues* (Virgil 1963) a collection of ten short pastoral poems written by Virgil about a decade before his *Georgics*, also contained a brief statement that has been occasionally interpreted as inferring that the canopy of the juniper had harmful qualities similar to that of the walnut (Bush 1854):

> The shade of this Juniper turns chill.
> Shade stunts a crop, and it's bad for a singer's voice. (Eclogue X, lines 75-76)

Classical agricultural writers were also aware that climbing plants, in particular the vine, had preferred trees or shrubs for support, but would fail when planted next to certain other trees (see Chapters 8 and 11). The use of elms or poplars in vineyards has been practised for centuries in Europe, and Virgil wrote:

> Then make ready and fit smooth reeds, poles of peeled wood,
> Ash stakes for the forked uprights,
> Upon whose strength your vines can mount and be trained to clamber
> Up the high-storied elm trees, not caring tuppence for wind. (Book II)

Yet another poetic work, by an unknown author, is the ode *Nux* (The Walnut Tree), written as an allegory, with an abused walnut tree representing Ovid, who was exiled, for no stated reason, to remote Tomis on the Black Sea in 8 A.D. by Augustus[17]. This short poem was formerly included among the works of Ovid (43 B.C. – 17 A.D.), but this is now considered incorrect. Nonetheless herein we find an early warning concerning the walnut:

> Lest I [the walnut tree] harm the crops, for I am even said to harm the crops, the furthest and extremist limit of the estate receives me.

---

[17] The poem is now judged to have been written by an imitator of Ovid, perhaps toward the end of the 1st century A.D. (Bramble 1982). Ovid never learned the reason for his exile, but assumedly he had somehow offended Augustus or learned some secret. Tomis is today the site of Constanta in Romania.

While it is usually Pliny the Elder who is cited with regard to Roman thoughts on allelopathy, he is preceded also by Columella (Lucius Junius Moderatus Columella). Little is known of the life of Columella, although it is thought he served in the Roman army and then became a farmer at Cadiz. His surviving works are the highly practical *De Rerum Rusticarum* (*c.* 64 A.D.) and *De Arboribus*, which is considered to be sole surviving remnant of an earlier version of his large work; in the former he wrote:

> For the ordinary oak, even if it has been cut down, leaves behind roots harmful to the olive grove, the poison from which kills the olive tree. (Book V, Chapter 8)

The Romans seemed to have had wide experience of "soil sickness", that is, declining yield on land cultivated repeatedly with the same crop, due to indeterminate reasons. Such problems were largely the result of poor agricultural practices, but were compounded, particularly in Roman times, by a deterministic attitude that like most things, the soil must deteriorate with age (Hughes 1975). This philosophy was stated to great effect by Lucretius (c. 98 – c. 55 B.C.) in his epic poem *De Rerum Natura*:

> For time transforms
> The whole world's nature, and all things must pass
> From one condition to another: nothing
> Continues like itself: all is in flux;
> Nature is ever changing and compelling
> All that exists to alter. For one thing
> Moulders and wastes away grown weak with age,
> And then another comes forth into the light,
> Issuing from obscurity. So thus Time
> Changes the whole world's nature, and the Earth
> Passes from one condition to another:
> So that what once she bore she can no longer,
> And now can bear what she did not before. (Book VI)

It has been suggested by some historians (Simkhovitch 1916, Huntington and Cushing 1924, Semple 1931) that the demise of the Roman Empire was in part linked to declining soil fertility, and its accepted inevitability[18]. Dissent was expressed by the practical Columella:

> You ask me, Publius Silvinus, and I have no hesitation in informing you at once, why in the preceding book I immediately at the start rejected the long-standing opinion of almost all who have discoursed on the subject of agriculture, and repudiated as mistaken the views of those who hold that the soil, wearied and exhausted by age-long wasting away and by cultivation now extending over a long period of time, has become barren. (Book II, Chapter 1)

Returning to Columella's ideas relating to allelopathy, we note his remarks:

---

[18] This is a controversial topic that seems to have been popular at this time, and apparently was first entertained by Liebig. Dissenting views were given by Rostovtzeff (1926) in *Social and Economic History of the Roman Empire*, and Salvemini (1939) in a lecture "Soil exhaustion and the decline of the Roman Empire", of which notes are held by the Water Resources Center Archives, University of California, Berkeley.

But when you have taken off a crop of it [barley], it is best to let the ground lie fallow for a year; or if not, to saturate it with manure and drive off all the poison that still remains in the land. (Book II, Chapter 9)

The idea of crops poisoning the soil is also found with regard to other crops:

Of those legumes, too, which are harvested by pulling, Tremelius[19] says that the poisons of the chickpea and of flax are most harmful to the soil, the one because of a salty nature; the other because of its burning qualities. (Book II, Chapter 13)

and,

And before considering the soil itself, we think it is a matter of very first importance that land hitherto untilled, if we have such, should be chosen in preference to that upon which there has been a crop of grain or a plantation of trees and vines. As to vineyards which have become worthless through long neglect, it is agreed by all authorities that they are worst of all if we wish to replant them, because the lower soil is imprisoned in a tangle of roots, as if caught in a net, and has not yet lost that infection and rottenness of old age by which the earth is deadened and numbed as by some poison or other. (Book III, Chapter 11)

Further to this, Columella wrote in *De Arboribus*:

You should plant your vineyard on ground which has lain fallow; for where there has been a vineyard, anything which you plant sooner than the tenth year will only take root with difficulty and will never attain to any strength. (Book III)

It is due to Caius Plinius Secundus (23-79 A.D.; Figure 2.6), known commonly as Pliny the Elder, that we find many references which may be interpreted as concerning allelopathy. After a career in the military, during which he travelled widely in Roman Europe, Pliny effectively retired in about 59 A.D. to pursue his scholarly studies. Pliny was a great encyclopaedist, and his passion for knowledge led to his famous death while investigating the eruption of Mount Vesuvius in 79 A.D. While many of his writings are lost, his great legacy is the compendial *Historia Naturalis* (Natural History), completed in about 77 A.D., and which has remained in print continuously since 1469.

Not surprisingly, Pliny reiterated and amplified Theophrastus's statement concerning the vine and cabbage:

The nature of some plants though not actually deadly is injurious owing to its blend of scents or of juice - for instance the radish and the laurel are harmful to the vine; for the vine can be inferred to possess a sense of smell, and to be affected by odours in a marvellous degree, and consequently when an evil-smelling plant is near it to turn away and withdraw, and to avoid an unfriendly tang. This supplied Androcydes with an antidote against intoxication, for which he recommended chewing a radish. The vine also abhors cabbage and all sorts of garden vegetables, as well as hazel[20], and these

---

[19] This appears to be GnaeusTremelius Scrofa, a contemporary of Varro and an estate owner, whose works are lost.

[20] The supposed antipathy of hazelnut (filbert) trees to other plants, especially vines, which becomes often repeated in later literature, e.g. Albertus Magnus, Konrad von Megenburg, Heresbach, etc., seems to originate with Pliny, but seems to have little basis. It is possible that the addition of *Corylus* to the list of antipathetic plants is due to an error in translating the Latin or Greek term for walnut. The Latin noun *nux*, as with the Greek χαρνα, may refer specifically to a walnut, or indeed to any reasonable sized nut, such as a hazelnut.

unless a long way off make it ailing and sickly; indeed nitre and alum and warm sea-water and the pods of beans or bitter vetch are to a vine the direst poisons.
(Book XVII, Chapter 37)

Pliny wrote extensively on trees, and some trees were regarded with dread, for example, the yew (*Taxus baccata*):

Sextius says that the Greek name for this tree is *milax*, and that in Arcadia its poison is so active that people who to sleep or picnic beneath a yew-tree die. Some people also say that this is why poisons were called 'taxic,' which we now pronounce 'toxic', meaning used for poisoning arrows. (Book XVI, Chapter 20)

Pliny extended this concept[21], in the case of the walnut tree, to include the noxious effects on other plants:

We turn now to certain special properties of the shade of trees. That of the walnut is heavy, and even causes headache in man and injury to anything planted in its vicinity; and that of the pine-tree[22] also kills grass."

Figure 2.6. A representation of Caius Plinius Secundus from Les Vrais Portraits et Vies des Hommes Illustrés *by André Thévet (1584). All images of Pliny are regarded as fictitious, as no images of him from his era are known.*

---

[21] The harmful effects of the shade of certain, but unnamed, trees were also stated earlier by Lucretius in his *De Rerum Natura*, Book VI.
[22] André (1964) in his French edition of Pliny notes that this advice is contrary to that of Theophrastus (*De Causis Plantarum*, Book III, Chapter 10). André suggests that Pliny was referring to the litter layer within groves of pine.

> Even this department of knowledge is not to be despised, nor put in the last class, inasmuch as to each kind of plant shade is either a nurse or else a step-mother - at all events for the shadow of a walnut tree or a stone pine or a spruce or a silver fir to touch any plant whatever is undoubtedly poison. (Book XVII, Chapter 18)

As with the cabbage and the vine, Pliny addressed issues of antipathy in trees:

> The oak and the olive are parted by such inveterate hatred that, if one be planted in the hole from which the other has been dug out[23], they die, the oak indeed also dying if planted near the walnut. Deadly too is the hatred between the cabbage and the vine; the very vegetable that keeps the vine at a distance itself withers away when planted opposite cyclamen or wild marjoram. (Book XXIV, Chapter 1)

Pliny continued, and remarked upon the deleterious effects when trees produce leaf runoff, but it is not clear whether the effects were construed as physical or chemical[24]:

> The question of raindrops falling from trees can be settled briefly. With all the trees which are so shielded by the spread of their foliage that the rainwater does not flow down over the tree itself the drip does cruel injury. (Book XVII, Chapter 19)

This issue of so-called "soil sickness" or "replant problem" was also addressed by Pliny, who wrote regarding trees, notably fruit trees:

> Nature has also taught the art of making nurseries, as from the roots of many trees there shoots up a teeming cluster of progeny, and the mother tree bears offspring destined to be killed by herself, inasmuch as her shadow stifles the disorderly throng – as in the case of laurels, pomegranates, planes, cherries and plums; although with a few trees in this class, for instance elms and palms, the branches spare the young suckers. But young shoots of this nature are only produced by trees whose roots are led by their love of sun and rain to move about on the surface of the ground. All of these it is customary not to put in their ground at once, but first to give them to a foster-mother and let them grow up in seed-plots, and then change their habitation again, this removal having a marvelous civilizing effect even on wild trees, whether it be the case that, like human beings, trees also have a nature that is greedy for novelty and travel, or whether on going away they leave their venom behind when the plant is torn up from the root, and like animals are tamed by handling. (Book XVII, Chapter 12)

There was a subtle reference to companion planting, involving the onion and the herb savory (*Satureja hortensis*)[25]:

> In addition, they recommend digging over the ground three times and weeding out the plant-roots before sowing onions; and using ten pounds of seed to the acre, with savory mixed in, as the onions come up better. (Book XIX, Chapter 32)

---

[23] Elsewhere (Book XVII, Chapter 30), Pliny said that planting olive trees in holes resulting from the removal of an oak in not advisable, as oak roots are a source of "worms" which will then attack the olive roots.

[24] There is a related statement in Book XVII, Chapter 18: "Very heavy raindrops fall from the pine, oak and holm-oak, but none at all from the cypress, which throws a very compact shadow around it."

[25] This was repeated subsequently by the 16th century Flemish botanist, Rembert Dodoens, in the later posthumous editions of his *Cruydt-Boeck*. Funke (1943) was intrigued and tested this theory, but his experiments showed a negative effect between savory and onion.

Pliny is also credited with recording the earliest instances of using natural herbicides or applied allelopathy:

> Bracken dies in two years if you do not let it make leaf, the best way to kill it is to knock off the stalk with a stick when it is budding, as the juice trickling down out of the fern itself kills the roots.
> Democritus has put forward a method of clearing away forest by soaking lupin-flower for one day in hemlock juice and sprinkling it on the root of the trees.[26]
>
> (Book XVIII, Chapter 8)

While Pliny clearly embraced the concepts of antipathy and sympathy, as developed by the Greeks, he did so fairly unquestioningly, with little concern for any mechanistic explanations (Gaillard-Seux 2003). Furthermore, Pliny had little regard for magical or occult explanations. For Pliny, and likely most people of the era, the compatibility and incompatibility of things were simply part of the natural order of things. At some point well into the writing of his *Natural History*, it appears that the concepts of antipathy and sympathy crystallised for him, as in Book 20 (of a total of 37)[27] he provided a remarkable, veritable philosophical statement, that really became the touchstone for such matters over the next 1600 years:

> Herein will be told of Nature at peace or at war with herself, along with the hatreds and friendships of things deaf and dumb, and even without feeling. Moreover, to increase our wonder, all of them are for the sake of mankind. The Greeks have applied the terms "sympathy" and "antipathy" to this basic principle of all things: water putting out fire; the sun absorbing water while the moon gives it birth; each of these heavenly bodies suffering eclipse through the injustice of the other. Furthermore, to leave the more heavenly regions, the magnetic stone draws iron to itself while another kind of stone repels it; the diamond, the rare delight of Wealth, unbreakable and invincible, by all other force, is broken by goat's blood. Other marvels, equally or even more wonderful, we shall speak of in their proper place. I only ask pardon for beginning with trivial though healthful objects. First I shall deal with kitchen-garden plants.[28] (Book XX, Chapter 1)

As mentioned previously, with regard to the Athenian Solon, a surprising commentator on matters relating to allelopathy was the Greco-Roman historian Plutarch. Another reference appeared in his rather chaotic *Moralia* wherein he reiterated Pliny's reference on the effects of the shade of the walnut tree [29], but perhaps more interestingly, he related the origin of the Greek word for walnut tree χαζύα to its

---

[26] The Democritus mentioned was Bolos Demokritos of Mendes, discussed previously. The pesticidal effects of crushed lupins in ridding vines of ants was mentioned by Columella (*De Arboribus*, Book XII, Chapter 14)

[27] As noted by Gaillard-Seux (2003), the terms "sympathy" and "antipathy" make their first appearance in Book XX, although clearly Pliny embraced the concepts under different terminology in the earlier chapters. Thus the botanical entries discussed above did not actually use the terms "sympathy" and "antipathy".

[28] Thereafter follows a lengthy section on the uses of various plants as curatives.

[29] The Loeb edition of Plutarch's *Moralia* (volume VIII) gives this as hazel, which seems clearly to be wrong (see note 20), but amplifies the point as to how the hazel later came to be identified as antipathetic to other plants (e.g. see Chapter 5).

harmful effects, a circumstance that has occurred in several languages (see Chapters 5 and 6). He gave insight into Roman understanding of antipathy and sympathy, and provided an early account of what are phytoncidal effects (in the broad sense), today marketed as aromatherapy:

> And Bacchus was counted a physician not only for finding wine, the most pleasing and most potent remedy, but for bringing ivy, the greatest opposite imaginable to wine, into reputation, and for teaching his drunken followers to wear garlands of it, that by that means they might be secured against the violence of a debauch, the heat of the liquor being remitted by the coldness of the ivy. Besides, the names of several plants sufficiently evidence the ancients' curiosity in this matter; for they named the walnut-tree χαρυα, because it sends forth a heavy and drowsy (χαρωιχου)spirit, which affects their heads who sleep beneath it; and the daffodil, νάρχισος, because it benumbs the nerves and causes a stupid narcotic heaviness in the limbs, and therefore Sophocles calls it the ancient garland flower of the great (that is, the earthy) Gods. And some say rue was called πήγυου from its astringent quality; for, by its dryness proceeding from its heat, it fixes (πήγνυσι) or coagulates the seed, and is very hurtful to great-bellied women. But those that imagine the herb fumes free passage to exhale, and those that are moderately cold repel and keep down the ascending vapors. Of this last nature are the violet and rose; for the odors of both these are prevalent against any ache and heaviness in the head. The flowers of the privet and crocus bring those that have drunk freely into a gentle sleep; for they send amethyst (ἀμέθυστος), and the precious stone of the same name, are called so because powerful against the force of wine are much mistaken; for both receive there names from their color; for its leaf is not of the color of strong wine, but resembles that of weak diluted liquor. And indeed I could mention a great many which have their names from their proper virtues. But the care and experience of the ancients sufficiently appears in those of which they made their garlands when they designed to be merry and frolic over a glass of wine; for wine, especially when it seizes on the head, and strains the body just at the very spring and origin of the sense, disturbs the whole man. Now the effluvia of flowers are an admirable preservative against this, they secure the brain, as it were a citadel, against the effects of drunkenness; for those that are hot open the pores and give the fumes free passage to exhale, and those that are moderately cold repel and keep down the ascending vapors. Of this last nature are the violet and rose; for the odors of both these are prevalent against any ache and heaviness in the head. The flowers of privet and crocus bring those that have drunk freely into a gentle sleep; for they send forth a smooth and gentle effluvia, which softly takes off all asperities that arise in the body of the drunken; and so all things being quiet and composed, the violence of the noxious humor is abated and thrown off. The smells of some flowers being received into the brain cleanse the organs and instruments of sense, and gently by their heat, without any violence or force, dissolve the humors, and warm and cherish the brain itself, which is naturally cold. (*Moralia*: Symposiacs[30], Book III, Question 1)

The damaging effects of growing related species successively in crop rotation, a notion that was fundamental to de Candolle's theory of crop rotation in the nineteenth century (see Chapter 7), was mentioned by the lexicographer Sextus Pompeius Festus, who was active likely in the second or third century A.D. Festus

---

[30] This is sometimes given as *Questiones Conviviales* or "Table-talk".

compiled an abridged version of the now lost work of Marcus Verrius Flaccus, *De Significatu Verborum*[31], and Festus recorded the following:

> A field must be allowed to rest, when it is grown for two years in succession with grain wheat, that is awned, lest it happen, as occurring when farms are leased.

Another overlooked, but important, agricultural work is *De Re Rustica* by Palladius (Rutilius Taurus Æmilianus Palladius), believed to have been written between 371 and 395 A.D. Little is known about Palladius, but his work drew heavily on Columella, and offered a very practical almanac. It has been largely neglected by historians, as there is no English translation readily available[32], but it is accessible in French translation. There are a few excerpts concerning allelopathy, and the first offered perhaps the earliest description of the allelopathic effects of a leaf leachate:

> This tree [walnut] prefers deep holes, because of its height, and should be planted at great distances, because the water which drips from its leaves, is harmful to those which neighbour it, even to those of its own species. (Book II, Chapter 15)

Also:

> You can also plant it [olive] in a soil which would have carried arbutus bushes or ilex, but the cherry or the oak when cut leave in the soil noxious roots of which the juice kills the olive. (Book III, Chapter 18)

Palladius, amongst others, described, likely in error, that there was antipathy between horseradish[33] and the vine:

> Horseradishes, as well as cabbages, do not like vines; sown around a vine they avoid it, according to antipathy. (Book IX, Chapter 14)

The works cited above represent more or less the entire body of agricultural and botanical writings surviving from the Greek and Roman eras. Many important works have become lost, and/or were destroyed accidentally or intentionally. An example is the often cited encyclopaedic work on agriculture by Mago the Carthaginian[34]. It is staggering to realise that an author such as Columella cited 55 other writers in his

---

[31] While the original work of Flaccus (fl. *c*. 10 B.C.) is completely lost, even that of Festus survives only through a single fragment, and through a further abridged version by Paulus Diaconus (8th century A.D.).

[32] There is a very rare translation into English by Thomas Owen (1807), who was also the translator of the only English edtion of the *Geoponika*. There is also one manuscript translation which dates from about 1420 or earlier, which was published in 1873 (Lodge 1873), and has recently been reprinted, but as this obscure translation is in Early English and has been rendered into verse, it is more of curiosity than of utility. For example, in Book II, verse 52 is found the following (equivalent to the quote above from Book II, Chapter 15):

> In delves deepe is sette thair appetite
> Thaire magnitude a larger lande requireth.
> Eke to noo tree thaire dropping is delite,
> Her brere thorne and her own kynde it ireth.

[33] The original Latin was *raphanus*, commonly translated as radish, but the translation into French gave *raifort*, or horseradish (see note 5). This error appears also in Gerard (see Chapter 5).

[34] Mago was the brother of the famous Carthaginian commander, Hannibal Barca.

work on agriculture, and Pliny in his encyclopedic *Natural History* named nearly 4000 authors. Pliny himself was the author of about 100 volumes, of which only the 37 books of *Natural History* have survived.

BYZANTINE GREECE AND ROME

The capital of the Roman Empire had become Constantinople (formerly Byzantium) in 330 A.D, and the focus of power shifted eastward. For the period from about the fifth century through to the twelfth century, the region which equated to the East Roman Empire is known commonly as the Byzantine Empire. Due to Byzantine influence, a number of agricultural treatises appeared with the title *Geoponika*. These were essentially compendia of earlier agricultural works by Greek and Latin authors, such as Theophrastos, Varro, and Columella. It is believed that the first of the Geoponikas appeared in the fourth century due to Vindonius Anatolius, and there were successive revisions and translations by Didymos, Sergios, and Cassianos Banos. The best version was produced by order of Emperor Constantine VII (also known as Porphyrogenitus) in Constantinople in about 950 A.D., and numerous editions of this subsequently appeared in Latin with the advent of printing[35] (Figure 2.7). It is little known that an English translation by Thomas Owen appeared under the title *Agricultural Pursuits*[36] in 1805-1806. The *Geoponika* is idiosyncratic in that while it obviously drew heavily from well-known agricultural writers such as Varro and Columella, it was embellished with credits to an array of historical figures, many of whom are not known to have authored any agricultural works. Nonetheless, the *Geoponika* does shed some new light on classical knowledge about agriculture.

The status of the supposed allelopathic antipathy between the vine and cabbage becomes very curious, when one reads the *Geoponika*. According to the *Geoponika*, Nestor[37], in a horticultural treatise, explained that the antipathy of the cabbage and the vine has a mythological basis:

> The cabbage is an emblem of the tear of Lycurgus; for, says he [Nestor], Bacchus being afraid of him, went under the sea, and Lycurgus being bound with the vine, shed a tear, and he says that from the tear sprang the cabbage, and that on this account the cabbage and the vine have an antipathy to each other. (Book XII, Section 17)

This view was endorsed by de Gubernaitis (1882), who also added that being tied to a vine was Lycurgus' punishment for having destroyed vines belonging to Dionysus (Bacchus). In another passage, sympathy between plants was described:

---

[35] Many of the early editions are abridged. The best available edition is judged to be the critical edition by Beckh (1895), which has been reprinted. Editions exist also in French and in Russian (Lipshits 1960).

[36] This edition is exceedingly rare; however, an electronic version is available courtesy of Missouri State University.

[37] According to Lipshits (1960), this was Nestor Larandeus (flourished 222-235 A.D.)

Democritus[38] also says that the pomegranate and myrtle betray an affection for each other, and that, when planted near each other, they will bear plentifully, and that their roots become mutually implicated, although they may not be very near. (Book X, Section 29)

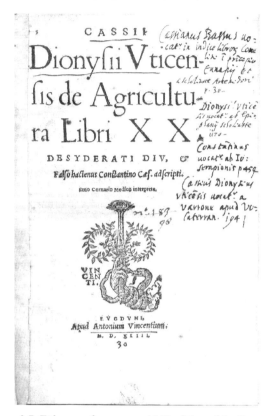

Figure 2.7. *Title page from a rare 1543 edition of the* Geoponika.

The *Geoponika*, being a product of Byzantine Constantinople, is of considerable interest as it is a work at the crossroads, both in terms of geography and of time. While the *Geoponika* draws heavily on the classical sources of Greece and Rome, it also appears to incorporate some practices, such as the soaking of seeds in various extracts, that are more in common with Asian lands, such as revealed in the *Vrikshayurveda* by Surapala (see Chapter 4) which appeared in India during roughly the same era. In time, the *Geoponika* bridges the period of classical Greece and Rome and the period dominated by subsequent Islamic authors such as Ibn Wahshiya,

---

[38] See note 10.

Ibn al-Bassal, and Ibn al-Awwam (see Chapter 3). Indeed recent research has suggested that the *Geoponika* and the seminal book of Islamic agriculture, *The Book of Nabathean Agriculture* by Ibn Wahshiyya, appear to share many similar sources (Carrara 2006).

REFERENCES

Aliotta, G. and A. Mallik. 2004. Agroecology: the allelopathic approach in ancient and modern times. In: R. Zeng and S. Luo (eds), *Proceeding of International Symposium on Allelopathy Research and Application, 27-29 April 2004, Shanshui, Guangdong, China*, pp. 1-21. South China Agricultural University, Guangzhou.

al-Awwam, Ibn. 1864-1867. *Le Livre de l'Agriculture d'Ibn al-Awwam. (Kitab al-Felahah)*. Volumes 1-2. Translated by J. J. Clément-Mullet. A. Franck, Paris.

Aristotle. 1936. On plants. In: *Minor Works: On Colours, On Things Heard, Physionomics, On Plants, On Marvellous Things Heard, Mechanical Problems, On Divisible Lines, Situations and Names of Winds, On Melissus, Xenophanes, and Gorgias*, pp. 142-233. Translated by W.S. Hett. William Heinemann Ltd., London.

Aristotle. 1937. On the parts of animals. In: *Parts of Animals, Movements of Animals, Progression of Animals*, pp. Translated by A.L. Peck and E.S. Forster. William Heinemann Ltd., London.

Athenaeus. 1927-1941. *Deipnosophistae*. Volumes 1-7. Translated by C.B. Gulick. William Heinemann, London.

Beckh, H. 1895. *Geoponica, sive Cassiani Bassi Scholastici De Re Rustica Eclogae*. B.G. Teubner, Leipzig.

Billiard, R. 1928. *L'Agriculture dans l'Antiquité d'après les Georgiques de Virgile*. E. de Boccard, Paris.

Bramble, J.C. 1982. Minor figures. In: E.J. Kenney and W.V. Clausen (eds), *The Cambridge History of Classical Literature. II. Latin Literature*, pp. 467-494. Cambridge University Press, Cambridge.

Bush, G. 1854. *Illustrations of the Holy Scriptures*. Lippincott, Grambo & Co., Philadelphia.

Carrara, A.A. 2006. *Geoponica* and *Nabatean Agriculture*: a new approach into their sources and authorship. *Arab Sciences and Philosophy* **16**: 103-132.

Cato. 1954. De Agri Cultura. In: *Marcus Porcius Cato on Agriculture, and Marcus Terrentius Varro on Agriculture*, pp. 1-157. Translated by W.D. Hooper. William Heinemann, London.

Chou, C.-H. 1999. Roles of allelopathy in plant biodiversity and sustainable agriculture. *Critical Reviews in Plant Sciences* **18**: 609-636.

Cicero. 1896. *De Natura Deorum (On the Nature of the Gods)*. Translated by F. Brooks. Methuen, London

Columella. 1941-1955. *De Re Rustica (On Agriculture)*. Volumes 1-3. Translated by H.B. Ash, E.S. Forster and E.H. Heffner. William Heinemann, London.

Columella. 1955. De Arboribus. In: *Lucius Junius Moderatus Columella On Agriculture X-XII, and On Trees*, pp. 341-411. Translated by E.S. Forster and E.H. Heffner. William Heinemann, London

Dioscorides. 1934. *The Greek Herbal of Dioscorides;Illustrated by a Byzantine, A. D. 512; Englished by John Goodyer, A. D. 1655, Edited and first printed A.D. 1933 by R.T. Gunther*. Oxford University Press, Oxford.

Festus. 1913. *De Verborum Significatu quae Supersunt cum Pauli Epitome*. Edited by W.M. Lindsay. B.G. Teubner, Leipzig.

Funke, G.L. 1943. The influence of *Satureja hortensis* L. on *Allium cepa* L. *Blumea* **5**: 294-296.

Gaillard-Seux, P. 2003. Sympathie et antipathie dans l'Histoire Naturelle de Pline l'Ancien. In: N. Palmieri (ed.), *Rationnel et Irrationnel dans la Médecine Ancienne et Médiévale: Aspects Historiques, Scientifiques et Culturels*, pp. 113-128. Publications de l'Université de Saint-Etienne, Saint-Etienne.

Gubernaitis, A. de. 1882. *La Mythologie des Plantes, ou Légends du Règne Végétale. Tome second, botanique spéciale*. Paris.

Hughes, J.D. 1975. *Ecology of Ancient Civilizations*. University of New Mexico Press, Albuquerque.

Huntington, E. and S.W. Cushing. 1924. *Principles of Human Geography*. John Wiley, New York.

Kohli, R.K. 1998. Allelopathic interactions in forestry systems. In: K. Sassa (ed.), *Environmental Forest Science*, pp. 269-284. Kluwer Academic Publishers, Dordrecht.

Lipshits, E. 1960. *Geoponiki, vizantiiskaya sel'skokhozyaistvennaya entsiklopediya X veka*. Akademiya Nauk SSSR, Moscow.
Lodge, B. 1873. *Palladius on Husbondrie, edited from the unique ms. of about 1420 A.D. in Colchester Castle*. N. Trübner & Co., London.
Lucretius. 1937. *De Rerum Natura*. Translated by R.C. Trevelyan. Cambridge University Press, Cambridge.
Olybrius. 1934. The cabbage and the vine. *Notes and Queries* **167**: 227.
Ovid. 1929. Nux. In: *The Art of Love, and Other Poems*, Translated by J.H. Mozley, pp. 293-307. William Heinemann, London
Owen, T. 1805-1806. *Geoponika. Agricultural Pursuits*. W. Spilsbury, London
Owen, T. 1807. *The Fourteen Books of Palladius, Rutilius Taurus Milianus, On Agriculture*. J. White, London.
Palladius. 1843. *L'Économie Rurale de Palladius Rutilius Taurus Emillianus*. Translated by Cabaret-Dupaty. C.L.F. Panckoucke, Paris.
Pease, A.S. 1927. The loves of the plants. *Classical Philology* **22**: 94-98.
Petriccione, M. and G. Aliotta. 2006. Ethnobotany and allelopathy of the Persian walnut (*Juglans regia* L.). *Acta Horticulturae* **705**: 297-300.
Pliny. 1950-62. *Natural History*. Volumes 1-10. Translated by H. Rackham, W.H.S. Jones and D.E. Eichholz. William Heinemann, London.
Pliny. 1964. *Histoire Naturelle*. Livre XVII. Translated by J. André. Société d'Éditions "Les Belles Lettres", Paris.
Plutarch. 1878. *Plutarch's Morals*. Volume 3. Little, Brown, and Company, Boston.
Plutarch. 1910. Solon. In: *Plutarch's Lives: the Dryden Plutarch revised by Arthur Hugh Clough*. Volume I. J.M. Dent, London.
Rice, E.L. 1983. *Pest Control with Nature's Chemicals: Allelochemics and Pheromones in Gardening and Agriculture*. University of Oklahoma Press, Norman.
Robin, L. 1928. *Greek Thought and the Origin of the Scientific Spirit*. Routledge and Kegan Paul, London.
Semple, E.C. 1931. *The Geography of the Mediterranean Region*. Henry Holt, New York.
Simkhovitch, V.G. 1916. Rome's fall reconsidered. *Political Science Quarterly* **31**: 201-243.
Spencer, E. 1836. *Sketches of Germany and the Germans, with a Glance at Poland, Hungary, & Switzerland, in 1834, 1835, and 1836*. Second edition.. Whittaker, London.
Terian, A. 1981. *Philonis Alexandrini de Animalibus: The Armenian Text with an Introduction, Translation, and Commentary*. Scholars Press, Chico.
Theophrastus. 1916-1926. *Enquiry into Plants, and Minor Works on Odours and Weather Signs*. Volumes 1-2. Translated by A. Hort. William Heinemann, London.
Theophrastus. 1976-1990. *De Causis Plantarum*. Volumes 1-3. Translated by B. Einarson and G.K.K. Link. William Heinemann, London.
Varro. 1954. Rerum Rusticarum. In: *Marcus Porcius Cato on Agriculture, and Marcus Terrentius Varro on Agriculture*, pp. 159-543. Translated by W.D. Hooper. William Heinemann, London.
Virgil. 1940. *The Georgics*. Translated by C. Day Lewis. Oxford University Press, Oxford.
Virgil. 1963. *The Eclogues*. Translated by C. Day Lewis. Oxford University Press, Oxford.
Wellmann, M. 1921. Die Georgika des Demokritos. *Abhandlungen der preussischen Akademie der Wissenschaften, philosophisch-historisch Klasse* **4**: 1-58
Willis, R.J. 1985. The historical bases of the concept of allelopathy. *Journal of the History of Biology* **18**: 71-102.
Wright, M.R. 1981. *Empedocles: the Extant Fragments*. Yale University Press, New Haven.

CHAPTER 3

ARABIC WORKS

> Possibly because some of their ancestors had come from the desert, and in most places where they lived the desert was still near at hand, sometimes within eyesight, ever ready to encroach on the land that had been claimed from it, the inhabitants of the early Islam world were, to a degree, that is difficult for us to comprehend, enchanted by greenery.
>
> *Agricultural Innovations in the Early Islamic World.*
> A. Watson (1983)

Following the demise of the Roman Empire, most of Europe slid into decline, into the Mediaeval period, and what is sometimes referred to as the Dark Ages. In marked contrast to this, the Arab-dominated world, which at its height in about 900 A.D. included northern Africa, Asia Minor, the Middle East, and Iberia, was ascendant. There were great advancements in mathematics, the physical sciences, astronomy, geography, and medicine, although there was an enormous debt to Greek and Roman scholarship, which the Islamic world had conserved through copies and translations. Indeed during the period spanning from the 9th through to the 12th century, more books were written in Arabic than in any other language. Despite significant advances made in the sciences noted above, there was only modest progress in botany and agriculture, which were still heavily reliant on Greek and Roman works.

From our point of view here, interest ultimately centers around one important work, which was Andalusian in origin. Andalusia (Islamic Spain), perhaps because of its relative remoteness, was less constrained by religious orthodoxy, and was thus a centre for science and practical arts. The book under discussion is the twelfth century treatise on agriculture, *Kitāb al-Filāha* (The Book of Agriculture)[1] by Ibn al-Awwam[2]. The *Kitab al-Filaha* is an amalgam of material from numerous sources, some of which are now known only through this work. One of the principal sources is The *Book of Nabathean Agriculture*, the name given to an enigmatic work by Ibn

---

[1] This title is common to the works of several authors.

[2] Like many Islamic authors, Ibn al-Awwam is known by different names. Firstly the spelling is sometimes given as al-Awam. According to manuscripts his full name was Abu Zakariya Yahya b. Muhammad b. al-Awwam, and the first part of this name is sometimes also used.

Wahshiya[3], which was known to European scholars over seven hundred years ago (known to 13th century St. Thomas Aquinas), but which remained essentially suppressed until the nineteenth century because of its supposed occult content. Until recently, it was known only from a number of Arabic manuscript copies, of which a facsimile of the Topkapi Sarayi Library, Instanbul copy has been reproduced in book form (Sezgin 1984). In 1995-1998, a critical edition edited by Toufic Fahd was finally published by the Institut Francais de Damas; however, the text still remains available only in Arabic.

Interest in the Book of Nabathean Agriculture was revived during the nineteenth century as manuscript copies of *Kitab al-Filaha* or The Book of Agriculture by Ibn al-Awwam (Figure 3.1) were rediscovered and were translated, firstly into Spanish in 1802, and then into French in 1864-1867. The *Kitab al-Filaha* was a work of unquestionable agricultural merit, and it cited the Book of Nabathean Agriculture frequently as a source of information.

Figure 3.1. Leaf (folio 138a) from a manuscript copy of Kitab al-Filaha *by Ibn al-Awwam, (Leiden Codex Org. 346) held by Leiden University (reproduced with permission).*

---

[3] The spelling varies: Wahshiya, Wahshiyah, Wahšiyya, etc.

Despite the fact that no translation of *The Book of Nabathean Agriculture* has appeared in any European language, it has generated considerable controversy over the past 150 years. The debate about it ranges from whether this book was in fact the legacy of one of the world's oldest works on agriculture, possibly dating back to several centuries B.C., or, at the other extreme, whether the work was a medieval fraud or forgery. Very recently the question of Ibn Wahshiya's integrity has come to the fore again, with the rediscovery[4] that Ibn Wahshiya was indeed skilled in ancient languages; for example, he was able to read some of the Egyptian hieroglyphs (El Daly 2005). Furthermore, Ibn Wahshiya wrote a detailed treatise on poisons, a work that demonstrated his skill in languages and his familiarity with ancient works from the Arabic, Greek and ancient Indian worlds (Levey 1966). Many scholars, notably the nineteenth century Russian Chwolson who translated the *Book of Nabathean Agriculture* into German (however the translation was not published), have studied the content and origins of the work. However, Chwolson erred in believing that the book was of Babylonian origin, for which he was harshly criticised by several writers including Renan (1862), in particular the German scholar Alfred von Gutschmid, and others (see Hämeen-Anttila 2004). The damage to Chwolson's credibility coupled with claims, albeit unsubstantiated, that the book was a forgery made the *Book of Nabathean Agriculture* a literary pariah, a situation which lasted for well over a century.

Recent research (Fahd 1996, El-Faiz 1995, Hämeen-Anttila 2006) has reinstated the authenticity of the *Book of Nabathean Agriculture*, and established that the version known today was written in the year 904 A.D., and that it was indeed a translation into Arabic from Syriac, a Babylonian language, by Ibn Wahshiya, who lived in rural Mesopotamia (present-day Iraq). The Syriac version was authored possibly in the third century[5] A.D. by a Babylonian, Qutama, of whom no details are known apart from textual inferences, and the original title was the "Book of Cultivation of the Soil, the Improvement of Seeds, Trees and Fruits, and their Protection against Disease." However, there is good reason to believe that Qutama compiled his work from two earlier sources.

The *Book of Nabathean Agriculture* has been largely ignored in scientific circles until recently, assumedly because of the damaging claims surrounding its authenticity, suppression due to its astrological and occult content, as well as the issue of language accessibility. However, despite this, there is much interesting and useful information. In Islamic culture there was great interest about understanding the natural world and in cataloguing the nature of things. As noted by Nasr (1968), this activity was not motivated merely by curiosity, but was part of the quest for an understanding of the deity through "signs of God" or *Vestigia Dei*. There are many references to various aspects of antipathy and sympathy in the *Book of Nabathean Agriculture*, although they are generally couched within the framework of opposition or similarity

---

[4] Ibn Wahshiya's manuscript on hieroglyphs was published by Hammer-Purgstall in 1806, but was regarded at the time as dubious.

[5] Hämeen-Anttila (2004) believes the Syriac version was compiled likely in the 6th century by Qutama, which is possibly a pseudonym, or by a small group of translators.

of the four basic elemental properties: hot, cold, wet and dry. It is difficult at first glance to comprehend these notions of antipathy and sympathy, as knowledge of the intrinsic properties of things, as understood at that time, is alien to us, and often the words in translation do not suffice. For example, Ibn Wahshiya discussed at some length the properties of the walnut, both in regard to the tree and the fruit. The stated property that characterised all aspects of the walnut was "heat", but this embodied a host of characteristics that had nothing to do with physical temperature, in somewhat the same way that pepper is still described as "hot". The following passage perhaps illustrates this:

> The walnut fruit has benefits as well as harms. The harms are greater. Therefore, on this basis, the walnut can be judged to be harmful and not beneficial. For something is judged by its majority characteristics. It is harmful because the heat of its fruit is intense and the heat causes pimples and black marks. Thus it has few nutrients for the body which takes it. If a person eats a large amount of walnut, due to its intense heat, his mouth gets scalded and makes him sleepless. This is because its intense heat upsets his nature and does not allow his nature to resettle. Because of these [harms] its nutritional value to the body is little and of poor quality.

It is tempting here to suggest that the many of the references to "heat" relate to the bitterness found in walnut plant parts, notably due to the juglone content.

Examples of antipathies among plants that are cited include those of the vine and the cabbage, and the *mutar syit* and tamarix. Sympathies cited include those of terebinth and myrtle, and the water-melon and several trees. Also according to the *Book of Nabathean Agriculture*, there was animosity between the trees of At-Tarfa (in present-day Oman) and Mwtrsyyt[6] (p. 1250, lines 15-18).

There is a number of unpublished Arabic agricultural and botanical manuscripts that are known in European or Middle Eastern libraries. One of interest to the discussion here was noted by Fahd (1996) and is held by Cambridge University: it is the *Julasat al-ijtisas fi ma'rifat al-qiwa wa-l-jawass* by an Andalusian, Ibn ar-Raqqam (1226-1315) of Granada[7]. The author states that he attempted to produce an abridgement of the *Book of Nabathean Agriculture*, without the heretical and heathen elements contained in that work. In this work, Chapter 14 concerns the sympathies and antipathies of trees (and other plants):

> Among those trees that enjoy harmony one can list the grapevine and the lotus tree[8], which always prosper well when together and are much more fruitful when planted in the same area. A great combination is also the grapevine and the olive tree, provided the latter is planted in the parameters of the field where the vines are. Equally the pumpkin[9] makes for a perfect combination with grapevines: they help each other grow and thrive. Grapevines, sugarberry[10] and the jujube trees[11] are in good harmony with the orange tree

---

[6] Place name is unknown.

[7] Ibn ar-Raqqam was a later figure, active in the early fourteenth century, and is known also as the author of a work on astronomy. A similar manuscript (Mingana MS. no. 933), perhaps an eighteenth century copy, is in the Mingana Collection, University of Birmingham

[8] Several plants have been interpreted as the "lotus tree" of antiquity (Smith 1882).

[9] This could mean any of the cucurbits: zucchini, calabash, gourd, etc. – not the American pumpkin.

[10] Also given as hackberry (*Celtis* sp.).

[11] = *Ziziphus jujuba*.

and also help one another thrive, just like the apple, the pear and the orange trees do. So do the pomegranate and the myrtle on the one hand and the walnut, the fig and the mulberry trees are suitable for one another on the other. Mint, narcissus, the lily of the valley and the mint[12] are another great combination, and one might also add the combination of the melon and the eggplant, that of the olive and pomegranate trees, and that of the sea onions, mulberry and pomegranate trees.

As to those which are in disharmony with each other, one could list the eggplant and the cucumber, the black and the white grapes, as well as the bay leave tree and the grapevine. In that category one need mention the disharmony of the walnut tree with the great majority of the other trees. Also in disharmony is the cabbage and the grapevine; the fenugreek is a known enemy of both the cabbage and the spinach. Grapevines and sumac trees[13] are unsuitable for one another and the latter (the sumac) is a known enemy of the apple tree, just like lupine is unsuited for the grapevine and fig tree: it dries the latter and really causes harm to many a tree as do the lentil and the broad bean. Equally harmful are the rue (herb of grace) and oreganum, particularly to the orange trees. What can harm the orange tree is all that has a sharp odour such as wild thyme[14]. Grapevines can really be harmed by the fig tree and the palm tree and the juniper alike. Cauliflower and wild cabbage are like a poison to the vine and can kill it. It is said that the fig tree is only a danger to the grapevine in the hot climates but not in the cool ones. Travellers across different lands are now lending weight to this argument. It is also now widely reported that turnips, radish and watercress do harm the grapevine. One need also mention that cedar[15] trees ought not be planted close to apple, pomegranate, pear, peach or palm trees as they cause the taste of their fruits to be either sour or gripping the tongue a bit in other cases. Parsnips ought not to be planted in a field in which flax was previously grown, while cane is totally against the violet and kills it. (folios 27-28)

Most of the extant mediaeval Arabic agricultural treatises originate from Andalusia, the name given to the southern part of Spain that had become dominated by the Moors commencing in the eighth century. During the medieval period, the cultural center of western Europe was indeed Andalusia. It also should not be forgotten that Spain has had a long and significant agricultural history, as it possessed favourable climate and soils; what is arguably the best and most practical of the Latin agricultural treatises, *De Rerum Rusticarum* by Columella (see Chapter 2), originated from Cadiz, during Roman domination of the region. During the eleventh to fourteenth centuries, there originated a corpus of agronomic works in Arabic from cities such as Seville, Toledo, Cordoba and Granada, and many of the manuscripts survive, and have been the focus of research in recent decades (Bolens 1981; see Table 3.1).

The best known and most important of the Andalusian agronomic treatises is that by Ibn al-Awwam (sometimes known as Abu Zakariya), *Kitab al-Filaha*, written during the second half of the twelfth century in Seville. This great work was the most comprehensive agricultural treatise to its date, and while it incorporated information from the well-known Greco-Roman works, it also drew heavily from Ibn Wahshiya, and other Arabic works such as that by Ibn Bassal. It remains the most accessible of the Arab agricultural works, as it was translated and published in Spanish

---

[12] = *Mentha sativa* L.

[13] = *Rhus* sp.

[14] = *Thymus serpyllum*.

[15] The term "cedar" in the Mediterranean region likely refers to either *Juniperus* spp. or *Cedrus* spp.

Table 3.1. Arabic agricultural works of the mediaeval era.

Author	Place	Date	Arabic Title transliterated	Title in translation	Editor or Translator	Notes
Ibn Qutayba	Iraq	c. 880	Uyun al-Akhbar	Book of Useful Knowledge	Kopf and Bodenheimer (1949)	Natural history section
Ibn Wahshiya	Iraq	c. 900	Al-Filaha al-Nabatiyya		Fahd (1995-1998)	In Arabic; Vol. 3 of Fahd edition contains commentary, mainly in French
Abu Hanifa al-Dinawari	Andalusia	c. 890	Kitab al-Nabat	Book of Plants	Bauer (1988)	
Anonymous (Ibn Abi l-Yawad?)	Andalusia	c. 1000	Kitab fi tartib awqat al-girasa wa-l-magrusat		Lopez y Lopez (1990)	
Ibn Wafid	Toledo	c. 1050	Maymu al-Filaha	Tratado de Agricultura	Cuadrado Romero (1997)	Original Arabic manuscript is lost; translation is based on an extant Catalan fragment.
Ibn al-Hayyay	Seville	c. 1056	Al-Muqni' fi l-filaha		Yirar and Safiyya (1982)	In Arabic
Abu l-Jayr al-Ichbili	Seville	c. 1100	Kitab al-Filaha	Tratado de Agricultura	Carabaza (1991)	
Ibn Bassal	Seville	c. 1150?	Kitab al-Qasd wa-l-bayan	Libro de Agricultura	Millas Vallicrosa and Aziman (1955)	Extant copy in Catalan, itself based on an abridged Arabic version of a large work, of which both are lost. Edited facsimile edition published in 1995.
Ibn al-Awwam	Granada	c. 1180	Kitab al-Filaha	Libro de Agricultura / Le Livre de l'Agriculture	Banqueri (1802) / Clement-Mullet (1864-1867)	Spanish facsimile edition 1988. French facsimile edition 1985 (Tunis), modern edition 2000 (Paris). Fahd (1996) cites Turkish and Urdu translations. University of Cambridge manuscript cited by Fahd (1996)
Ibn ar-Raqqam	Granada	c. 1310				
Ibn Luyun	Granada	c. 1348	Kitab al-Filaha	Tratado de Agricultura	Ibanez (1988)	Original Arabic is in verse; translation is in prose

(Banqueri 1802) and French (Clément-Mullet 1864-7; Figure 3.2), of which reprints of the rather rare original editions are now available. Little is known of the life of Ibn al-Awwam, except that he was born *c.* 1150 in Seville.

The following excerpts from *Kitab al-Filaha* are sourced and translated from the French edition, *Le Livre de l'Agriculture d'Ibn al-Awam*, by Clément-Mullet. There are a few scattered passages concerning the sympathy and antipathy of plants in the main body of the text, e.g. concerning walnut (Figure 3.3):

> All plants planted in its [walnut] vicinity show antipathy to it, with the exception of the fig which is found to have several points of similarity with it. (Volume 1, p. 275)

Then, concerning the orange we find:

> One must prevent oneself from planting in the vicinity of cedar and orange any rue (*Ruta graveolens* Linn.), or plantain, or lemon-balm, or euphorbia, or any plant exhaling a penetrating odour; the trees suffer from these. (Volume 1, p. 300)

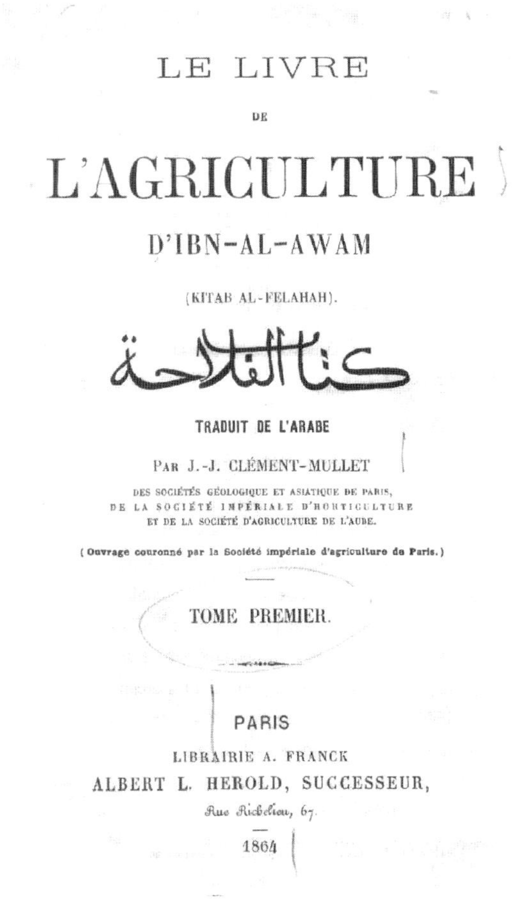

LE LIVRE

DE

L'AGRICULTURE

D'IBN-AL-AWAM

(KITAB AL-FELAHAH).

كتاب الفلاحة

TRADUIT DE L'ARABE

PAR J.-J. CLÉMENT-MULLET

DES SOCIÉTÉS GÉOLOGIQUE ET ASIATIQUE DE PARIS,
DE LA SOCIÉTÉ IMPÉRIALE D'HORTICULTURE
ET DE LA SOCIÉTÉ D'AGRICULTURE DE L'AUBE.

(Ouvrage couronné par la Société impériale d'agriculture de Paris.)

TOME PREMIER.

PARIS

LIBRAIRIE A. FRANCK
ALBERT L. HEROLD, SUCCESSEUR,
Rue Richelieu, 67.

1864

Figure 3.2. *Title page from Volume 1 of* Le Livre de l'Agriculture d'Ibn al-Awam*, the French translation by Clément-Mullet (1864-1867) of* Kitab al-Filaha.

Figure 3.3. *A leaf depicting the walnut tree from a 12th-13th century Andalusian version of* De Materia Medica *by Dioskorides. (Manuscrit Arabe 2850, folio 131v, courtesy of the Bibliothèque Nationale de France)*

However, most of the information concerning sympathy and antipathy is assembled in Volume I, Chapter XII, Article 2, which in addition to discussing tree interactions also provides details of occult sympathetic charms for benefitting trees, of which I have included a sample here.

> One reads in *Nabathean Agriculture*, that everything that has an analogy for form (among plants) helps one another, is protected (reciprocally), and that everything that is a different or contrary form is also antagonistic, in which it tends to be weakened and debilitated. One reads again that in *Nabathean Agriculture* that there is a sympathy between the vine and the jujube tree, especially in nature (habitats); such that every time the vine is found planted in the vicinity of the jujube, from one to the other, a sort of sympathy like that which a man feels for a beautiful woman; he is attached to her and he loves her with passion, and the breath of one gives strength to the other by virtue of its vicinity. Also *Nabathean Agriculture* says that, when one has planted an olive tree in the vicinity of the vine; that is advantageous for them both. Nevertheless the olive should be maintained at some distance from the vine, for it is useful for this; that was at least the opinion of most of the ancients. According to the same *Nabathean Agriculture*, there is a sympathy of convenience between the gourd and the vine, and each of them lends assistance to its ally.

One reads in the book by Hadj of Granada, that there exists between the white *nachem*, called almis, an elm (?)[16] which is a tree with a black round fruit, in which inside one finds a kernel; the upper part (the pulp) is sweet; there is (as we were saying between this tree) and the vine, a sympathy and an affection which acts advantageously from one tree to another, and always such that the vine is associated with the elm, its yield is most abundant, and it is protected from all troublesome accident.

Cassius says: every time that an apple tree is planted in the vicinity of the *idjac*, that is the pear tree, or of the cedar, there is established between one and the other an affection which is of a reciprocal utility. Macarius says that between the pomegranate and the myrtle there is established, by their proximity, sympathy and friendship; thus, when one plants a myrtle near a pomegranate, the yield of the latter is more abundant, and it gains a large advantage. According to Kastos, the roots are intertwined, and consequently, the fruiting is larger, but the advantage does not become apparent before (the mixing of the roots). It is the same between the walnut tree and the fig tree or the mulberry tree.

It is said that the olive tree and the wild pomegranate[17] lend each other a mutual advantage if they are planted near each other, because of the affection that exists between them. It has been said that the olive tree loves the vine, and that the apple tree is a friend of both. Further, it has been said that if one plants bulbs of sea-onion (*Scilla maritima*) around an olive tree at its base, it is highly useful for the tree which becomes very productive.

According to *Nabathean Agriculture*, there is an antipathy between the white grape and the black grape: they cannot live planted together, or in the vicinity of one another; one avoids pressing their grapes together, because the must that would be provided would spoil quickly. It has been said that one of the curious things in the nature of the laurel is that if one plants a turnip beside it, and if the turnip remains in its entirety for two seasons of the year, the fruit of the laurel acquires a certain acridity and a disagreeable smell.

Hadj of Granada says that the walnut is antipathetic to most trees that one would plant in its vicinity, except the fig and the mulberry, because the walnut is of an excessive heat and dryness, which is pernicious for everything that comes too close to it, and that is not sympathetic to it; similarly it destroys everything that grows underneath it, except certain winter herbs, or forage plants, that one can plant underneath its branches, when it has lost its leaves; when one wants to associate climbing vines on it, they do not succeed and they fall at the last limit of enfeeblement. It has been said again, that if one plants cabbages in the vicinity of the vine, its shoots neither proceed nor extend themselves beside the cabbage, but on the contrary, they are directed to the opposite side. According to Kastos, there is no plant more harmful to the vine; there is nothing that is more harmful to anything than the cabbage. If it happens that one plants cabbage under a vine, it dies; it is the same when the wind carries the refuse from a cabbage planting onto a vine (see Geoponica, XII, 17 and V, 11, Palladius, Aug., V, 3). It is said that if one plants fenugreek in the vicinity of the vine, both plants die; they make each other sickly, they turn away from each other, and they look to go to another side. It has been said also that if one plants sumac in the vicinity of the vine, it languishes and becomes dry. It is said also that the cabbage was the enemy of the apple tree. When lupin is planted under a vine, it makes it dry…[18] When a peach tree loses its fruit before ripening, one must hang a bone whatever it would be; the pubic bone and the skull bone of a dog are those that are preferable; the tree then becomes fertile, and the fruits do not fall any more. One obtains the same result if one attaches to the tree a red cloth or a rag found in a heap of manure; in this case, the fruit does not fall, divine will helping.

---

[16] Perhaps *Celtis*.

[17] Clément-Mullet uses *balaustier* here, which also provides the English term balaustine, an astringent preparation from wild pomegranate flowers.

[18] According to Clément-Mullet, there follows here an indecipherable passage.

> Abou'l-Khaïr and others say that if the peach tree is sterile, one must strip the root, and make a slit in which one drives a plug of juniper, completely fresh and of pleasant odour; next one puts back the soil on top, and the tree becomes fertile, God willing. It would be the same for apricot, almond, cherry and plum trees. When, after having made a hole in the base of a peach tree, one drives a plug of *ghirab*, which is the willow, the walnut diminishes the size (see *Geoponica*, X, 16). The service tree can be made fertile through means of gold of good quality, in this way; one pierces at the base of the tree a hole on four sides; in the biggest root one inserts a little piece of gold of weight of about an eighth of a dinar, that one buries in the wood; this operation is done when the tree is in flower. One takes the excrement of a dog of which the eyes are no longer open; one buries it in the roots of the service tree at the time of flowering; and the flower will not fall at all (will not become sterile), divine will helping. (Volume 1, pp. 518-522)

Ibn al-Awwam provided more information, particularly concerning plant antipathies:

> Each tree has its antipathetic object (literally its enemy). Thus, when one plants in the vicinity of the bitter orange tree (Seville orange), beans, rue, oregano, euphorbia, or any type of plant which has a strong odour, it will do badly. The antipathy of the palm and the red cedar is a common thing and well known; it is the same for pitch. According to Nabathean agriculture, the vine suffers if it is in the vicinity of peas, and naphtha[19] [which it forms], as it suffers in the vicinity of the palm. The fig and the cabbage are completely harmful to the vine; it is a type of poison to it, like the euphorbia and the pythuse[20] and other similar plants. The common cabbage and the cauliflower are particularly noxious to the vine. It is said that the fig is harmful to the vine only in warm lands; for in the cool areas as in the lands of the Romans, Greece and other regions where snow falls, the proximity of the fig is useful to the vine; according to some, it is the same for the olive. Iambouschad claims that the beet, radish and kale are harmful to the vine. (Volume 1, p. 542)

There is a number of other less known Andalusian agricultural works. In Toledo one is attributed to the physician Ibn Wafid (1008-1074), employed by the Sultan of Toledo, and was titled *Maymu'a fi l-filaha* (*c*. 1068). However, it survives only as a fragment translated into Catalan, which has recently been rendered into Spanish (Cuadrato Romero 1997).

A *Kitāb al-Filāha* (Book of Agriculture) is believed to have been written in about 1080 by Ibn Bassal of Toledo, gardener to the Sultan. This work was cited by Ibn al-Awwam, and is of special interest, as it appears to be based largely on original ideas. The text by Ibn Bassal is accessible, although somewhat indirectly, as only fragments of the original Arabic text survive, but a surviving Catalan translation has been rendered into Spanish by Millás Villacrosa and Aziman (Ibn Bassal 1955).

> It is noticed that the fruit trees that are to the side of the walnut tend to disappear, because the vicinity of the walnut kills and destroys them, by effect of the warm nature of this tree; only the fig tree tolerates its company and both can coexist. (Chapter V)

> The walnut prospers especially in very cool soil, in which the cold surpasses the humidity, and it is explained accordingly that this cold air counters the heat that has the nature of the tree and, therefore, this shifts and it benefits; as far as warm soil is concerned, the walnuts do not last in them. When the walnut has grown, it is reckoned

---

[19] Naphtha, an Arabic word, refers to a flammable volatile substance, which usually originates as a distillate from the soil or rocks.

[20] The meaning of pythuse is unknown.

that this is significant, but if the walnut is small this does not concern us. The walnut is not susceptible to be grafted nor can it serve as graft for another tree, and it is thus because of its heat and of its strength, according to what we said before. (Chapter V)

Ibn Luyun was active in Granada in the twelfth century, and composed an agricultural work in verse, comparable to the Virgil's Georgics; this has been translated into Spanish by Eguaras Ibañez (1988). However, the poetry has been translated as prose and has become fairly inelegant; Section 64 deals with sympathies and antipathies of plants, including the pomegranate (Figure 3.4):

> Some fleshy fruit trees demonstrate a certain inclination towards another three species; thus we have the orange tree is inclined toward the olive tree, and the oleander toward latex-bearing trees[21], and equally apparent is the propensity of the pistachio toward the resin-bearing trees. The walnut-tree, on the other hand, occurs alone, because its shade, with its drowsiness, is intensely damaging to the plants which occur underneath, which thus reach the point of dying, except for the vine and the fig-tree, which are never harmed.
>
> There are other plants which are mutually repulsive, as occurs certainly between the palm and the juniper according the report by Abu Hanifa in his treatise on plants – because there is something to hold to account between the vine and the cabbage. On the other hand, in contrast those that follow all sympathise, as occur with myrtle and pomegranate or with poplar and the vine. Experiment is to be seen. The wild pomegranate is beneficial for the olive, so therefore, agree to plant this around it.

An important book, antecedent to the work of Ibn al-Awwam, is the *Kitab al-Filaha* of Abu l-Jayr, which is believed to originate from Seville in the late eleventh century. This work is extensively quoted by Ibn al-Awwam, and has only become available comparatively recently in translation (Carabaza 1991). Firstly Abu l-Jayr summarised the causes and treatment of failing plants, and therein stated the fundamental ancient belief that plant decline was associated with injurious substances in soil:

> In summary, most of the plant diseases must be due to the influence of the four factors: water, air, manure, and soil, and what is suitable is when one factor concurs with another one and even with a third. In order to improve the state of all the trees and protect them from injurious substances, the soil over its roots is uncovered, non-salty alpechin[22] mixed with fresh water is mixed with them, and the soil replaced. If straw has been laid down on these roots, the fruit increases and improves in its condition. People with experience agree then that which has suited fruit trees, spilled on their roots, serves to them as antidote against different ailments. When the tree has been in a poor and not any cure is known, you will have to take sheep and human dung, dissolve it in water and water with this; or excavate deeply its roots, or make in them a hole that traverses them, introduces into it a stake of ash wood and soon throw in aged urine. (Carabaza, p. 248)

Regarding the walnut, Abu l-Jayr wrote:

> You should know that the walnut is an enemy to all trees and is incompatible with them, due to the heat of its vapours and the power of its aroma. When any tree is planted beside it, it will remain rigid and perhaps in time will soon wither. Of all the tree species, none agrees more than the fig tree in that both are similar in the heat of their

---

[21] The word given here is *lechosos*, which in other contemporary works, e.g. that by Ibn al-Awwam, refers to plants such as the fig, mulberry, etc.

[22] Alpechin is the term given to the aqueous residue that results from the extraction process for olive oil.

vapours. When it has finished budding, the bases of the tree and the roots are uncovered and most of them are decorticated, the soil is quickly replaced and it is watered, and thus their bark will shortly return to their previous state. (Carabaza, p. 263)

The orange tree was at antipathy with other aromatic plants:

> It [the orange tree] is suitable to black soil, reliable, rough and sandy, and it grows in vicinity with neither rue, oregano or lemon, or anything that has strong inhalations, because they do harm to it. (Carabaza, p. 271)

Although the *Kitab al-Filaha* by Ibn al-Awwam provided the most comprehensive summary of agricultural knowledge to its time, it had surprisingly little impact on subsequent periods. This is for a variety of reasons. Firstly the Andalusian works were written in Arabic, although a few were translated later in Catalan, and thus they were limited in their accessibility to other European cultures. Arabic works were often feared or their use prohibited in Christian cultures, because of their supposed cabbalistic content. A fact that is overlooked too often is that works prior to the

Figure 3.4. *A manuscript leaf depicting the pomegranate from a 14th century Egyptian or Syrian work by Masâlik-al-absâr al-Umari (Manuscrit Arabe 2771, folio 263v, courtesy of the Bibliothèque Nationale de France)*

second half of the fifteenth century generally were distributed exclusively through manuscript copies which were both laborious and expensive to produce; consequently, in most cases, books were rare during their own time, and original copies are naturally even rarer today[23]. In the case of Andalusian works, the major cities of southern Spain, with the exception of Granada, were heavily damaged in the thirteenth century due to the ravages of the Crusaders in the name of Christianity[24]. Thus, overall, few subsequent authors were aware of *Kitab al-Filaha*, although it was used as a source by the sixteenth century Spanish agricultural author Gabriel Alonso de Herrera, and to a lesser extent by de Crescenzi and Estienne.

An early work in the Arab literature that ensued the decline of the Greco-Roman era was the encyclopaedia *Uyun al-Akhbar* (The Choice of Transmitted Information) by Ibn Qutayba (828-889). This work clearly has drawn from Greek or Byzantine sources, as we find:

> Between the cabbage and the grapevine there is enmity. If cabbage is planted in the vicinity of a grapevine, one of the two will die.

A similar statement was written by Ali ibn Rabban al-Tabari[25] in his *Firdaws al-Hikmah* (Paradise of Wisdom), a medical encyclopaedia that appeared about 850 A.D.

The concepts of antipathy and sympathy among plants is regarded as a minor derivative of ideas that were once much more important in human life. Particularly in Medieval times there was a strong fusion of science and the occult. An interesting example which touches on the subject matter at hand is the *Ghayat al-hakim* (Aims of the Sage), often known in the West as *Picatrix*, an anonymous work, sometimes attributed to an Andalusian author of the eleventh century. In any case, there is a brief passage which at first glance one could regard as plant antipathy, but actually concerns plant parasitism:

> The orobanche[26] destroys all the trees and all the plants around it; no tree or no plant can grow in the place where the orobanch occurs. (Book IV, Chapter 7, Section 51)

---

[23] The first book printed in Europe with moveable type was the so-called Gutenberg bible of 1454. It is estimated that by the end of the fifteenth century up to 30000 different book titles were printed, with a total of about nine million copies.

[24] Just as the Crusaders destroyed Islamic libraries in Andalusia and elsewhere, it is reckoned, for example, that Omar, in the name of Islam, wiped out a large proportion of Greco-Roman culture in destroying the library in Alexandria during the seventh century. This in part explains why many works cited by Pliny, Cassianos Banos, and others are no longer extant.

[25] Also known as Sahl al-Tabari.

[26] The term "orobanche" is difficult in the classical literature, as in both Theophrastus and Pliny it is translated as dodder, which usually refers to the genus *Cuscuta* (Convolvulaceae), a twining photosynthetic parasitic plant, whereas true *Orobanche* (Orobanchaceae) lacks chlorophyll and is a root parasite.

REFERENCES

Abu l-Jayr. 1991. *Kitab al-Filaha.* *Tratado de Agricultura.* Translated by J.M. Carabaza. Agencia Española de Cooperacion Internacional. Madrid.
Bakhouche, B., F. Fauquier. and B. Pérez-Jean. 2003. *Picatrix: un Traité de Magie Médiéval.* Brepols, Turnhout, Belgium.
Bolens, L. 1981. *Agronomes Andalous du Moyen-Age.* Librairie Droz, Geneva.
El Daly, O. 2005. *Egyptology: The Missing Millenium. Ancient Egypt in Medieval Arabic Writings.* UCL Press, London.
Fahd, T. 1996. L'Agriculture nabatéene en Andalousie. In C. Álvarez de Morales (ed.), *Ciencias de la naturaleza en el-Andalus. Textos y Estudios IV*, pp. 41-52. CSIC, Grenada. (reprinted in Fahd, T. 1998. *L'Agriculture Nabatéene.* Tome III. Institut Français de Damas, Damascus).
Hämeen-Anttila, J. 2004. The *Nabatean Agriculture*: authenticity, textual history and analysis. *Zeitschrift für Geschichte der arabisch-islamischen Wissenschaften* **15**: 249-280.
Hämeen-Anttila, J. 2006. *The Last Pagans of Iraq: Ibn Wahshiyya and his Nabatean Agriculture.* Brill, Leiden.
Hammer-Purgstall, J. von. 1806. *Ancient Alphabets and Hieroglyphic Characters Explained.* W. Bulmer, London.
Ibn al-Awwam. 1802. *Libro de Agricultura.* Translated by J.A. Banqueri. Imprenta Real, Madrid.
Ibn al-Awwam. 1864-1867. *Le Livre de l'Agriculture d'Ibn-al-Awam (Kitab al-Felehah).* Translated by J.-J. Clément-Mullet. A. Franck, Paris
Ibn ar-Raqqam. c. 1310. *Julasat al-ijtisas fi ma'rifat al-qiwa wa-l-jawass.* Manuscript held by Cambridge University.
Ibn Bassal. 1955. *Libro de Agricultura, editado, traducido y anotado por José M. Millás Vallicrosa y Mohmed Aziman.* Instituto Muley El-Hasan, Tetuán. (A facsimile edition with an added preliminary section was also published in 1995 by the Government of Andalusia).
Ibn Luyun. 1975. *Tratado de Agricultura.* Translated by J.E. Ibanez. Patronato de la Alhambra, Granada.
Ibn Wafid. 1997. *Tratado de Agricultura. Traduccion Castellana (Ms.S.XIV).* Edited by C. Cuadrado Romero. Analecta Malacitana, Malaga.
Ibn Wahshîya, Abu Bakr. 1984. *al-Filâha al-nabatîya / The Book of Nabatean Agriculture. Parts I-VII.* Edited by F. Sezgin. Institut für Geschichte der Arabe-Islamischen Wissenschaften an der Johann Wolfgang Goethe Universität, Frankfurt.
Ibn Wahshiya. 1995-1998. *L'Agriculture Nabatéenne. al Filaha al-Nabatiyya. Tomes 1-3* (edited by T. Fahd). Institut Français de Damas, Damascus.
Kopf, L. and F.S. Bodenheimer. 1949. *The Natural History Section from a 9th Century "Book of Useful Knowledge", The Uyun al-Akhbar of Ibn Quyayba.* Collection de Travaux de l'Académie Internationale d'Histoire des Sciences no. 4. E.J. Brill, Leiden.
Levey, M. 1966. Medieval Arabic toxicology. The *Book on Poisons* of Ibn Wahshīya and its relation to early Indian and Greek texts. *Transactions of the American Philosophical Society, New Series* **56**(7): 1-130.
Nasr, S.H. 1968. *Science and Civilization in Islam.* Harvard University Press, Cambridge.
Rabban al-Tabari, Ali ibn. 1928. *Firdaws al-Hikmah fi al-Tibb.* Sonne Press, Berlin.
Renan, E. 1862. *An Essay on the Age and Antiquity of the Book of Nabathean Agriculture.* Trübner & Co., London.
Smith, J. 1882. *A Dictionary of the Popular Names of Plants which Furnish the Natural and Acquired Wants of Man, in all Matters of Domestic and General Economy.* Macmillan and Co., London.

CHAPTER 4

ANCIENT INDIA, CHINA AND JAPAN

> Thus more and more the science and learning of the Sages of East and West will fuse into one.
> *Si Xue Fan*[1].
> Guilio Aleni (1623)

The following accounts owe little more than a related geography and a common paucity of information for their inclusion together in one chapter. As far as I know, there has been no substantial treatment in English of the early Asian literature relating to allelopathy, and doubtless much has been omitted here. Particularly in the case of China, there is a very rich body of literature of which very little has been translated into European languages. Ironically, the survival of some of these ancient documents is due to the fact that copies have found their way into Western libraries where they have been preserved, whereas native copies have perished, as for examples appears to be case with the work of the Indian writer Surapala.

ANCIENT INDIA

Information concerning plants and agriculture in India dates back thousands of years, with the earliest writings originating from the early Vedic period (*c.* 1500 B.C.). The sacred texts from the Vedic period are known as the Vedas, or books of knowledge. These works contain a mixture of information about all things, and amongst the wealth of information, there are numerous sections relating to plants: their uses in ceremony and medicine, methods of propagation and cultivation, and protection from disease (Raychaudhuri 1964, Kansara 1995, Pandey 1996).

In these early works, which are arguably the first records concerning botany, there is nothing that relates directly to the concept of allelopathy. However, it is worth observing that there was indication of an understanding of plant interaction.

---

[1] This statement, which appears in Needham *et al.* (1986), was written and published originally in China by one of the early Italian Christian missionaries in China, Guilio Aleni ((1582-1649). Few copies of the original work survive, and there is an Italian translation by Pasquale d'Elia (1950).

The *Rgveda* and *Yajurveda* indicate that crop rotation was practised in ancient India, likely long before it was advocated in Europe. A later Vedic text, *Taittirīya Samhitā* (*c.* 1000 B.C.) stated, for example, that rice grown in summer was alternated with pulses grown in winter. Similarly there was mention that ryegrass and clover were grown in rotation with wheat, barley or oats, as was beans with peas.

The *Brhatsamhita* by Varahamihara (*c.* 500 AD) is perhaps the earliest known Indian work to suggest allelopathy. It was recommended that sesame be planted, chopped down, and turned into the soil before a particular crop was sown (Bhat 1981). Perhaps the virtue of this was to reduce weeds.

The most important surviving early work from India that mentions what can be interpreted as allelopathic phenomena is the Sanskrit work *Vrikshayurveda*[2] (or *Vrkṣāyurveda*) by Surapala (Sadhale 1996). The exact date of this work is uncertain, but while completely unrelated, it is somewhat analogous in style and period with the *Geoponika*. Surapala is believed to have been the court physician to Bhimapala, which places him as a figure in Uttar Pradesh in the twelfth century. *Vrikshayurveda*, which translates as "The Science of Plant Life", is a compendial work, which draws on earlier material, such as *Brhatsamhita* by Varahamihara, but much of the material seems to be original. Until comparatively recently, the *Vrikshayurveda* by Surapala was known only by repute, but a copy is now known from the Bodleian Library, Oxford (see Figure 4.1), and this has been translated into English (Sadhale 1996).

Vrikshayurveda was written in verse form, as were most early Indian works, and two passages, in particular, stand out as inferring knowledge of allelopathy. The first of these, verse 63, concerns again sesame:

> The seedling then should be planted in beautiful, even ground, on which sesame or black gram [*Vigna mungo*] is not grown earlier and which is strewn over with heaps of flowers.

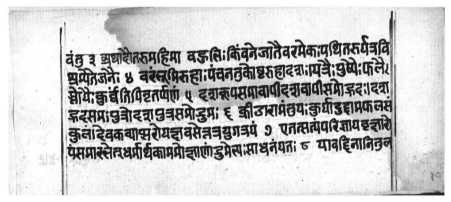

Figure 4.1. Leaf 2 of the manuscript of the Vrikshayurveda *by Surapala, MS. Walker 137a (courtesy of Bodleian Library, Oxford University).*

---

[2] The term *Vrkṣāyurveda*, came into use in about 300 B.C., and means simply a text on plant science, and thus denotes when botany became recognised as individual study.

The second of these is within verses 274 and 275:

> The coconut tree is destroyed if fed by water used for cleaning rice. Cotton tree immediately perishes if fed by water treated with the leaves of neem tree [*Azidirachta indica*]. A stick of the hingu [*Ferula asa-foetida*] tree kept at the root of the plantain tree [banana] destroys it.

In *Vrikshayurveda*, there is also some indication of familiarity with the phenomenon now known as "soil sickness". In Verse 208, one reads:

> If [trees] dried due to bad soil, the original soil from the root should be removed and it should be replaced by healthy soil and milk-water should be sprinkled over it.

Similarly, in Verse 222:

> Plants which are not cured by any one of the various above-stated remedies should be transplanted at other special sites.

Throughout *Vrikshayurveda*, there are numerous references to the beneficial effects of applying various plant decoctions to seeds, diseased plants, and soil, and as noted by Nene (1996), such information offers pointers for investigating the stimulatory and biocidal effects of otherwise little researched plants, e.g. *Embelia ribes*, although the modern reader would likely dismiss most of Surapala's directives as shamanistic.

Early Indian texts are slowly becoming available to Western readers, largely through the efforts of the Asian Agri-History Foundation, which in the past ten years has published five early agricultural works in English. The second publication in the series was *Krishi-Parashara* (Agriculture by Parashara) translated by Sadhale (1999). The exact authorship of this work is uncertain, as Parashara is a relatively common Indian name, but Sadhale thinks it is likely a work from the fourth century A.D., although Nene (1999) believes, in view of content, for example the scant knowledge of plant protection, that the original book could predate the fourth century B.C. It is also possible that the work dates from about the first century A.D., as there is another book by an author of the same name, the *Vṛkṣāyurveda* of Parāśara (Sircar and Sarkar 1996, Sadhale 1999), believed to date from the First Century A.D. Until recently, this Parāśara was merely known as a historical figure, assumedly from about 2000 years ago. When the translation of this *Vṛkṣāyurveda* appeared in 1996, many believed the Parasara manuscript, in Sanskrit, was a fraud, but a second translation has seemingly consolidated its authenticity. What is remarkable about this latter work are the astonishingly modern viewpoints concerning plant morphology, biogeography, germination and plant systematics based on floral structure, that seem in strong contrast to the primitive views of *Krishi-Parashara*. There are rudimentary ideas concerning a dual vascular system, and photosynthesis in the leaf, in which its coloured matter (chlorophyll), is involved in the assimilation of food (albeit from the roots), with the aid of energy (light) and air.

While the *Krishi-Parashara* and another Sanskrit agricultural text, the *Kashyapiyakrishisukti* by Kashyapa (c. 700-800 A.D.) contained little of allelopathic interest, the *Vishvavallabha* (Dear to the World) by Chakrapani (translated by Sadhale 2004), a Sanskrit work believed written during the sixteenth century is more relevant, and more in the style of Suarapala's *Vrikshayurveda*.

Firstly there was an allusion to the harmful effects of plants such as *Sesamum*:

> A field from which ripe grass (or *harital* (?)), *masha* and *tila*[3] are completely cut off and which is plowed repeatedly is suitable for a garden. (Chapter IV, Verse 8)

The *Vishvavallabha* contains a number of recommendations that one may construe as "companion planting" and/or biological control. For example,

> If one suspects existence of termites etc., one should plant *ajagandha* and *shatapushpi*[4] (fennel) in that place. (Chapter IV, Verse 8)

> In between smaller trees, the wise planter should plant densely in the field *shatapushpa* and *kuberakshi*[5] as a result of which he can get rid of insects. (Chapter VI, Verse 5)

> When *palasha* tree, planted in between other trees, bears fruit, it prevents the water related diseases and water-borne insects from infecting other trees as does *ashoka*[6].
> (Chapter VIII, Verse 45)

As in Surapala's *Vrikshayurveda* there are many recommendations concerning the treatment of seeds, soil, and plants with various and often seemingly bizarre mixtures of animal and plant products, particularly with regard to disease prevention and treatment; the use of neem (*Azadirachta indica*) is to be noted (Chaper VIII, Verses 21, 43). The following more closely approximates what might be considered indicative of allelopathic potential:

> Never sprinkle trees with water mixed with the decoction of *kulattha*. That ruins its flowers and fruits. Sprinkling with water mixed with salt, bark of *arjuna*, *karkarika* and *kimshuka*[7] also acts similarly. (Chapter IX, Verse 23)

The last Indian work that I wish to mention here could equally be treated in Chapter 3, as it emanates from Moghul India, that is the northern part of India that came under Islamic influence in the sixteenth century. The work in question was written in Arabic and is the *Nuskha Dar Fanni-Falahat* (The Art of Agriculture) likely by the seventeenth century scholar Dara Shikoh of the Delhi region. The book (translated in 2000 by Razia Aktar) is of interest as it provides early examples of companion planting:

> Those basil plants that are grown near brinjal (eggplant) will be stronger and better.
> (265)

> As the soil becomes vigorous when garlic plants are grown in it, they are grown in between other plants. (293)

> It is said that if pomegranate and guava trees, which are "very friendly" with each other are planted near each other, both will yield more fruits. (302)

---

[3] The three Sanskrit plant names here refer to *Cynodon dactylon*, *Vigna mungo* and *Sesamum indicum* respectively.

[4] *Cleome gynandra* and *Anethum sowa* respectively.

[5] *Anethum sowa* and *Caesalpinia crista* respectively.

[6] *Butea monosperma* and *Saraca asoca* respectively.

[7] Sanskrit names refer to *Dolichos uniflorus*, *Terminalia arjuna*, unknown, and *Butea monosperma* respectively.

Similarly, if pistachio and almond trees are planted near each other, they will yield more fruits. (303)

If a rose is grown near a peach tree, the fruits of this peach tree will be sweet-smelling.
(38)

ANCIENT CHINA

It is regarded that the first "books" in China appeared nearly 5000 years ago, although writings about agriculture did not appear until much later, possibly about 500-300 B.C. (Figure 4.2). Influential works such as *Guan Zi – Zhi Guo* by Guan Zi, and *Xun Zi – Fu Guo* by Xun Kuang appeared during the declining late Zhou Dynasty (see Table 1). Thus, the history of agricultural writings in ancient China has striking parallels with that in ancient Greece and Rome, both in regard to the time frame and the topics explored.

In China, as in much of Asia, the land holdings were commonly small, and consequently it became necessary to grow a variety of crops for subsistence, and evidently trial-and-error led to certain crop combinations that proved most advantageous. The climate of much of China, unlike Europe, is amenable to two or more harvests during the year. Consequently, in the Chinese agricultural literature, there has been far great emphasis on recommendations for intercropping (simultaneous growth of usually two different crops) or multiple cropping (multiple harvests of different crops within a year) than there has for crop rotation. A fallow field was anathema to a Chinese farmer, and the renowned American soil scientist F.H. King (1911) noted that in some regions of China, farmers actually went so far as to rotate the soil, that is, periodically they laboriously stripped soil from mulberry orchards and transported it to rice paddies, and vice versa. The combinations of crops were apparent in crop rotation where different types of crop succeeded one another on the same ground, and intercropping where one crop was planted in between rows of another crop. Chinese agriculture differed also from European agriculture in far less emphasis on animals for food, and consequently minimal use of land for grazing.

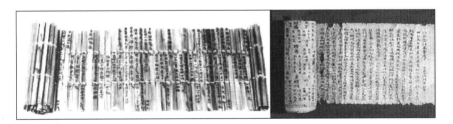

Figure 4.2. Examples of: (left) zhu jian *(bamboo strip book) – the text is read from right to left;* (right) scroll *– this is an example of a copy of* Pen Ts'ao Ching *by Shen Nung, a herbal of which the original text is reputed to date from about 2700 B.C.*

Table 4.1. Chinese dynasties and their major subdivisions and dates.

Dynasty and subdivisions	Years
Mythical	2953-2357 B.C.
Patriarchs (Tao Tang, You Yu)	2357-2205 B.C.
Xia	2205-1766 B.C.
Shang	1766-1131 B.C.
Zhou (Eastern Zhou, Chun Qiu, Warring States)	1121-221 B.C.
Qin	221-206 B.C.
Han	206 B.C.-220 A.D.
(Western Han)	206 B.C.-25 A.D.
(Eastern Han)	25-220 A.D.
Three Kingdoms	220-280 A.D.
(Wei)	220-265 A.D.
(Shu)	221-265 A.D.
(Wu)	222-280 A.D.
Jin	265-420 A.D.
(Western Jin)	265-317 A.D.
(Eastern Jin)	317-420 A.D.
Song	420-479 A.D.
Six Dynasties	386-589 A.D.
(Qi)	479-502 A.D.
(Liang)	502-557 A.D.
(Chen)	557-589 A.D.
(Wei)	386-557 A.D.
(Northern Qi)	550-589 A.D.
(Northern Zhou)	557-589 A.D.
Sui	589-618 A.D.
Tang	618-907 A.D.
Five Dynasties and Ten Kingdoms	907-960 A.D.
Song	960-1280 A.D.
(Northern Song)	960-1127 A.D.
(Southern Song)	1127-1280 A.D
Yuan (Mongol)	1280-1368 A.D.
Ming	1368-1644 A.D.
Qing (Manchu)	1644-1911 A.D.

Access to the early Chinese literature is extremely difficult as the books are rare, use obsolete characters, are not indexed because of the enormous number of different characters, and seldom have been translated. Although very many of the early Chinese works have been lost, there is still a surprisingly good record as there have been various historical compilations which have provided, more or less verbatim, ancient texts. The two most useful general references in English to the old Chinese botanical literature are Bretschneider (1882, 1885-1895) and Needham's volumes on botany and agriculture (Needham *et al.* 1986, Bray 1984), but none of these splendid sources sheds much light on early writings relating to allelopathic ideas.

In view of the preceding, I am particularly indebted to a recent review of the ancient Chinese literature that pertains to allelopathy (Zhou 1998), upon which much of the following is based.

Allelopathy

One of the earliest works to describe a phenomenon that might be classed as allelopathic is *Lin Hai Yi Wu Zhi* (Records of Strange Things Occurring near the Sea) by Shen Ying, an author active during the period of the Three Kingdoms (Wei or North Kingdom, 220-265 A.D.). Shen Ying described what sounds like a strangler fig, in that a sort of vine plant killed other trees by winding tightly around the tree trunks, and he ascribed its fatal effect to the secretion of "evil juices" to facilitate the more rapid rotting of the tree trunks. With the demise of the tree, the vine grew into the size of a large tree (Zhou 1998). There is a number of epiphytic figs native to China, and they are found principally in the subgenus *Urostigma*.

The most notable plant to be described as allelopathic in the early Chinese literature is sesame (*Sesamum indicum*), and in the third century, Quan Yang in *Wu Li Lun* wrote of the inhibitory effects of sesame on weed growth. It was commonly recommended that *Sesamum* be used early in the exploitation of virgin agricultural soil, as subsequent weed growth would be diminished. According to a Tang Dynasty work of the ninth century, *Si Shi Zuan Yao* (Main Points for the Four Seasons):

> When making land for farming, burn the grass first, then plough the soil, sow sesames in the first year. Sesames will destroy the roots of grasses and bushes, farmers can get rid of their worries on weeding. Every farm should know this.

During the Ming Dynasty, shifting cultivation was also practiced extensively. For example, Chinese fir (*Cunninghamia lanceolata*) plantations were prepared by planting sesame (*Sesamum* spp.) for weed control in the preceding year; this was then followed by intercropping the fir with millet (*Setaria italica*) or wheat (*Triticum* spp.) in successive years (Wu and Zhu 1997).

Similar advice appeared in later Ming or early Qing texts such as *Yang Yu Yue Ling* (Monthly Recommendations for Farming) by Dai Xi (1640, Ming Dynasty) and *San Nong Ji* by Zhang, Zong-Fa (1760, Qing Dynasty). According to Zhou (1998), two similar works *Nong Pu Liou Shu* (Six Books for Farming) by Zhou Zhi-Yu, and *Zhi Fu Qi Shu* (A Wonder Book for Getting Wealthier) by Chen Mei-Gong, explained that "drops of rain and dew drained from sesame leaves made other plants withered, one should never grow sesames near flowers and fruit trees" It is remarkable that these seventeenth century books are contemporary with European works such as Worlidge's *Systema Horticulturae*, which also presented an early statement of allelopathy through leaf leachates.

A second plant mentioned in the *San Nong Ji* as being useful as a pioneering crop in having a deleterious effect on weeds is *Perilla frutescens*, a member of the Lamiaceae used as a culinary herb and in traditional Chinese medicine.

Other indicators of potential allelopathic interactions were included in statements from the early twelfth century (Song Dynasty) text *Fen Men Suo Sui Lu* (Record of

Classified Activities) by Wen Ge (supplemented by Chen Hao), that "water lily was afraid of tung oil[8]; Cassia[9] killed grasses and woody plants".

The *Nong Sang Ji Yao* (Main Points for Agriculture and Mulberry Production) by Si Nong-si (1273, Yuan Dynasty) recorded that "mulberry was not suitable for intercropping with millets" and "reeds harmed bamboos".

The *Zhong Yi Bi Yong Bu Yi* (Supplement to the Essential Book for Plant Cultivation) by Zhang, Fu (late 13th century, late Yuan Early/ Ming Dynasty) advised that "red beans badly harmed cotton". A work from the Qing Dynasty, *Hua Jing*, stated that "weeds did not grow on lands covered with chips of cassia plant" Chen, Hao-zi (1688). Similarly, according to Minakata (1913), the pharmacopoeia *Pen Ts'ao Kang Mu* by Li Shih-Chin (1578) stated that bamboo could by killed with a decoction of the brown algal seaweed *Ecklonia bicyclis*, and that the grapevine could be killed by puncturing the stem with a peg made from licorice root.

References to the positive interactions of plants through their secretions in the early literature are rare, but in *Chen Fu Nong Shu*, by Chen Fu (1149, Southern Song Dynasty) there is mention that the intercropping of mulberry (*Morus* spp.) and ramie (*Boehmeria nivea*, Urticaceae) was mutually beneficial because of secretions released by both the roots and above-ground parts.

Crop Rotation and Intercropping

As in Europe, and indeed other areas such as Mexico, it appears that the virtues of cropping systems such crop rotation and intercropping were discovered pragmatically long ago. As indicated above, the growth of a crop, or fallowing, strictly for the purpose of improving the soil was a luxury few land-holders could afford, especially if soil fertility could be maintained through manuring and labour. Nonetheless in *Fan Sheng-Zhi Shu* (*c.* 100 B.C) is found the advice that: "If a field gave a poor crop in the second year, fallow it for one year." Similarly in *Qi Min Yao Shu* (6th century) is found an early description of what might be interpreted as soil sickness in hemp:

> Hemp needs good ground, one should not plant it on the same soil repeatedly – but no doubt it can be done; however, there will be stems and the problem is that the leaves will die early. (Chapter 2, § 8)

Until a few decades ago, agriculture in China was practised much as it had been for centuries, and thus relatively contemporary observations of traditional Chinese agricultural likely reflected ancient practise (see Figure 4.3). In reference to the allelopathic effects of sorghum stubble, Breazeale (1924) noted that it was common practise in China for farmers to remove all the sorghum stubble and burn it prior to replanting, and hence it was possible to maintain sorghum on the same land without difficulty.

---

[8] Various trees, notably dipterocarps, yield a product described as tung-oil or wood-oil, but in China the source is likely to be the seeds of the tree *Aleurites cordata* (Euphorbiaceae).

[9] The name cassia is applied to two unrelated plants – those of the genus *Senna*, and *Cinnamomum cassia*, a type of cinnamon.

Figure 4.3. *A Han Dynasty tomb tile depicting an agricultural scene (Collection of the author).*

Rotation cropping of Bananas and sugar canes not only promoted each other's growth but also improved each others' quality. According to *Quang Dong Xin Yu* (early Qing Dynasty), they smelled and tasted better than those produced in other fields. This crop rotation experience was widely adopted in Guangdong during the Ming and Qing Dynasties.

The virtues of intercropping were manifold. Firstly, agreeable species could improve each others' growth, perhaps through nitrogen supplementation or possibly through stimulatory chemicals released into the soil. There are many examples of this where crops from the Fabaceae are involved, although the chemical basis of the benefits of legumes was not discovered until the nineteenth century. An early example of this is recorded in *Fan Sheng Zhi Shu* by Fan Sheng Zhi (*c.* 100 B.C.), where it was stated that the growth of beans was beneficial for other plants as it made the soil softer and more fertile.

Secondly certain species could afford protection against various pests, diseases or weeds; for example the *Fan Shen Zhi Shu* also stated that the growth of alliaceous plants such as garlic and chives amongst melon plants was recommended, ostensibly to reduce the damage caused by insects and pathogens. However, it was earlier, in the Yuan Dynasty (13th to 14th centuries) that the first ecological and biological interactions among trees and associated intercrops were observed. It was reported, for example, that proso millet (*Panicum miliaceum*) grown under mulberry could promote the growth of both trees and crops, but that foxtail millet (*Setaria italica*) could have a negative effect by stimulating the occurence of pests (Wu and Zhu 1997). A much later example of this appears in *Nong Sang Jing* (Scripture for Crops and Silk Worms Farming) by Pu Song-Ling (1705, Qing Dynasty), where the interplanting of hemp or mustard plants was advised to reduce insect damage to bean plants. In other instances, a second crop was used as a nurse species, in order to protect the first species from damaging physical elements such as wind, too much sunlight or low temperature; for example, in the case of the latter, the interplanting of hemp was recommended to protect paper mulberry (*Broussonetia papyrifera*) from freezing during winter.

While intercropping was described in works from the Zhou Dynasty, its use proliferated later, and culminated in the writing of a work known as *Qi Min Yao Shu* (6th century A.D.) by Jia Si-xie. This important agricultural encyclopaedia recommended the use of beans as a nurse crop for melon plants, where the bean plants were killed and the residues were left to fertilise the melon plants. *Qi Min Yao Shu* reported that during the sixth century, the Chinese scholar tree (*Sophora japonica*), a legume, was planted with hemp in order to increase hemp growth and to improve the form of trees for future road-side plantations (Wu and Zhu 1997).

Finally, it was viewed that in some cases one species could modify the taste of the other species. An example of the latter was given in *Zhong Shu Shu*[10] by Guo Tuotuo, where he stated that if a vine was planted so close to a jujube tree (*Ziziphus jujuba*) that the roots of the plants came into contact, the grapes of the vine would assume the flavour of the jujube (Bretschneider 1882). It was widely believed that certain plants growing amongst other plants prized for their scent or flavour could degrade or improve their quality. Fragrant plants, such as plums, magnolia, pines, and bamboo orchids[11], and chrysanthemums that were grown between tea shrubs were believed to improve the taste of the tea leaves (*Wu Ben Xin Shu* (New Book for Practical Agriculture), late Song or early Yuan Dynasty; and *Explanation for Tea Production*, Luo Lin (1609, Ming Dynasty))[12].

Much advice was given concerning interplanting with mulberry trees (*Morus alba*), a plant which was esteemed for its variety of uses, including of course, the production of silk through silkworms. It is interesting that one author (*Hu Zhou Fu Zhi* (Records of the Facts in the Hu Zhou Fu District), Qing Dynasty) warned against the use of mulberry leaves from trees grown amongst barley, and that "silkworms were sick after being fed with leaves from mulberry trees grown in wheat fields." It is possible that these interactions were based on phytochemicals produced by certain grain crops.

Grafting

While the subject of grafting does not come directly within the confines of the subject of allelopathy, there are parallels in that grafting does concern plants that are

---

[10] This work is not known from extant copies. It is known through *Ku Chin T'u Shu Chi Ch'eng*, published in 1726. This latter work, a compendium of Chinese knowledge, is one of the most monumental publishing tasks ever achieved. It was produced in moveable type and is reputed to contain 852,408 pages (Giles 1911). Unlike western encyclopaedias, it is not a synthesis; rather, it is a collation of previously published work, and, for example, contains the work by Guo Tuotuo.

[11] Likely the terrestrial orchid *Arundina graminifolia*, although also used with some *Dendrobium* spp.

[12] I guess it is possible that a very small amount of terpenoid material could be become dissolved in the cuticle of tea leaves, and slightly affect the flavour.

viewed as sympathetic or antipathetic. Furthermore, there are remarkable similarities within this topic between Roman writers, such as Pliny, and ancient Chinese writers. Today grafting is generally viewed as only practicable between congeneric species, or at best species within the same family. Yet, Pliny[13] described the grafting of heterogeneous species, such as vine, fig, walnut, olive, pomegranate; he regarded certain species of *Platanus* and *Quercus* as being capable of accepting almost any graft. Similarly in a seventh or eighth century work by Guo Tuotuo[14] entitled *Zhong Shu Shu* (The Cultivating of Trees) is found the description of grafts: between plum and pear on mulberry, which was reckoned to yield sweeter fruit; peach on *Diospyros kaki* (oriental persimmon), which was alleged to produced more golden fruit; pomegranate on *Osmanthus fragrans* (sweet olive); and *Prunus* on *Melia azadarach* (chinaberry). Little is known of this book, as the work is lost and is known only through portions which have been reproduced in later works, such as *T'u Shu Tsi Cheng* (Book on the Art of Planting Trees). Kuo T'o t'o was a villager farmer who lived near the city of Ch'ang during the Tang Dynasty; however, his name was a pseudonym.

ANCIENT JAPAN

Little is known of Japanese agricultural or botanical literature prior to the unification of Japan and establishment of the Tokugawa Shogunate in the early seventeenth century.

Banzan Kumazawa (1619-1691) was born into a privileged family. He achieved fame in Japan as a statesman, and at the age of 37 he sought the secluded life of a scholar. However, his ideas caused him to have numerous enemies, often for petty reasons, and he lived the second half of his life in virtual exile (Fisher 1933). Amongst numerous books attributed to him, the most famous is *Dai Gaku Wakumon* (A Discussion of Public Questions in the Light of the Great Learning), believed to have been written during the closing years of his life. In reality, it was a manifesto for government reform, and it touched on many potential areas of reform, including agriculture and forestry. Regarding forestry, Banzan wrote:

> If a good plan is set up, after a while heavy forests will mature and provide plenty of firewood for generations to come. Meanwhile the shortage of fuel can be met by giving up some cultivated fields to fast-growing pine trees. The soil where pines have grown is bad for rice, but pines will thrive on soil too poor for vegetables and other trees. It is wise to bear present loss for ultimate gain. In following the foregoing plan, thin out the young pines before they get deeply rooted and other trees will spring up and make a mixed forest.
>
> Rain and dew wash down a poison from pines so that underbrush and grass will not grow beneath them, and such water is bad for crops. (Chapter X)

---

[13] *Natural History* Book XVII, section 26.

[14] Most older Chinese names have changed following the reforms in pinyin spelling. For example the old name for Guo Tuotuo is Kuo T'o t'o. See Shang Dai Zhong Shu Shu , Taibei Shi: Dong Fang Wen Jua Shu Ji, (c. 1981).

According to the extraordinary Japanese scholar Kumagusu Minakata (1913a), the Japanese philosopher and naturalist Kaibara Ekken (1630-1714) wrote in his *Yamata Honzō* (Plants of Japan) that there was antipathy between the white and red-flowered varieties of the lotus, *Nelumbo nucifera* (formerly known as *Nelumbium speciosum*), when planted together in the same pond (Kaibara Ekken 1709). Minakata (1913b) also noted that in the 1880's, orange growers in the province of Kii were officially instructed to plant onions under each orange tree in order to protect it from black mould, but the origin of this practise was unknown. There seems little else recorded in ancient Japanese writings that concerns allelopathy, although *Lycoris* spp. were thought to have allelopathic effects in ancient Japan (Y. Fujii, pers. comm.)

SOUTHEAST ASIA

Botanical knowledge in the regions that include today Indonesia, Papua New Guinea and Malaysia is also undoubtedly ancient, but was not documented until the arrival of Europeans, chiefly the Portuguese, followed by the Dutch, in the sixteenth and seventeenth centuries. The German botanist Georg Eberhard Rumpf (1627-1702,) known as Rumphius, lived the last fifty years of his life on the island of Ambon, in present-day Indonesia. His *magnum opus* was the *Herbarium Amboinense*, of which the first manuscript copy was lost at sea in 1692. Although blind by this stage, he managed to organise the rewriting of the entire work, but then publication was suppressed by the Dutch, because of the detailed content, until well after Rumphius' death. Rumphius cited some examples of traditional knowledge which can be regarded loosely as allelopathic in nature. He stated in *Herbarium Amboinense* (Rumphius 1741) that:

> There is great antipathy between the Durians and the leaf of the Siri plant[15], so much so that if one single leaf is placed in a prahu[16] full with Durians, then all of them are supposed to spoil.

Similarly he recorded the use of opium to affect other plants:

> The natives are also known, from maliciousness, to play a trick on one another, and do so by drilling a hole in someone else's Durian-tree and putting some Opium or Amphium[17] in it, that causes all Durians, be they ripe or not, to fall down a short time later. And even when this is found out, the perpetrator is still rewarded, because the tree will perish from it.

Rumphius was also largely responsible for giving substance to tales about the dreaded upas tree (*Antiaris toxicaria*) of the East Indies, but this matter remains better related in the next chapter.

---

[15] Likely *daun sirih*, or *Piper betle*, of which the leaves are used commonly in traditional medicine.
[16] A boat.
[17] Assumedly this refers to a toxic preparation from certain salamanders (e.g. genus *Amphiuma*).

REFERENCES

Akbar, R. 2000. *Nushka Dar Fanni-Falahat (The Art of Agriculture). Agri-History Bulletin No. 3.* Asian Agri-History Foundation, Secunderabad.
Ayachit, S.M. 2002. *Kashyapiyakrishisukti (A Treatise on Agriculture by Kashyapa). Agri-History Bulletin No. 4.* Asian Agri-History Foundation, Secunderabad.
Bhat, M.R. 1981. *Varahamihira's Bṛhat Sahitā. Part 1.* Motilal Banarsidass, Delhi.
Bray, F. 1984. *Science and Civilisation in China. Volume 6 Biology and Biological Technology. Part II: Agriculture.* Cambridge University Press, Cambridge.
Beazeale, J.F. 1924. The injurious after-effects of sorghum. *Journal of the American Society of Agronomy* **16**: 689-700.
Brettschneider, E. 1882. *Botanicon Sinicum. Notes on Chinese Botany from Native and Western Sources.* Trübner & Co., London.
Brettschneider, E. 1892-1895. *Botanicon Sinicum. Notes on Chinese Botany from Native and Western Sources. Parts II and III.* Kelly and Walsh, Shanghai.
D'Elia, P. 1950. Le "Generalità sulle Scienze Occidentali"; *Hsi Hsueh Fan* [*Si Xue Fan*] di Giulio Aleni. *Revista di Studi Orientali* **25**: 58-.
Fisher, G.M. 1933. Banzan Kumazawa: his life and ideas. *Transactions of the Asiatic Society, Second Series* **16**: 221-258.
Giles, L. 1911. *An Alphabetic Index to the Chinese Encyclopaedia.* Oxford University Press, Oxford.
Gopal, L. 2000. *Vṛkṣāyurveda in Ancient India.* Sundeep Prakashan, New Delhi.
Kaibara Ekken. 1709. *Yamata Honzō.* Kyoto. A collected edition of Kaibara Ekken's work was published as *Ekken Zenshū* in 1910-1911, and *Yamata Honzō* was also republished in 1936.
Kansara, N.M. 1995. *Agriculture and Animal Husbandry in the Vedas.* NAG Publishers, Delhi.
King, F.H. 1911. *Farmers of Forty Centuries or Permanent Agriculture in China, Korea and Japan.* Mrs. F.H. King, Madison.
Kumazawa Banzan. 1933. Dai Gaku Wakumon. *Transactions of the Asiatic Society, Second Series* **16**: 261-356.
Minakata, K. 1913a. Botany. *Notes and Queries* **11 S.VII**: 72-73.
Minakata, K. 1913b. Onions planted with roses. *Notes and Queries* **11 S. VII**: 516.
Needham, J., Lu, G.-D., Huang, H.-T. 1986. *Science and Civilisation in China. Volume 6 Biology and Biological Technology. Part I: Botany.* Cambridge University Press, Cambridge.
Nene, Y.L. 1996. Ailments. In: N. Sadhale (1996), *Surapala's Vrikshayurveda (The Science of Plant Life by Surapala). Agri-History Bulletin No. 1.* Asian Agri-History Foundation, Secunderabad.
Jayasekera, S.J.B.A. 2005. Ancient agricultural crop production and protection practises in Sri Lanka. In: Y.L. Nene (ed.), *Agricultural Heritage of Asia. Proceedings of the International Conference.*, pp. 204-211. Asian Agri-History Foundation, Secunderabad.
Pandey, L.P. 1996. *History of Ancient Indian Science Volume 1, Botanical Science and Economic Growth. A Study of Forestry, Horticulture, Gardening and Plant Science.* Munshiram Mancharlal Publishers Pvt. Ltd, New Delhi.
Raychaudhuri, S.P. 1964. *Agriculture in Ancient India.* Indian Council of Agricultural Research, New Delhi.
Rumphius, G.E. 1741-1750. *Herbarium Amboinense.* Franciscum Changuion, Joannem Catuffe, Hermannum Uytwerf, Amsterdam.
Sadhale, N. 1996. *Surapala's Vrikshayurveda (The Science of Plant Life by Surapala). Agri-History Bulletin No. 1.* Asian Agri-History Foundation, Secunderabad.
Sadhale, N. 1999. *Krishi-Parashara (Agriculture by Parashara). Agri-History Bulletin No. 2.* Asian Agri-History Foundation, Secunderabad.
Sircar, N.N. and Sarkar, R. 1996. *Vṛkṣāyurveda of Parāśara (A Treatise on Plant Science).* Sri Satguru Publications, Delhi.
Wu, Y and Zhu, Z. 1997. Temperate agroforestry in China. In: Gordon, A.M and Newman, S.M. (eds.). *Temperate Agroforestry Systems*, pp. 19-179. CAB International Publishing, Wallingford.
Zhou, Z. 1998. *Zhongguo Zhi Wu Sheng Li Xue Shi.* Guangdong Gao Deng Jiao Yu Chu Ban She, Guangzhou.

ANCIENT CHINESE REFERENCES

Fan Sheng Zhi Shu (Fan Sheng-shi's Book) [c. 100 B.C.] by Fan Shen Zhi. Available in English as *On "Fan Shêng-chih Shu", an Agriculturalist Book of China written by Fan Shêng-Chih in the First Century B.C.* by S.-H. Shih (1963), Science Press, Peking.

Lin Hai Yi Wu Zhi (Records of Strange Things Occurring near the Sea) [c. 220-265 A.D.] by Shen Ying. Recent edition published: I Wen, Taipei (c. 1965)

Wu Li Lun (On Weeds?) [3rd century, *c*. 265] by Quan Yang. Several annotated Chinese editions have been published.

Qi (Chi) Min Yao Shu (Important Arts for the People's Welfare) [6th century A.D.] by Jia (Xia) Si Xie. Several annotated Chinese editions have been published. A summary is available in English as *A Preliminary Survey of the Book Ch'i Min Yao Shu, an Agricultural Encyclopaedia of the 6th Century. Second Edition* by S.-H. Shih (1962), Science Press, Peking. It is also scheduled to be published in English by Foreign Language Press in "Library of Chinese Classics Series"

Si Shi Zuan Yao (9th century) by Han E. Several annotated Chinese editions have been published.

Si Min Yue Ling (Essential Arts for the People) [25-220 A.D.] by Cui Shi. Several annotated Chinese have been published.

Zhong Shu Shu (Book on Planting Trees) [7th or 8th century A.D., Tang Dynasty] by Guo Tuotuo. No extant versions; known only from later encyclopaedias, e.g. *Gu Jin Tu Shu Ji Cheng*. See *Shang Dai Zhong Shu Shu*, Taibei Shi: Dong Fang Wen Jua Shu Ji, (c. 1981).

Fen Men Suo Sui Lu (Record of Classified Activities) [early 12th century] by Wen Ge. Recent edition published c. 1995.

Chen Fu Nong Shu (1149, Song Dynasty) by Chen Fu. A Chinese annotated edition was published in 1981.

Bei Yuan Bie Lu (1186, Song Dynasty) by Zhao Ru-li. A Chinese edition was published in 1965 and is often found amongst a collection of treatises under the title *Cha Lu*.

Wu Ben Xin Shu (New Book for Practical Agriculture) [Song and Yuan Dynasty] author unknown.

Nong Sang Ji Yao (Fundamentals for Agriculture and Mulberry Production) (1273, Yuan Dynasty) by Si Nong-Si. Recent edition published by I. Wen, Taipei (1969).

Zhong Yi Bi Yong Bu Yi (Supplementary to an Essential Book for Plant Cultivation) (late 13th century, Yuan Dynasty) by Zhang Fu. Recent edition published c. 1994.

Bian Min Tu Zuan (A Convenient Book of Diagrams for Farmers) (1493, Ming Dynasty) by Kuang Fan. Several recent editions have been published.

Pen Ts'ao Kang Mu (1578, Ming Dynasty) by Li Shih-Chen.

Hu Zhou Fu Zhi (Records of the Facts in the Hu Zhou fu District) by Zhang Duo (1542, Ming Dynasty).

Qun Fang Pu (Beauty List) (1604, Ming Dynasty) by Wang, Xiang-Jin. Several editions have been published.

Cha Jie (Explanation for Tea Production) (1609, Ming Dynasty) by Luo Lin.

Yong Zhuang Ziao Pin (c. 1609-1621) by Zhu Guo-Zhen.

Nong Zheng Quan Shu (A Complete Book for Agricultural Management) (1628, Ming Dynasty) by Xu Guang-Qi. Recent edition published in 2002. To be published in English by Foreign Language Press in "Library of Chinese Classics Series".

Yang Yu Yue Ling (Monthly Guidance for Farming) (1640, Ming Dynasty) by Dai Xi. Edition published in 1956.

Nong Pu Liou Shu (Six Books for Farming). [Ming Dynasty] by Zhou Zhi-Yu.

Hua Jing. [*c*. 1670, Qing Dynasty] by Chen, Hao-zi. Edition published in 1956

Zhi Fu Qi Shu (A Wonder Book for Getting Wealthier). [late 17th century, Qing Dynasty] by Chen Mei-Gong.

Quang Dong Xin Yu (Cantonese New Words and Phrases). [late 17th century, Qing dynasty] by Qu, Da-jun.

Nong Sang Jing Jiao Zhu (Scripture for Crops and Silk Worms Farming) by Pu Song-ling (1705, Qing Dynasty). Edition published in 1982.

Gu Jin Tu Shu Ji Cheng (Chinese or Imperial Encyclopaedia). [*c*. 1726, Qing Dynasty].
Known with various spellings (e.g. T'u Shu Tsi Ch'eng or *Thu Shu Chi Chheng*), this was a massive compilation from earlier works. It consists of about 10000 books and apparently has 852408 pages. As such it is about three to four times the size of a western encyclopaedia (e.g. 11th edition of *Encyclopaedia Britannica*) based on a word count. See Giles, L. (1911). *An Alphabetical Index to the Chinese Encyclopaedia*. Oxford University Press, Oxford.

San Nong Ji (1760, Qing Dynasty) by Zhang, Zong-Fa. Edition published in 1989.

CHAPTER 5

MEDIAEVAL PERIOD AND RENAISSANCE

> Oates, rie or else barlie, and wheat that is gray,
> Bring land out of comfort, and soone to decay;
> One after another, no comfort betweene
> Is crop upon crop, as quickly will be seene,
> Still crop upon crop many farmers do take,
> And reape little profit for greedinesse sake.
>
> *Five Hundred Points of Husbandrie*
> Martin Tusser (1534)

INTRODUCTION

The decline of the Roman Empire was followed by a lengthy period of about a thousand years during which few advancements were made, particularly in botany, and many classical works were likely lost or forgotten.

The notable exception to this was with the flourishing of Byzantine and Islamic culture in Asia Minor and in Andalusia respectively. An important achievement in Byzantine agriculture was the compilation of the *Geoponica*, successive editions of which are believed to have been compiled from the sixth century through to the tenth century, and which is dealt with, more appropriately in Chapter 2. In Andalusia, Islamic patronage led to the production of several agricultural and botanical works of which one is moderately well known today: *Kitab al-Filaha* (Book of Agriculture) by the twelfth century writer Ibn al-Awwam, which is based largely on an earlier and controversial work, the *Nabathean Agriculture* by Ibn Wahshiya. These and related works are described in Chapter 3.

For the rest of Europe, the period is one of slender botanical advancement, but nonetheless, there is a surprisingly rich body of literature, which has essentially borrowed the classical writings, and commonly immersed them in the occult. The era was also one that witnessed the European discovery of distant lands. This, coupled with the invention of printing in Europe, led to the rapid diffusion and demand for information of things foreign, and much that was printed related to the fabulous.

It is very difficult to pinpoint in time the Mediaeval Period or so-called Middle Ages, and indeed the Renaissance, as they were defined largely by social conditions, and they occurred at different times in different places. Also, when one speaks of a renaissance in botany, for example, even in one place, such as Italy, this occurred at a very different time from the renaissance in art. Very broadly, the Mediaeval Period or Middle Ages represents that period when there was little advancement, or even a regression in knowledge, and in much of Europe it may be said to span the period from about 500 to 1500 A.D. (Orlob 1971). The renaissance in botany was selective, and was largely led by German herbalists of the sixteenth century who recognized that the plants in northern Europe were often dissimilar from those described by Theophrastus, Pliny and Dioskorides. The renaissance in understanding how plants function followed much later, commencing with the seventeenth century work of Malpighi and Ray, which relied heavily on the technological advancement of microscopy.

Having recognised the above constraints, we can now address the information concerning allelopathic phenomena that emanated from the period, under three main headings: basic botanical lore, antipathy and sympathy, and myths and travelers tales.

BOTANICAL AND AGRICULTURAL LORE OF THE MIDDLE AGES

The knowledge in the botanical sciences increased little during the Middle Ages, with the exception of contributions noted above, and that of a very small number of exceptional individuals.

For example, there was the 12th-13th century English scholar Alexander Neckam[1] (1157-1217), who is known chiefly for a work entitled *De Naturis Rerum. Libri Duo*, which was succeeded years later by a poetic work with similar content, *De Laudibus Divinae Sapientiae*. In the former, Neckam wrote concerning the walnut (Figure 5.1):

> The walnut, placed among dangerous herbs and fungi, expels and extinguishes whatever is poisonous in them. The walnut-tree is injurious to all other trees growing under or adjacent to it, contrary to the pine, under which all plants flourish.
>
> (Book II, Chapter 81)

Perhaps the most significant individual of the era was Albertus Magnus, who was born Albrecht de Groot in Lauingen, Suabia (present day Bavaria), possibly in 1206 or earlier[2], and who died in Cologne in 1280. He trained as a Dominican monk, and served as a theologian, teacher and administrator, eventually acquiring the title of Bishop of Ratisbon. He was a prodigious writer, and his collected works are truly encyclopaedic, covering every discipline, including botany. The substantial volume on botany, *De Vegetabilibus et De Plantis*, is not simply a compilation of Greek and Roman writings, but is a creditable and largely original account of the plants that would have been known to Albertus in northern Europe. Thus, with reference to allelopathy, Albertus mentioned that the "indwelling extreme toxic bitterness" of the shade of walnut was harmful to surrounding plants (Book VI, Tract I, Chapter XXVII, § 147. Similarly he warned of the planting, at the same time, hazelnut[3] or cabbage near vines, as well as helleborus or scammony near other plants, and zizania

---

[1] Sometimes given as Neckham.
[2] Some authorities give the year of birth as early as 1193.
[3] See Chapter 2, note 12.

*Figure 5.1. The walnut tree (*Juglans regia*) from* Cosmographia *by Sebastian Münster (1552).*

next to wheat (Book VII, Tract I, Chapter IX, § 75). Another interesting statement suggestive of an understanding of soil sickness appeared in Book VII, Tract I, Chapter VI, § 50: ??:

> Not any practise, as indicated, can cure soil which has become barren through continuous scorching and dryness. Its looseness does not hold introduced moisture, its dryness prevents plants from using manure and its inherent bitterness interferes with the nutrition of seeds and plants.

Also, it is seldom appreciated that it was Albertus Magnus who first recognised the existence of germination inhibitors; he realised that the flesh of fruits such as apples and pears was inhibitory to the seeds contained in the fruit (Köckemann 1936). Albertus Magnus' contribution to botany has been largely overlooked, as his work preceded the invention of printing in Europe, and achieved little distribution in his time. Furthermore, when his work was published, it was never translated from the original Latin, and thus has remained largely inaccessible.

A lesser known encyclopedia of nature was *Das Buch der Natur* written by Konrad of Megenberg (1309-1374) in *c.* 1350. It was the first natural history book written in the German language. It too was an amalgam of classic lore and local knowledge. For example, Konrad von Megenberg recorded that the cabbage and the hazelnut tree (*Corylus* sp.) were harmful to grape vines.

Another highlight was that of the Bolognese, Pietro di Crescenzi (*c.* 1233-*c.*1320), who compiled an agricultural encyclopedia in the thirteenth century based largely on his own experiences in Italy, supplemented by the Latin agricultural texts of Cato, Varro, Columella and Palladius. This work, *Ruralium Commodorum* (known in later editions as *De Agricultura Vulgare*), is believed to have been written during the period 1304-1309, at the request of Charles II, King of Sicily. It is notable as comprising one of the earliest secular books to have been printed, having been first printed in Augsburg (Bavaria) in 1471. Like Albertus Magnus, di Crescenzi was judicious in his borrowing of material from the Roman authors, and from an allelopathic point of view, he mentioned the harmful nature of walnut trees. *Ruralium Commodorum* has been translated into most European languages, but a full edition has never appeared in English.

Yet another work, which was likely written during the fourteenth century, was the *Pelzbuch* of Gottfried von Franken (*c.* 1350), which was first printed in 1530. Gottfried's book was concerned largely with the disorders of trees, and as such it mentioned the concepts of antipathy of sympathy.

BOTANICAL AND AGRICULTURAL LORE DURING THE RENAISSANCE

Botany during the Renaissance was expressed in a number of ways:

1) the surviving Latin and Greek texts, in particular Pliny's *Natural History*, were widely published in printed form and often in modern languages;
2) new handbooks of plants, or herbals, began to appear – these were not simply reiterations of classical works, but often incorporated descriptions of local plants and fresh observations; and
3) experimentation and critical observation became important tools.

Allelopathy was a benefactor of these advancements. Apart from the reappearance of Pliny's work and the so-called "*Scriptores Rei Rustica*", which were compilations of the works of Cato, Varro, Columella and Palladius, new natural history or agricultural works appeared, albeit strongly based on these predecessors.

The advent of printing had a profound effect on European culture and science. Books previously had only been available through the tedious, expensive, and sometimes erratic process of manual copying. Printing allowed hundreds of copies of a book to be available within a comparatively short period of time, and at a price which many individuals could afford. The first printed book in Europe, the so-called Gutenberg Bible, was printed in 1454 in an edition of 300 copies. It is reckoned that by 1500, approximately 30000 separate titles or editions, with an estimated total of nine million volumes, had come into circulation; yet, these *incunabula* are extraordinarily rare today. Included among such books is the agricultural tract by di Crescenzi, of which the 1471 edition would fetch well over $100,000 today.

It is suffice to mention that many of the classic texts of the Greek and Roman authors were amongst those that came into print in the early years of printing. However, I will only provide detail of new works, particularly those that offered some variant or innovative perspective on allelopathy.

An important early work is *Obra de Agricultura* by Gabriel Alonso de Herrera (c. 1460 – 1530?), which was first published in 1513, and despite the fact it appeared in over thirty editions, it is rare today. Herrera was based in Toledo in central Spain, and his was likely the first printed book to incorporate the work of Ibn al-Awwam. This point is not absolutely certain, as, interestingly, Herrera did not directly cite Ibn al-Awwam, whereas he cited extensively from the classical Greek and Latin authors, and even the Bible. However, there are considerable similarities in the organisation of the two works, and there are similar citations of Abencenif (Ibn Sina) who was held in respect as a physician and was better known under the Latin name Avicenna. Simply, it may have been dangerous at that time, in Roman Catholic Spain, to cite an Arabic work that had some occult content. Herrera's work drew on the full range of classical and later works, such as de Crescenzi, that were available to him, and he also incorporated local knowledge. It is regrettable that Herrera's book has never been translated into languages other than Italian or French.

The Italian, Girolamo Cardano (1501-1576), known primarily for his contributions to mathematics, was another author of an omnibus work. In reference to antipathy, he wrote (Cardan[4] 1550):

> It is fairly widely known that the plants have hatreds between themselves.... it is said that the olive and the vine hate the cabbage; the cucumber flies from the olive.... Since they grow by means of the sun's warmth and the earth's humour, it is inevitable that any thick and opaque tree should be pernicious to the others, and also the tree that has several roots.

Numerous encyclopaedic works appeared, commencing in the sixteenth century, and had titles such as "Spectacle de la Nature", or "Théâtre de la Nature". For example, antipathies of plants were discussed in *La Théâtre de la Nature Universelle* by Jean Bodin (1597).

A contemporary of Herrera was Jean Ruel (1479-1537), known as Ruellius, who trained as a physician and became dean of the medical faculty in Paris. His most important work was a huge compilation of classical natural history lore, *De Natura Stirpium Libri Tres*, first published in 1536, but virtually none of the content was original. For example, the material regarding antipathy and sympathy (*discordia* and *concordia*; see Book I, Chapter 22) was drawn more or less verbatim from Pliny (Book XXIV, Chapter 1).

The German botanist Hieronymus Bock (1498-1554), also known as Tragus, wrote a work entitled *Neu Kreutterbuch* which, along with that of Leonard Fuchs, marks the very beginning of the Renaissance for descriptive botany. The works of Fuchs and Bock are known for their departures from the botanical works of Greece and Rome, and the accuracy and naturalness of the illustrations (Figure 5.2); however, much anecdotal material from the classical sources remained. For example, Bock reiterated the lore concerning the antipathy of the fern and the reed (Boch 1560).

Konrad Heresbach (1496-1576), who lived most of his life in what is now the western parts of Germany, wrote an agricultural compilation which employed the usual Roman and Greek sources, as well as local sources, as suggested by the reference to the antipathy of hazelnut, attributable to Albertus Magnus and Konrad von Megenberg. The work was first published in Latin in 1570 as *Rei Rusticae*, but achieved its greatest popularity through the several English editions by Googe entitled *Foure Bookes of Husbandrie*, published firstly in 1577. In Book II, Heresbach quaintly blended Roman agricultural lore with sixteenth century practice:

---

[4] Names were often altered according to the language used, and in French editions, Cardano is known as Cardan.

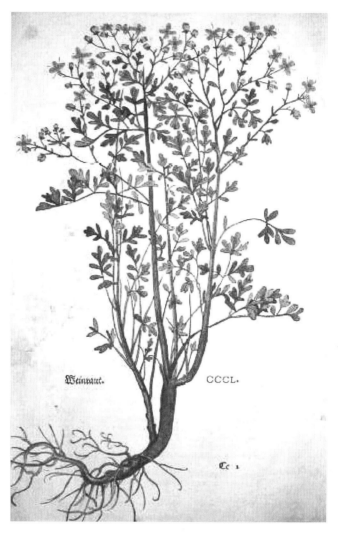

Figure 5.2. *Illustration of rue (*Ruta graveolens*) from the 1543 edition of the* Neu Kreuterbuch *which was the German edition of* De Historia Stirpium *by Fuchs.*

And because there is naturall freendshippe and love betwixt certayne trees, you must set them the nearer togeather, as the Vine and the Olyve, the Pomegranate and the Myrtel. On the other side, you must set farre a sunder, such as have mutual hatred among them, as the Vine with the Filberte & the Bay. There are some of them, that desire to stand two and two togeather as the Chestnut: the droppinges also doo hurt of all sortes, but specially the droppings of Okes, Pinetrees, and Mastholmes[5]. Moreover, the shaddowes of divers of them are hurtful, as of Walnut tree, whose shaddowe is unholsome for men, and Pine tree that kylleth young springes: yet they both resist the winde, and therefore are best to be set in the outer sides of the Orchardes, as hereafter shalbe sayde.

---

[5] *Quercus ilex*

> Neyther doo they [vines] like all manner of trees, for they hate the Nuttree, the Bay, the Radishe, and the Coll[6]: as agayne they love the Poplar, the Elme, the Willowe, the Figge, and the Olyve tree.

It may be surprising to some, but perhaps less surprising to others, that the lore of antipathy and sympathy became embedded in religious and didactic works. This seems to have occurred firstly in the work of the great Dutch reformationist writer Desiderius Erasmus (1466-1536), whom one does not usually associate with natural history; however, up until the end of the mediaeval period, it was still possible for a great mind to have familiarity with most of recorded knowledge. Erasmus summarised the largely Plinian lore on antipathy and sympathy to illustrate the bases of natural aversions and friendships in human life. In one of his "Colloquies", entitled *Amicitia*, or friendship, (written in 1513), Erasmus used a dialogue between two characters, Ephorinus and John, in which Ephorinus remarked:

> What you have heard, as to that matter, is no Fiction. But, not to mention Democrital[7] Stories, do we not find by Experience, that there is mighty Disagreement between an Oak and an Olive-Tree, that they will both die if they be planted into the Ground of each other? And that an Oak is so opposite to a Walnut-Tree, that it will die tho' it be set at a good Distance from it; and indeed a Walnut-tree is hurtful to most sorts of Plants and Trees. Again tho' a Vine will twine its Sprigs round all other Things else, yet it shuns a Colewort; and, as tho' it were sensible of it, turns itself another Way, as if another Person gave the Vine Notice that his Enemy was near at Hand. The Juice of Coleworts is a Thing contrary to Wine, and they are used to be eaten against Drunkenness; But the Colewort has its enemy too; for if it be set near the Herb called Sow-Bread, or wild Marjoram, it will wither presently. There is like Disposition between Hemlock and Wine; as Hemlock is poison to Man, so is Wine to Hemlock. What secret Commerce is there between the Lily and the Garlick, that growing near to one another, they seem, as it were, mutually to congratulate one another? The Garlick is the stronger, but the Lily-Flower smells the sweeter[8].
>
> (*The Colloquies of Erasmus, Volume II*, pp. 311-313)

Similarly, St Francis of Sales (1567-1622) of Geneva used the cabbage-vine antipathy as a religious metaphor in his *Treatise on the Love of God*, Chapter 11.

> We have but one soul, Theotimus, and an indivisible one; but in that one soul there are various degrees of perfection, for it is living, sensible and reasonable; and according to these different degrees it has also different properties and inclinations by which it is moved to the avoidance or to the acceptance of things. For first, as we see that the vine hates, so to speak and avoids the cabbage, so that the one is pernicious to the other; and, on the contrary, is delighted in the olive:—so we perceive a natural opposition between man and the serpent, so great that a man's fasting spittle is mortal to the serpent: on the contrary, man and the sheep have a wondrous affinity, and are agreeable one to the other. Now this inclination does not proceed from any knowledge that the one has of the hurtfulness of its contrary, or of the advantage of the one with which it has affinity, but only from a certain occult and secret quality which produces this insensible opposition and antipathy, or this complacency and sympathy.

---

[6] Old term for *Brassica* spp.

[7] Demokritos of Mendes (see Chapter 2). Johnson, in his notes to the 1878 English edition of *Colloquies* wrongly states that this refers to Democritus of Abdera.

[8] This appears to be one of the earliest statements regarding "companion planting".

As mentioned above, one of the features of the era was that classical knowledge was viewed more critically. It was still regarded mostly with respect, but was supplemented with local knowledge. An erstwhile example of this is the important early English herbal by William Turner (1562). While this work relied heavily on earlier writers, there was almost no mention of antipathy and sympathy, which evidently were held in low esteem by Turner. On the other hand, Turner provided an interesting observation regarding the walnut tree, which accords well with contemporary knowledge regarding the distribution of juglone in the plant:

> The walnut tree, both in his leaves and buddes, hath a certayne bindinge, but the bindinge is most evidently perceived in the utter huskes, both moyst and drye, and therefore fullers dorse them.[9]

A French work that enjoyed great popularity, commencing in the late sixteenth century in the seventeenth century, was *l'Agriculture et Maison Rustique* by Estienne and Liebault, which was published in many editions. Like most similar works it was an amalgam of local knowledge and practice coupled with the often embellished lore of the Roman and Greek authors. The section that deals with walnut trees is instructive, as it highlights these points, and observes also the connection between terms for walnut such as *noyer* and words meaning "harmful", such as *nocif*[10].

> The walnut is a species common enough in all parts, & known to bear such a name [*noyer*] because it is noxious to others which are neighbouring, in the places where it is planted, & and to people & even in babies, all the more one sees by experience, that if a man sleeps below, he will wake up with great heaviness of the head and so stunned that he nearly cannot move himself. And its shade is bad that nothing good can grow underneath there, & that also the roots are of marvelous extent, which spoils all the land where this tree is situated & planted. Thus it should not be planted in a workable field, and especially in those which are richest and most fertile, rather towards the north, on the side of roads or elsewhere, where there are no other fruits which can receive damage from this tree. To place a tree of another species among them, it is not any more useful than putting a little artisan among two great lords; for walnut trees which are naturally great miners with their large roots, remove its food even in a trench, & the cover from its above blocks the sun, & the liberty of the air also: but because the things of this world are thus composed, there is not anything that would not have some adversaries, one must not house the walnut, even plant, or transplant near the oak, not even place it in a trench where an oak has been planted before, because these two trees have a natural hatred for each other, & cannot grow together. (Estienne and Liebault 1689)

A very important work in the history of allelopathy is *Horticultura*, published firstly in 1631 (Figure 5.3), and authored by Peter Lauremberg (1585-1639), a multifaceted professor at the University of Hamburg, and then Rostock. The title provided the origin of the term "horticulture". However, what is important here, is

---

[9] "Binding" means causing constipation; the meaning of "dorse" is obscure (variant of endorse?).

[10] The link between the Castilian words for "walnut" and "harm" is also made by Herrera (1513). It is now generally regarded that the similarity is coincidental, as words for "walnut" such as *nogal* (Spanish), *noyer* (French), and *noce* (Italian) are likely derived from the Latin *nux* for nut. This etymological confusion also occurs in English with the word "noxious". Curiously, Fuchs (1550) claimed that the unrelated Greek word for walnut, *karyon*, was derived from a word meaning "causing headache"; this actually originates with Plutarch in his *Moralia*.

Figure 5.3. The title page from Horticultura *by Peter Lauremberg (1631).*

that this work described what is arguably the first experiment concerning allelopathy. Lauremberg was an eminently practical man, and was very sceptical of the claims concerning antipathy and sympathy among plants. He recorded how he tested the alleged antipathy between cabbage and the vine (see Figure 5.4):

> Firstly, what pertains to the occult discord concerning that between the vine and the cabbage, although it may be worthless, I learned first by testing it in my garden, and then from the records of other diligent researchers of the natural sciences. A few years ago, I sowed cabbage, both the common sort and Savoy, quite densely around and close to two hundred vine cuttings, which I had cut from the vine in the month of March as is customary and struck the roots of the plants. Not only did the cabbage grow most luxuriantly, but also the cuttings, one and all, sprouted successfully and grew to a great height. After three years and with the vines able to be seen fruiting, I again planted out cabbage in great abundance: it did not prevent at all the vines in my nursery from producing a bountiful yield of grape bunches according to my desire. (pp. 65-66)

Lauremberg also noted that he had similarly tested the alleged sympathy between rue and the fig, but he found no improved growth.

A practical view also emanated from John Worlidge (*c.* 1630-1693), a Hampshire gentleman of whom little is known, but who wrote some important agricultural works, prized for their fresh and practical information. One of these, *Systema Horti-Culturae*,

Figure 5.4. Illustration of harvesting cabbages, which gives an idea of the mediaeval cabbage plant. This illustration originates from a Bibliothèque Nationale de France copy of Tacuinum Sanitatis, *a 15th century Latin translation of* Taqwim al Sihhah *(Tables of Health) by Ibn Butlan (d. 1066). The illustration is from BNF MS Latin 9333, fol. 20).*

first appeared only with the authorship of "J.W." in 1677, but later editions, e.g. the third edition (1688) named the author as J. Woolridge (*sic*). In this work, Worlidge wrote:

> There is a sympathy and antipathy in Plants. And many fabulous traditions there are concerning them, but this is certainly observed that some Trees will not thrive under the shade or drip of another, as the drip of a Walnut tree and of a Cherry tree are injurious to other Trees, because the leaf is bitter, and the drip destroyeth such Trees or Plants that are under it. The like doth the drip of the leaves of the Artichoke, and of Hemp, which destroyeth all other vegetables near it, those grounds being free from weeds where they grow, from that cause.

It is worth making a brief note about the topic of crop rotation, a practice which had been certainly known since classical times. While crop rotation became the focal point of allelopathic interest in the nineteenth century, there was little theoretical basis to it during the renaissance. It was acknowledged that it was unwise to grow certain crops in succession on the same ground (Lippay 1663).

POPULARISATION: SYMPATHY, ANTIPATHY AND THE OCCULT

The concepts of sympathy and antipathy originated with the Greek philosophers, including Demokritos, Empedocles, Plato and Aristotle (see Chapter 2). The Romans referred to these ideas as *concordia* and *discordia*, but their natural history writers, particularly those in agriculture, were far more concerned with the everyday practicalities of raising stock and growing crops, than esoteric theory.

Sympathy and antipathy gained a fresh impetus with the rise of Islamic culture, which was in part based itself on cultures, such as that of the Chaldeans, which had been permeated by occultism. The antiquity of those cultures contributing to Islamic works is controversial, and discussed elsewhere (Chapter 3). Nonetheless, from the tenth century onwards, the concepts of sympathy and antipathy became increasingly infused with a plethora of astrological and arcane lore, where objects, including plants, were allied to celestial bodies, and accorded a position within the tetragram of the elements. This essentially determined the object's relationship to various other objects, either in conjunction or in opposition. The assignation of plants to planets and zodiac signs was known to both the Greeks and Romans (see Ducourthial 2003), but seems to have had made only a limited impact with them, at least according to their few surviving botanical works. While the traditional classical teachings waned, the occult ideas, which had probably always maintained popularity among the masses, flourished and remained popular until the eighteenth century.

The fifteenth, sixteenth and seventeenth centuries inherited a hotch-potch of botanical lore, which was often assembled uncritically, with the exception of some of the herbalists, such as Fuchs, and anatomists, such as Malpighi and Ray. Another legacy of the Middle Ages was a strong leaning toward mysticism, and there was a revival of the ideas of antipathy and sympathy, which often incorporated an astrological basis. This commonly became the domain of the herbalists, as most of botanical lore was strongly tied to medicine, which also became permeated with occult practise.

Sympathy and antipathy, particularly involving plants and animals, became common knowledge amongst the populace, and the pervasiveness of these ideas has been underestimated. As noted by Foucault (1966), antipathy and sympathy were seen as important in preserving balance in the world; antipathy in particular was seen as essential in maintaining individuality and preventing coalescence into one monotonous harmonious whole.

Within the domain of plant pathology, which really had to await advances in microscopy for its birth as a science, the concepts of antipathy and sympathy have been regarded as restrictive to its development (Orlob 1971). However, while certain facets of allelopathy are sometimes included within the realm of phytopathology, the popularity of antipathy and sympathy served to maintain interest in some phenomena that may have had a real allelopathic basis. It should be added that the valuable work of Orlob (1973) is really the only other work which has addressed allelopathy, albeit somewhat haphazardly, within the framework of plant pathology, for the time period before 1500, inclusive of the Middle Ages.

One aspect of antipathy and sympathy that has been relatively well documented is the so-called Doctrine of Signatures, wherein the form or attribute of the object supposedly indicated its utility, especially for human health or benefit. This was championed in detail by Giambattista della Porta (1588) in his *Phytognomonica* (Figure 5.5). Many plants have derived their common name from this basis; for example, the liverwort, which on account of its lobed appearance, was alleged to be beneficial for the liver. Many relationships were extrapolated; for example, as cabbage was supposedly antipathetic to the vine, so cabbage was esteemed as a cure for a hangover. Outside of botany, sympathy and antipathy were invoked in many diverse areas, from medicine to warfare. There became widespread belief that certain objects, because of their antipathy or sympathy to other things, could serve as amulets, and have effects that could act over a distance.

A well-known figure associated with this era was the Swiss-borne Philippus Theophrastus Bombast von Hohenheim, known simply as Paracelsus (1493-1541). Although Paracelsus graduated with a least one university degree, he became disdainful of the traditional academic texts dominated by classic authors, and became determined to assimilate practical knowledge. Paracelsus wrote on a great number of matters including iatrochemistry, herbalism, astrology, and medicine, and his writings present

Figure 5.5. Illustration from Phytognomica *by della Porta (1588), showing the resemblance of a tuber (*Orchis *sp.), grass inflorescence (* Digitaria *sp.), and iris tuber (*Hermodactylus tuberosus*) to the human hand.*

an unsettling mixture of original observation, insight, quackery and arrogance[11], which may also reflect the fact that many of the writings credited to Paracelsus may have been written by his students (Browne 1944). Although it is all too easy to take his writings out of context, his diverse writings included a couple of statements that are of allelopathic interest. He was first to have made a clear statement about the fact that almost any substance can be poisonous if sufficiently abundant or concentrated; this concept, particularly in pharmacology is known today as hormesis. Hormesis has been increasingly recognized in today's world as an important concept in both medicine and ecology, and it addresses the fact, in dose-response relationships of almost any substance with physiological activity, that the relationship is biphasic, with stimulation at low concentration and inhibition and/or toxicity at higher concentrations (Calabrese and Baldwin 2003, Belz et al. 2005). Paracelsus (1538)[12] in his *Sieben Defensiones* expressed the basis of this concept quite clearly almost 500 years ago:

> In all things is a poison, and there is nothing without a poison. It depends alone on the dose whether a poison is a poison or not.

Secondly, Paracelsus is noted as having written concerning the concept of companion plants. In his *Buch der Natur* (1525), he observed that St Johnswort (*Hypericum perforatum*) grew larger and more prolifically when associated with other plants.

Many of the works that mention plant antipathies deal with what we would regard as occult phenomena. A prime is example is *De Occulta Philosophia* by Heinrich Cornelius Agrippa von Nettesheim (c. 1486-1535), who was born in Cologne, and taught widely in Germany, France and Italy. *De Occulta Philosophia* was an ambitious compendium of esoteric knowledge first written when Agrippa was only twenty-three years old. It first appeared in print in Latin in 1533, and it has remained in print more or less continuously since that time. The quotes here are derived from the first English edition of 1651, and they provide a taste of the pervasive influence of the concepts of the virtues of things, and of antipathy and sympathy (see Figure 5.6):

> In the next place, it is requisite that we consider that all things have a friendliness, and enmity amongst themselves, and everything hath something that it fears & dreads, that is an enemy, and destructive to it; and on the contrary something that it rejoyceth, and delighteth in, and is strengthened by. So in the elements, Fire is an enemy to Water, and Aire to Earth, but yet they agree amongst themselves. (Book 1, Chapter XVII)

> The Vines love the Elme, and the Olive-tree, and Myrtle love one the other: also the Olive-tree and Fig tree. (Book 1, Chapter XVII)

> Also Origanum is contrary to a certain poisonous fly, which cannot endure the Sun, and resists Salamanders, and loaths Cabbage with such a deadly hatred, that they destroy one the other. (Book 1, Chapter XVIII)

Giambattista della Porta (c. 1535-1615) was born in Naples, and was something of a prodigy, an image which he unhesitatingly promoted. He is noted as founding

---

[11] This is simply exemplified in his chosen names of Paracelsus, or "beyond Celsus".

[12] None of Paracelsus' works were actually published during his lifetime, and the date given is that of the manuscript according to the collected edition by Sudhoff.

Figure 5.6. *The use of certain sympathetic trees in the culture of grape vines. This damaged illustration is from a Bibliothèque Nationale de France manuscript copy of* Taquinum Sanitatis, *a Latin translation of* Taqwim al Sihhah *(Tables of Health) by Ibn Butlan (d. 1066). The illustration (Nouvelle Acquisition MS Latine 1673, fol. 2) dates from about 1390-1400.*

Europe's first scientific association, the Accademia Secretorum, the predecessor of the Accademia Lincei, of which Galileo was its most famous member. Della Porta published treatises in many disciplines, but our interest centres on a work known as *Magia Naturalis*, which was first published in its entirety in 1589 in Latin, and which became very popular. The definitive English edition did not appear until 1658, under the title *Natural Magick*. Della Porta's status in the scientific community was high, despite his obvious interest in the occult, which indicates that acceptance of occult explanations was commonplace. Della Porta wrote:

> By reason of the hidden and secret properties of things, there is in all kinds of creatures a certain compassion, as I may call it, which the Greeks call Sympathy and Antipathy, but we term it more familiarly, their consent, and their disagreement. For some things are joined together as it were in a mutual league, and some other things are at variance and discord, among themselves; or they have something in them which is terror and destruction to each other, whereof there can be rendered no probable reason: neither will any wise man seek after any other cause hereof but only this, That it is the pleasure of Nature to see it should be so, that she would have nothing to be without his like, and that, amongst all secrets of Nature, there is nothing but hath some hidden and special property; and moreover, that by this their Consent and Disagreement, we might gather many helps for the uses and necessities of men, for when once we find one thing at variance with another, presently we may conjecture, and in trial so it will prove, that one of them may used as a fit remedy against the harms of the other: and surely many

things which former ages have by this means found out, they have commended to their posterity, as by their may appear. There is deadly hatred, and open enmity betwixt Coleworts and the Vine; for whereas the Vine windes it self with her tendrels about every things else, she shuns Coleworts only: if once she come neer them, she turns her self another way, as if she were told that her enemy were at hand: and when Coleworts is seething, if you put never so little wine unto it, it will neither boil nor keep the colour. By the example of which experiment Androcides found out a remedy against wine, namely, that Coleworts are good against drunkennesse, as Theophrastus saith, in as much as the Vine cannot away with the savour of the Colewort. And this herbe is at enmity with Cyclamine or Sow-bread; for when they are put together, if either of them be green, it will dry up the other: now this Sow-bread being put into wine, doth increase drunkennesse, where as Coleworts is a remedy against drunkennesse, as we said before. Ivy, as it is the bane of all Trees, so it is most hurtful, and greatest enemy to the Vine; and therefore Ivy also is good against drunkennesse. There is likewise a wonderful enmity betwixt Cane and Fern, so that of them destroyes the other. Hence it is that a Fern root powned, doth loose and shake out the darts from a wounded body that were or cast out of Canes: and if you would not have Cane grow in a place, do but plow up the ground with a little Fern upon the Plough-shear, and Cane will never grow there. Strangle-tare or Choke-weed[13] desires to grow amongst Pulse, and especially among Beans and Fetches, but it choaks them all: and thence Dioscorides gathers, That if it be put amongst Pulse, set to seethe, it will make them seethe quickly. Hemlock and Rue are at enmity; they thrive each against other: Rue must not be handled or gathered with a bare hand, for then it will cause Ulcers to arise; but if you do chance to touch it with your bare hand, and so cause it to swell or itch, anoint it with the juice of Hemlock.

(Book I, Chapter VII, pp. 8-9)

It is surprising that a seminal figure concerning a scientific view of the sympathy and antipathy of plants was the Jesuit, Athanasius Kircher (c. 1601-1680). The Jesuits had a tradition in scientific investigation, which was encouraged within their Order, and other notable Jesuit works which dealt with sympathy and antipathy were those of Juan Eusebio Nieremberg (1635).and Antoine Mizauld (1689), the former of which was an influence on Kircher. Kircher is best known for his works on magnetism and geology. However, while the subject of magnetism concerned attraction and repulsion, for Kircher, this also embraced the allied phenomena of sympathy and antipathy. Thus Kircher proposed what he regarded as an original theory of the sympathy and antipathy of plants, although in essence, it differed little from the ideas of Theophrastus. He suggested that plants through their vapours or exhalations generated a sphere of influence, which could either be harmful or beneficial. Thus, for example, is explained the enmity of the cabbage and the vine, known to the ancients, and other similar interactions, such as the cabbage and the cyclamen, and the fern and the reed (Kircher 1641).

The most extreme form of occult botany was astrological botany, in which not only were plants assigned a position within the tetragram of the elements, but they were also aligned with one or more of the major celestial bodies. Thus plants that were soothing, such as mint, were reckoned to be influenced by the planet Venus (which at the time was regarded as cool and moist), whereas spicy plants were often aligned with fiery Mars, the red planet. The aims of astrological botany were

---

[13] Strangle-tare and chokeweed were terms for *Orobanche* spp.

distinctly anthropocentric, in allowing the reasoned prescription of various remedies. A nicety for the herbalist was that a remedy might be deemed to work because it was antipathetic against an ailing organ or because it boosted an organ through sympathy. Not surprisingly, the assignations of plants varied considerably among practitioners. Astrological botany gained a foothold in Europe, and attained its greatest influence and complexity in the seventeenth century (e.g. Cardilucius 1686).

As on the Continent, England too gradually became enveloped in the sway of astrological botany, and there was often acrimony between the traditional medical practitioners and those invoking the occult, especially as the latter often enjoyed considerable popularity. The early English works generally reiterated the traditional lore of antipathy and sympathy; however, by the seventeenth century, the most popular botanical work in England was that of the astrological botanist Nicholas Culpeper (*q.v.*).

Amongst the orthodox contributions was that of the English physician Thomas Cogan (*c*. 1545-1607), who in his *Haven of Health* (Cogan 1584) provided the following:

> The Vine and the Coleworts be so contrarie by Nature that if you plant Coleworts neare to the rootes of the Vine of it selfe it will flee from them, therefore it is no maruaile[14] if Coleworts be of such force against drunkennesse.

Another contribution came from Thomas Hill, an English writer on gardening and horticulture, who wrote *The Gardener's Labyrinth*, originally pseudonymously under the punned name Didymus Mountain in 1577. Beginning at about this point in time in the English literature, we find that the enemy of the vine can be a plant other than colewort or cabbage, an error which likely arose out of similarity of the Greek or Latin words for the various related plants:

> This no doubt is a secrete very marvellous that the radish in no wise agreeth to be planted or grow nigh to the vine, for the deadly hatred between them, insomuch that the vine nere growing, turneth or windeth backe with the branches, as mightily disdaining and hating the radish growing fass by: if we may credite the learned Plinie, Galen and the Neapolitaine Rutilius, which seem to have diligently noted the same.
>
> (Book II, p. 7)

> Athaenius writeth, that the colwort ought not in any case to be planted or sowne neare to the vine, nor the vine in like manner nigh to it, for such is the great enmitie between these two plants (as Theophrastus witnesseth) that being both in one plot together, these so hinder one another, that the vine in branches growing further, rather turneth or goeth back againe, from the colewort, then stretching toward it, and it yieldeth less fruite there through.

There was little original in these works, and the rendering, if not dubious, is quaint, as may be seen in excerpts from *Cornucopiae* by Thomas Johnson (1595):

> The Vine is greatly delighted with the Elme and yeeldeth more frute being placed together.

> The myrtle tree and the Olive tree love each other mutually, even so doth the Olive tree and the figge tree.

---

[14] = marvel.

The Olive tree so detesteth the Cowcumber[15] that being placed nere together, they will turne backe and growe hookewise lest they shoulde touch one another.

Another amusing contribution came from John Taylor, who was a ferryman by profession, but had a way with words, that he was able to exploit for profit. His often irreverent poems were commonly published by subscription. In 1620 he published such, as a raucous booklet entitled *The Praise of Hemp-seed*, in which appeared:

> Moreover, *Hempseed* hath this vertue rare
> In making bad ground good, good corne to beare,
> It fats the earth, and makes it to excell
> No dung, or marle, or mucke can do't so well :
> For in that Land which beares this happy seed
> In three yeares after it no dung will need,
> But sow that ground with barley, wheat, or rye
> And still it will encrease aboundantly ;
> Besides, this much I of my knowledge know
> That where *Hemp* growes, no stinking weed can grow,
> No cockle, darnell, henbane, tare, or nettle
> Neere where it is can prosper, spring, or settle,
> For such antipathy is in this seed,
> Against each fruitlesse undeserving weed,
> That it with feare and terror strikes them dead,
> Or makes them that they dare not shew their head.
> And as in growing it all weeds doth kill
> So being growne, it keepes it nature still,
> For good mens uses serves & still releives
> And yeelds good whips and ropes for rogues and theeves.

A thoroughly interesting compilation of the lore of antipathy and sympathy relating to plants, animals, minerals and humans was authored by the Scottish physician, Sylvester Rattray (1658) and was also included in *Theatrum Sympatheticum* (1662) a valuable German compendium of treatises relating to antipathy and sympathy. In some respects, Rattray's treatise has some ecological merit, as, the relationships of certain plants to particular environments, such as streams or sand, are recorded variously as sympathies or antipathies. Amongst the sympathies given was an early version of "garlic loves roses". The plants that Rattray recorded as being antipathetic to other plants are of especial interest, and one can find what are likely the earliest statements concerning allelopathy, where the harmful effects of a plant were attributed to a particular chemical substance:

> Cabbage[16], if planted close to cyclamen, withers.
> Cabbage through oregano, withers.
> Cabbage, and rue, occurring together, wither.
> Vines (grape), if planted with cabbage, wither & bend away.
> Even their juices are able to exert an antipathy, for cabbage reduces drunkenness from wine
> Also if cabbages boiled in a pot, they act weakly on the vines, which are not effected nor grow weak.
> Cucumber and the olive tree have an enmity.

---

[15] = cucumber.

[16] The term *Brassica* may translate as cabbage or colewort.

> The oak tree is opposed to the olive tree, thus if they occur together, they wither.
> The vine (grape) abhors laurel, and does not grow, if laurel is planted near it.
> Harmful medical plants are abhorrent to grain crops, for if water is sprinkled on grain crops, they grow rotten without delay.
> The reed wilts opposite ferns, & vice versa.
> The oak growing near to the walnut tree, withers.
> Hemlock wilts due to the vine (grape), & to the poison of hemlock, wine is effective as an antidote, if really offered hemlock, this shows the way.
> The nut[17] tree, by the shade of the sun is pernicious to neighbouring trees, and plants.
> Lavender is fairly inimical to neighbouring herbs.
> Mandrake is harmful to vines (grape), as it imparts to it a narcotic.
> *Chamaepitys*[18] is opposed to the vine (grape), and its shoots as a garland act against drunkenness.
> Colocynth [19] infuses all plants in its vicinity with poison, & renders them bitter.
> Basil poisons Cuscuta, growing on it.
> Willow is inimical to grain plants, for willow thickets do not grow alongside.
> Rue does not grow with basil.
> Yews do not allow the grafting of other trees.
> Cherry does not allow the grafting of peach or terebinth.[20]
> If wheat is placed in flour sacks, the fruits contracts blight, from its products.
> Incense trees taint other fragrant trees.
> Lupins are dangerous to sycamore, and they do not grow together.
> The rose abhors onion.
> Rue detests hemlock.
> Solomon's seal[21] is opposed to cabbage.
> Orobanche strangles all legumes, if growing among them.
> Aconite is adverse to rue.
> Ivy is thoroughly harmful to all trees.
> If residues of bean plant are placed near tree roots, they render the trees unproductive.
> Antithura[22] are inimical to aconite. (pp. 1-30)

One the most important contributions to early English botany was that of John Gerard (1545-1612). Typical of botanists of his era, Gerard was a physician with a love of plants. Gerard's *The Herball or General Historie of Plants* (Gerard 1597) was the standard English compendium on plant diversity and uses of plants, for the following fifty years. In this work we find a curious variation of the all too familiar vine antipathy:

> Divers think that the Horse Radish[23] is an enemie to Vines, and that the hatred between them is so great, that if the rootes hereof be planted neere to the vine, it bendeth backward from it as not willing to have fellowship with it.

---

[17] The original Latin word used is *nux*, which can refer to any of the nut trees, and is variously translated as the walnut tree, hazelnut tree, and almond tree, which explains the varied statements about the alleged antipathy of such trees.

[18] This term can refer to either *Ajuga* or *Hypericum*.

[19] A type of small wild melon.

[20] *Pistacia terebinthus*

[21] Rattray uses the term *sigillum st. mariae*, which Gerard (1597) refers to as Solomon's seal or *Polygonatum multiflorum*.

[22] Meaning is unknown.

[23] The horse-radish is *Armoracia rusticana*, which although in the Brassicaceae, is quite easily distinguished from the common radish. Part of the problem may due to translating the Greek term for

However, Gerard was critical of the supposed antipathy of the fern and the reed:

> [It] is vaine to thinke that it hapneth by any antipathetic or naturall hatred, and not by reason that this ferne prospereth not in moist places, nor the reed in dry.

Another English herbalist was John Parkinson (1567-1650), an apothecary and herbalist to the king. In *Theatrum Botanicum* (Parkinson 1640), he wrote:

> .... and even Galen himselfe applied the juice thereof, to the temples of them that had paines in their head caused by drunkenness; for as they say there is such an antipathy or enmity between the Vine and the Colewort, that one will die where the other groweth. (p. 271)

The most famous of the astrological botanists was the Englishman, Nicholas Culpeper (1616-1654). Culpeper spent some time at Cambridge University, but a series of misfortunes caused him to train eventually as an apothecary, where he gained familiarity with herbs, and he was also encouraged to study astrology by William Lilly. Culpeper's success as an apothecary led him to question the utility of the Royal College of Physicians. In 1649 Culpeper published an English lay version of *Pharmacopoeia Londonensis*, previously only available in Latin, which brought him into vehement conflict with the physicians. This occurred during the Cromwellian era, and consequently the Royal College of Physicians, with abolition of the Star Chamber, had no power to prosecute Culpeper who was also a staunch anti-royalist. It is relevant that during this period of civil war (1642-1649), much censorship was lifted with the disbanding of the royalist Company of Stationers established in 1603, and consequently works which had previously been banned, notably in subjects such as astrology, enjoyed immediate popularity. Occult ideas which would have been once censored, due to their offence to the Church, were cleverly manipulated by Culpeper, who maintained that astrology was endorsed by the Bible in Genesis (1: 15-18): "God made the Sun, Moon and Stars to rule over night and day…to be signs of things to come."[24] (Thulesius 1992). Culpeper was known especially for his populist herbal, which while drawing heavily from Gerarde and other authors, enjoyed unparalleled success. It was first published in 1652 with the title *The English Physitian, or an Astrologo-Physical Discourse of the Vulgar Herbs of this Nation*, and this work, with its numerous editions, has remained in print, more or less continuously, for over 350 years. Its contribution to the lore of allelopathy is indeed minimal, but the work has served to maintain the concepts of antipathy and sympathy very much in the public eye. Despite an injudicious embrace of astrology, the work of Culpeper epitomises a period of social and cultural upheaval in which traditional ways were challenged, and there was at least attempts to find new methods based on empirical information.

---

cabbage, which also translates as colewort, a general term for any plant resembling kale in appearance (see note 5, Chapter 2).

[24] This quote is given by Thulesius without reference. This biblical text differs substantially from normal versions, and one is tempted to speculate that Culpeper has taken considerable liberty to promote his own cause.

Of cabbages and coleworts, Culpeper wrote:

> They are much commended being eaten before meat, to keep one from surfeiting, as also from being drunk with too much Wine, or quickly make a man sober again that is drunk before. For (as they say) there is such Antipathy or enmity between the Vine and the Colewort, that the one will die where the other groweth.

The entry for fennel provides an interesting example of the astrologic basis of a plant's efficacy:

> One good old fashion is not yet left off, viz. to boil fennel with fish: for it consumes that phlegmatic humour which fish most plentifully afford and annoy the body with, though few that use it know it wherefore they do it. I suppose the reason of its benefit this way is, because it is an herb of Mercury, and under Virgo, and therefore bears antipathy to Pisces.

Francis Bacon (1561-1626) excelled in many disciplines. His prodigious career in law led ultimately to the position of Lord Chancellor of England, and honours of peerage, including acquisition of the title Lord Verulam in 1618. While his political career became unstuck with charges of corruption in 1621, his philosophical and scientific studies continued unabated. In botany he is remembered chiefly through a work published posthumously, firstly in 1627, *Sylva Sylvarum*. Bacon's approach in science was one of rigour, and he attempted to disencumber science from the shackles of the Greek and other dicta, and to make statements based upon experimentation, observation and the testing of hypotheses, often referred to today as the Baconian method. He had scant regard for the ideas of sympathy and antipathy in plants, as apparent in the following from *Novum Organum*, which was initially published under the grandiose title *Franciscus de Verulamio Summi Angliae Cancellaris Instauratio Magna* (Bacon 1620), a projected six-part magnum opus that saw completion of just two parts:

> But the inner consents and aversions, or friendships and enmities, of bodies (for I am almost weary of the words sympathy and antipathy on account of the superstitions and vanities associated with them) are either falsely ascribed, or mixed with fables, or from want of observation very rarely met with. For if it be said that there is enmity between the vine and colewort, because when planted near each other they do not thrive, the reason is obvious — that both of these plants are succulent and exhaust the ground, and thus one robs the other. If it be said that there is consent and friendship between corn and the corn cockle or the wild poppy, because these herbs hardly come up except in ploughed fields, it should rather be said that there is enmity between them, because the poppy and corn cockle are emitted and generated from a juice of the earth which the corn has left and rejected; so that sowing the ground with corn prepares it for their growth. And of such false ascriptions there is a great number. (From the 1863 edition of *The New Organon*, translated by J. Spedding *et al*. Aphorisms, Book II, Section L)

The same fundamental points were dealt with at length in *Sylva Sylvarum*:

> There are many Ancient and Received Traditions and Observations, touching the Sympathy and Antipathy of Plants: for that some will thrive best growing neere others; which they impute to Sympathy: and some worse; which they impute to Antipathy. But these are Idle and Ignorant Conceits; and forsake the true Indication of the Causes; as the most Part of Experiments, that concerns Sympathies and Antipathies doe. For as to Plants, neither is there any such secret Friendship, or Hatred, as they imagine; and if we should be content to call it Sympathy, and Antipathy, it is utterly mistaken; for their

> Sympathy is an Antipathy, and their Antipathy is a Sympathy: for it is thus: wheresoever one plant draweth such a particular Juyce out of the Earth; as it qualifieth the Earth; So as that Juyce which remaineth is fit for the other Plant, there the Neighbourhood doth good; Because the Nourishments are contrairie or severall: But where two Plants draw (much) the same Juyce, there the Neighbourhood hurteth; for the one deceiveth the other.
>
> First therefore, all Plants that doe draw much Nourishment from the Earth, and so soake the Earth, and exhaust it; hurt all Things that grow by them; as Great Trees, (especially Ashes,) and such Trees, as spread their Roots, neere the Top of the Ground. So the Colewort is not an Enemy (though that were anciently received) to the Vine only; But it is an Enemie to any other Plant; because it draweth strongly the fattest Juyce of the Earth. And if it be true, that the Vine, when it creepeth neere the Colewort, will turne away; This may because there it findeth worse Nourishment; For though the Root be where it was, yet (I doubt) the Plant will bend as it nourisheth.
>
> Where Plants are of severall Natures, and draw severall Juyces out of the Earth, there (as hath beene said) the One set by the other helpeth: As it set downe by divers of the Ancients, that Rew doth prosper much, and becommeth stronger, if it be set by a Figge-Tree; which (we conceive) is caused Not by Reason of Friendship, but by Extraction of a Contrairie Juyce: the one Drawing Juyce fit to result Sweet, the other bitter. So they have set downe likewise, that a Rose set by Garlick is sweeter[25]; Which likewise may be, because the more Fetide Juyce of the Earth goeth into the Garlick: And the more Odorate into the Rose. (1631 edition, Century V, pp. 121-122)

Bacon gave short shrift to those espousing ideas of sympathy and antipathy based on astrology, and had no time for contemporaries such as Paracelsus:

> Some of the Ancients, and likewise divers of the Moderne Writers, that have laboured in Natural Magick, have noted a Sympathy, between the Sunne, Moone, and some Principall Starres; And certaine Herbs, and Plants. And So they have denominated some Herbs Solar and some Lunar; and such like Toyes put into great words.
> (1631 edition, Century V, p. 124)

Given that allelopathy, in part, owes its origins in the twentieth century to the effects of ethylene (see Chapter 11), it is noteworthy that Bacon was likely the first to record the effect of one ripening fruit on another, evidently due to ethylene:

> Note, that all these were compared with another Apple of the same kind that lay of it selfe; and in comparison of that, were more sweet, and more yellow, and so appeared to be more ripe. (1631 edition, Century IV, Experiment 323, p. 83)

The Englishman, Ralph Austen, a proctor at Oxford University, wrote a critique of some of Bacon's experiments with plants, which quite fairly questioned Bacon's seeming support of sympathy between plants, for example, rue and fig, garlic and rose (Austen 1658). Austen could not support the idea that the soil contained myriad juices, which were selectively extracted by various plants, a notion which seemingly explained the diversity of plants growing in the same soil, and which persisted into the nineteenth century.

---

[25] Note that this is an early statement of the relationship widely known amongst modern gardeners through the book on companion planting, *Roses Love Garlic*, by L. Riotte (1985); see also note 5. Curiously the French equivalent to this title is *Le Poireau Préfère les Fraises* (The leek prefers strawberries) by Hans Wagner (2001).

Another important Englishman who shared Bacon's disdain for discourse about antipathy was John Evelyn (1664). Evelyn was a great proponent of planting walnut in England, and he pointed across the English Channel to Burgundy (Bourgogne) in central France, where walnut was planted in wheat fields[26]:

> It is so far from hurting the crop, that they look on them as great preservers, by keeping the ground warm.

Thomas Browne (1605-1682) trained as a physician, but is remembered chiefly for his verbose essays, on a wide range of often esoteric topics, that overflowed with classical references. One of his most curious works is titled *The Garden of Cyrus, or the Quincunciall Lozenge, or Net-work Plantation of the Ancients, Artificially, Naturally, Mystically Considered* (1658). The work delved into discovering the significance of diamond-shaped or rhomboidal patterns in nature, wherein any point would neighbour four others (hence five points per group). Thus the work touched on the subject of close-packing in plant structures, as found in the series of scales in pine cones, or florets in composite inflorescences. Pursuing this line of argument, Browne considered the spacing of plants, recalled Solon's law (see Chapter 2), and discussed the possible causes:

> Whereby they also avoided the peril of συνολεθρια [synolethria[27]] or one tree perishing with another, as it happeneth ofttimes from the sick effluviums or entanglements of the roots, falling foul with each other. Observable in Elmes set in hedges, where if one dieth the neighbouring Tree prospereth not long after.

> And as they send forth much, so may they receive somewhat in: For beside the common way and road of reception by the root, there may be a refection and imbibition from without; For gentle showrs refresh plants, though they enter not their roots; And the good and bad effluviums of Vegetables promote or debilitate each other. (Chapter IV):

The above passage raises the question as to what exactly are effluvia? There is no simple answer to this. This term, or the equivalent "exhalations", dates back to early Greek works. Amongst the fragments of the writings of Empedocles is a statement that decrees that all created (living) things have effluvia. However, these effluvia can include gases now associated with respiration, fluids associated with excretion, perspirants, exudates, and so forth. The nature and role of effluvia in plants was not at all understood, and those associated with the roots least of all. Swedenborg (1763) later advanced the naive idea that effluvia from plants can give rise spontaneously to insect pests.

Another citation of sympathy and antipathy in plants occurred in *Anatomy of Melancholy* by Robert Burton (*c.* 1620), who adopted the pen-name of Democritus Junior:

> No creature, S. Hierom concludes, is to be found, quod non aliquid amat, no stock, no stone, that hath not some feeling of love, 'Tis more eminent in plants, herbs, and is especially observed in vegetables; as between the vine and elm a great sympathy, between the vine and the cabbage, between the vine and the olive, Virgo fugit Bromium, between the vine and bays a great antipathy, the vine loves not the bay, nor

---

[26] However, see Stendahl (1830), Chapter 7.

[27] There is a word in English, synlethal, who means roughly the same as this coinage of Browne's.

his smell, and will kill him, if he grow near him; the bur[28] and the lentil cannot endure one another, the olive and the myrtle embrace each other, in roots and branches if they grow near.

These notions were reiterated by others, and it is quite surprising how widespread was the notion of antipathy and sympathy in literature. For example, the English poet Abraham Cowley (1618-1667), known chiefly for his romantic poems and his essays, in later life turned his hand to the study of medicine and consequently the study of plants. It is of interest that Cowley was a contemporary of Culpeper; however, Cowley was a royalist, and did not enjoy favour until the restoration of the monarchy. In 1662 he published the first two parts of a botanical work in elegaic verse, entitled *Libri Plantarum*. This Latin work was eventually completed in six parts, and an English translation appeared, well after Cowley's death, in 1689. Cowley wrote in his poems, on sow-bread[29] (*Cyclamen* spp.):

> See how with Pride the groveling Pot-herb swells[30],
> And sawcily the generous Vine repells:
> Her, that great Emperours oft in Triumph drew,
> A base, unworthy Colewort does subdue.
> But though o'r that the wretch victorious be,
> It cannot stand, puissant Plant! Near Thee
> For Meat to Medicines still must give the place,
> That feeds Diseases, which away these chase.
> You bravely Men and other plants outvie,
> Who no kind Office do, until they die;
> Thy virtues thou, yet living, do'st impart,
> And ev'n to thy own Garden Physick art.

Similarly, the noted English poet John Philips, in imitation of Virgil's *Georgics*, wrote a long piece entitled *Cyder* (1708), of which the following alluded to ideas on sympathy and antipathy borrowed from classical writers and Worlidge:

> The Prudent will observe, what Passions reign
> In various Plants (for not to Man alone,
> But all the wide Creation, Nature gave
> Love, and Aversion): Everlasting Hate
> The *Vine* to *Ivy* bears, nor less abhors
> The *Coleworts* Rankness; but, with amorous Twine,
> Clasps the tall *Elm*: the *Pæstan Rose* unfolds
> Her Bud, more lovely, near the fetid *Leek*,
> (Crest of stout *Britons*,) and inhances thence
> The Price of her celestial Scent: The *Gourd*,
> And thirsty *Cucumer*, when they perceive
> Th' approaching *Olive*, with Resentment fly
> Her fatty Fibres, and with Tendrils creep
> Diverse, detesting Contact; whilst the *Fig*
> Contemns not *Rue*, nor *Sage*'s humble Leaf,
> Close neighbouring: The *Herefordian* Plant

---

[28] Likely *Arctium* spp. or *Xanthium* spp.

[29] Sowbread was an old name for cyclamen, based on the fact that it was eaten by wild pigs.

[30] At this point there was a note in the original edition of the poem, which stated: "The Colewort is said to kill the Vine, and it self kill'd by this Herb." The term Pot-herb refers to a vegetable added to the pot in cooking, but especially cabbage or colewort.

> Caresses freely the contiguous *Peach*,
> *Hazel*, and weight-resisting *Palm*, and likes
> T' approach the *Quince*, and th' *Elder*'s pithy Stem;
> Uneasie, seated by funereal *Yeugh*,
> Or *Walnut*, (whose malignant Touch impairs
> All generous Fruits), or near the bitter Dews
> Of *Cherries*. Therefore, weigh the Habits well
> Of Plants, how they associate best, nor let
> Ill Neighbourhood corrupt thy hopeful Graffs. (Book I, pp. 16-17)

It is believed by some that the works by William Shakespeare, were actually written pseudonymously by someone such as Francis Bacon. This case is taken to the extreme by Bormann (1895) who claims that the sympathy and antipathy of plants as discussed in *Sylva Silvarum*, is represented symbolically in *The Taming of the Shrew*; however, to my mind, the connection is far-fetched.

In 1692 Richard Bentley, the leading English classical scholar of his time, was compelled to state:

> When Occult Quality, and sympathy and antipathy were admitted for satisfactory explication of things, even wise and vertuous men might swallow down any opinion that was countenanced by Antiquity.

By the beginning of the eighteenth century, the tide had begun to turn against the credibility of sympathy and antipathy. Following is a translation of a rather verbose summary given by Vallemont (1705), a French priest and collector of curiosities:

> The ancient philosophers said a lot of pith on mutual love, and reciprocal aversion of plants. It is true, that they had recourse to the pompous words of sympathy, and antipathy, like a special refuge to hide their ignorance. According to the naturalists, there are some plants, which seek one another, and which live together with every possible agreement: there are others which cannot tolerate each other, and of which the vicinity is equally fatal to one another. Bacon, Chancellor of England, mocked these supposed hatreds and imaginary friendships. Here, according to this great man, all is a mystery. Two plants, which are nourished by the same type of juice, harm each other utterly, when they are too close. The sharing of the food, which is available to both of them, emaciates one and the other: *obest viciniae, altera alteram fraudante*. That is antipathy. On the contrary, two plants, which need, for food, two very different juices, grow and flower together perfectly well. *Plantae indolis non unius, et succo diverso alendae amica conjunctione gestiunt*. That is sympathy. *Sylva Sylv.* Cent. V. n. 480 and 481.
>
> But the mystery will be revealed, by an explanation so simple, the philosophy becomes to the whole world.: its credit diminishes; and near the people, it loses the reverence which it deserves. What would it be: thus there is sympathy according to the principle of Bacon, between the fig and the rue. There is no argument about the food. The juice, which it transports to the rue, does not suit the fig. Their good intelligence will show that evermore.
>
> There is then sympathy between garlic and the rose. There must be an odorous juice from the rose, and a ill-smelling juice from the garlic. That being, nothing prevents the rose from growing in the same ground with the garlic; then the garlic does not at all vie with the rose to steal its food. When even the rose has garlic as a neighbour, it is in it, most beautiful, and odoriferous.
>
> On the contrary, there is antipathy between rosemary, lavender, laurel, thyme, marjory, which would only suffer together; because they need nourishing juices that are very similar. Thus, plants starve one and another, and visibly dwindle, when they are neighbours.

> There is a tremendous antipathy between the cabbage and the cyclamen; between hemlock and rue; between the reed and the fern. These plants endeavour so terribly of it, says P. Kirker the Jesuit, that they cannot live together, within the sphere of one another. Their struggles are so cruel, that one of the two must die, and often one and the other are consumed with pity, and die of sadness: *Adeo saevas luctas meunt ut utrumque viribus destitutum marcescens contabescat.* Art. Magnet. Lib. Iii. Cap. 2 pag. 494. There is that one calls an irreconcilable hatred. One would not have thought that there was such unruliness, and a discord so murderous in the family of plants. Perhaps that the philosophers pinch sometimes the cothurne of the poets, in order to enhance and swell their style. This savant Jesuit gives the reason of the demise of these plants, that exhaled from the body of certain plants is a vapour, an exhalation, a bad breath that does not please at all to others; and that when a delicate plant has the misfortune to find itself in the sphere of the strong odour of a foul-smelling plant, the other suffers, dwindles without cessation, and finally dies of disgust: *Plante enim, sive vapore, sive exhalatione certas quasdam sphaeras causantur, intra quas alia consituta alterant.* Thus it is explained the antipathy of certain plants. I would accommodate more willingly the physics of Bacon, who attributes the demise of this plant to theft, that its neighbour does for itself of a food which it needs. (pp. 168-173)

Similarly, Henry Curzon, in his *Universal Library*, dismissed the lore of antipathy as being ignorant explanations by ancient authors, and wrote (Curzon 1712):

> And that which is called Antipathy between the Vine and the Cabbage is as improper, for the reason of their not thriving when sown near to one another in the same Ground, is, because the Nutrient proper for the Growth of one, is also proper for the increase of the other, and the Vine draws away all that Aliment by Strength (as great Fishes devour less) which should nourish the Cabbage, whereby the latter droops and dies. The like may be said of many other Vegetables, which are accounted to their Antipathies. (p. 529)

Herman Boerhaave (1668-1738) is more deservedly treated as part of the eighteenth century, a seminal period in the history of allelopathy, as it appears his simple theory of root excretion set a basis for later workers (See Chapter 6). Boerhaave had a distinguished career in medicine, and he was an important teacher of chemistry. A rare publication entitled *An Essay on the Virtue and Efficient Cause of Magnetical Cures*, reminiscent of the writings of Kircher, has been attributed, albeit likely falsely, to Boerhaave (1743)[31]. Parts of its contents are more relevant to the subject at hand, and perhaps highlight the lingering influence of the theories of antipathy and sympathy:

> According to Pliny, there is no greater Poison for Trees than wild Parsnip, because, being near them, it taketh not only all the Substance away, but it boreth and pierceth Holes, like as with a Sword, through the very Roots of the trees.[32] (p. 13)
>
> I proceed to Vegetables, where equally we meet with an evident and notable Discord. For an Oak-Tree will not prosper in Places where Olives grow; and an Olive-Tree

---

[31] This is real doubt as to whether Boerhaave was the real author of this work. The occult content, inaccuracy, and the lack of editions in languages used by Boerhaave cast further suspicion on the matter.

[32] I cannot find any passage in Pliny that is similar. My best guess is that this refers to Pliny (Book XVIII, Chapter 8; see Chapter 2 of the present work) wherein hemlock is used to kill plants. Hemlock (*Conium maculatum*) has on occasion been mistaken by wild plant harvesters for the benign parsnip (*Pastinacea sativa*), both tap-rooted members of the Apiaceae, with dire results.

> leaveth in Groves of Oaks, such offensive Roots, which actually do kill the Oak-Trees. The same being planted near Wallnut-Trees, either dieth itself or remains always weak, or causeth the same Effects of the Wallnut.
>
> The Hatred between Colewort or Cabbage and Vines, is more visible. For a Vine with its crooked Tendrels doth tye and bind itself to every Thing that it doth catch, only to refuseth the same to Cabbage, and being near it, it bends to the opposite Side. Colewort or Cabbage being boiling, if only a few Drops of Vine are pour'd upon it, will immediately cease to boil, and the Cabbage will lose its Colour. The same will dry through and through, if *Cyclamen*, a Kind of Briony or Origan, or wild Marjorum is near it; and Vines will become worse in the Neighbourhood of Bay-Trees.

It should be stated in this discussion that an interaction described during this era as seemingly what we might describe as allelopathic, was likely to fall under the rubric of antipathy and sympathy, or something similar. However, mere description of an interaction as antipathetic does not necessary imply that it is allelopathic; it simply means that the species do not readily co-occur. For example, while the reed and the fern were often described as antipathetic, given their respective ecologies, it is hardly surprising that they are not found together.

MYTHS AND TRAVELLERS TALES

There is little doubt that myth and superstition have contributed to the lore of allelopathy, and likely vice versa. The alleged antipathy between the vine and the cabbage is a good example. However, other examples exist. The elder (*Sambucus nigra*) was commonly credited with being able to affect other species through its leaf leachate. The elder is a tree which is also strongly associated with superstition, and has reputations both good and bad. An early source of its power was its alleged association with the suicide of the apostle Judas. In medieval times, it was believed that Judas hanged himself from an elder tree. This possibly originated from a traveller's tale, as John Mandeville[33] wrote:

> ....faste by' the Pool of Siloam, the identical 'Tree of Eldre that Judas henge himself upon, for despeyr that he hadde, when he solde and betrayed oure Lord.

At about the same time during the middle of the fourteenth century, Langland in *Vision of Piers Plowman* wrote:

> Judas he japed with Jewen silver
> And sithen an eller hanged hymselve.

This story was later reinforced with lines from Shakespeare's Love's Labour's Lost: "Judas was hanged on an elder." Curiously, the elder did not grow in biblical regions. In any case, as a result, the elder is regarded as either lucky or unlucky, depending on your point of view. The trunk of the elder lacks heartwood, and therefore, the elder is regarded as heartless. Such melancholy attitudes to elder are echoed in Shakespeare's Cymbeline ("the stinking elder, grief", and Edmund Spenser's "Shepherd's Calender":

---

[33] *The Travels of Sir John Mandeville* was written in the 14th century, and was once highly esteemed for its details of exotic lands. It is now regarded as a largely fabricated work, cobbled together from travellers' tales, perhaps by a well educated Englishman living in France.

The water nymphe, that wont with her to sing and daunce,
And for her girlond olive braunches beare,
Nowe balefull boughes of cypres doen advaunce!
The Muses, that were wont greene bayes to weare,
Now bringen bitter Eldre braunches seare.

Given the intense research that has surrounded *The Bible*, and its panoptic view of the customs and lore of the ancient Near East, including that relating to natural history (e.g. Carpenter 1832), it is surprising that little, if anything seems to relate, to the ideas of plant antipathy or sympathy. Bush (1854) indicated that the story of Elijah in the Beersheba wilderness (1 Kings, Chapter 18) has been interpreted dubiously as concerning the noxious qualities of the juniper. Elijah, in despair, fled to the desert, and while resting under the shade of a juniper[34] tree, he beseeched God to take his life. This supposedly indicated that the shade of a juniper tree, like that of a walnut, was injurious to life, and the argument obscurely drew support from Virgil's *Eclogues* (see Chapter 2). However, as Bush noted, in ancient times, to lie beneath almost any tree, particularly at night, was regarded with some trepidation.

An outright hoax of the Middle Ages, recorded as early as the fifth century in the *Talmud*, was the so-called vegetable or Scythian lamb. It was reputed to be a plant, but with animal form which was attached to the roots by a stalk. It was said to be able to graze on the grass around itself (Figure 5.7). The fabled plant was reported by travelers to Asia, who were sometimes sold rootstocks (likely of the fern *Cibotium barometz*), which had been cunningly carved and shaped by locals to fool naïve travellers. Acceptance of these bizarre creatures was encouraged by a belief that these were a type of *lusus naturae*, or joke of nature, and that the Creator had a sense of humour.

Figure 5.7. Left: A woodcut of The Scythian Lamb as figured in the work by Claude Duret (1605). Right: An engraving of an actual prepared "Scythian Lamb" from the collection of Sir Hans Sloane (Sloane 1698).

---

[34] The translation of the original term *rothem* may be either a broom (*Genista* sp.) or a juniper.

Another plant surrounded by fable and extraordinary claims was the upas[35] or bohun-upas (*Antiaris toxicaria*), a tree of the family Moraceae, which is native to the East Indies (Figure 5.8). This tree first came to the attention of Europeans through the writings of the Italian, Friar Odoric (*c.* 1330), a missionary who travelled widely in East Asia, and John Mandeville. Early writers wrote of the use of the sap of the tree for poisoning the tips of darts and arrows. Over the following centuries, stories of the upas became more and more sensational, and the tree was reputed to kill anything, including animal and plant life, for kilometers around.

It was largely the lengthy description by the German-born Dutch botanist Rumphius (see Chapter 4), written in about 1685 which initially gave credibility to these tales (Figure 5.9), although the remoteness of the trees and his blindness later in life prevented him from ever seeing an actual tree:

> Up to now I have never heard of a more horrible and villainous poison coming from plants, than that which is produced by this kind of Milk-tree.
>
> Under this tree and for a stone's throw around it, there grows neither grass nor leaves, nor any other trees, and the soil stays barren there, russet, and as if scorched. And under the most pernicious ones one will find the telltale sign of bird feathers, for the air around the tree is so tainted that if some birds want to rest themselves on the branches, they soon find themselves get dizzy and fall down dead. (Volume 2, pp. 263-268)

Nonetheless, it was accounts such as these that helped to make the concept of allelopathy in more benign plants seem credible.

Perhaps the most remarkable account of the upas appeared in 1783 in *The London Magazine* (Foersch 1783), and it was alleged to be a translation of a report by a Dutch surgeon, N.P. Foersch[36], stationed in the East Indies. Foersch described in considerable detail how only condemned prisoners were used to collect the deadly upas latex, and the likelihood of surviving exposure to the tree was reckoned as one in ten. There has been much controversy about whether a person named Foersch did exist, and the article has frequently been regarded, with little foundation, as a very clever and elaborate piece of fiction written by the English writer George Steevens. According to Bastin (1985), research has now shown that a German-born naval surgeon, John Nichols Foersch, who had spent some years in the Dutch East Indies, did exist, was in London in 1783, and was even known to Joseph Banks. Foersch likely authored the largely fictitious story about the upas to create publicity, as he was intending to publish a book about the East Indies. It highlights the fact that there are certain phenomena that people want to believe. Even the acerbic German botanist J.M. Schleiden[37] regarded the various stories regarding the upas as based at least in part on truth (Schleiden 1848). The celebrated Erasmus Darwin (1731-1802)

---

[35] A very useful summary of lore concerning the upas may be found in *Hobson-Jobson, The Anglo-Indian Dictionary* by Yule and Burnell (1886) and in Beekman (1981).

[36] Confusion about Foersch was partly due to the fact that at the beginning of *The London Magazine* article, his name was given as N.P. Foersch, but at the close, the correct initials, J.N., were given.

[37] Jacob Matthias Schleiden is sometimes referred to as M.J. Schleiden largely due to error in English translations of his work.

Figure 5.8. A charming woodcut of the Bausor tree (assumedly the upas) from Hortus Sanitatis *(1491)*.

Figure 5.9. Engraving of the "Bohon upas, the Java poison tree, showing its alleged properties, from a rare German work on gambling by "A.Z." published in 1845.

was compelled to incorporate the story of the upas in his *Loves of the Plants*, part of his lengthy poem *The Botanic Garden* (Darwin 1789), although he was warned that the story might be spurious. It was Darwin's enthusiasm, as much as anything that helped maintain the legend of the upas during much of the nineteenth century:

> Fierce in dread silence on the blasted heath
> Fell Upas sits, the Hydra-tree of death.
> Lo; from one root, the envenom'd soil below,
> A thousand vegetative serpents grow;
> In shining rays the the scaly monster spreads
> O'er ten square leagues his far-reaching heads.

Today, it is acknowledged that the stories surrounding the upas were wildly excessive, although the sap of the tree is toxic. The alleged lack of life in certain areas inhabited by the upas has been attributed to suffocating volcanic gases (Sykes 1837). Generally, in its native habitat, the upas tree supports both wildlife and undergrowth. Marsden (1811) cited a report from a Dr. Charles Campbell, who had seen the trees in Sumatra, and who dismissed the alleged injury to undergrowth: "Every one who has been in a forest must know that grass is not found in such situations." The celebrity of the tree, especially during the nineteenth century, caused it to become an image used in poetry and drama for suffering, and many notable authors could not resist exploiting the upas story, despite its lack of veracity, e.g. Byron, Ruskin, Pushkin, Colman and Charlotte Bronte.

A tree with remarkably similar properties, occurs in the New World, mainly on coastal sands in the Caribbean region. The plant is the manchineel or poison guava (*Hippomane mancinella*), which is a shrub or small tree of the family Euphorbiaceae. Its poisonous qualities were known to native Americans, who used the latex as an arrow poison, and sailors on Columbus' second voyage in 1493 were the first Europeans to suffer from eating the toxic fruit. As with the upas, tales spread rapidly about the virulent qualities of the manchineel, and records of Columbus' voyage warned that to sleep beneath a manchineel tree was dangerous (de Herrera 1601-1615). As far as allelopathy is concerned, it was reputed that grass was unable to grow underneath the canopy of the manchineel tree (e.g. Lindley and Moore 1873). These tales are regarded as much exaggerated, although it must be said that contact with manchineel latex can cause severe skin or eye irritation. As with the upas, the poisonous manchineel has featured in many literary works, e.g. by Maturin, Melville and J.-P. Richter. The American novelist Herman Melville (1849) provided an eloquent description of what may be construed as allelopathy in *Mardi: and a Voyage Thither*:

> Near by stood clean-limbed, comely manchineels, with lustrous leaves and golden fruit. You would have deemed them Trees of Life; but underneath their branches grew no blade of grass, no herb, nor moss; the bare earth was scorched by heaven's own dews, filtrated through that fatal foliage. (Chapter 107)

Remarkably, a very similar description to those of the upas and manchineel originated from central Africa. The British explorer Verney Lovett Cameron, in his futile quest to find Dr. Livingstone, eventually became the first European to traversed central Africa, from east to west. The published account of his arduous expedition of 1873-1875 provided the following account of an unknown tree:

> Another story had a curious resemblance to that of the upas-tree. At a certain place in Urguru, a division of Unyamwési[38], are three large trees with dark green foliage, the leaves being broad and smooth. A travelling party of Warori on seeing them thought how excellent a shelter they would afford and camped under them; but the next morning all were dead, and to this day their skeletons and the ivory they were carrying are said to remain there to attest their sad fate.
>
> Jumah assured me that he had seen these trees, and that no birds ever roosted on their branches, neither does any grass grow under their deadly shade' and some men who were with him when he passed them corroborated his statement in every particular. (Volume II, pp. 88-89)

Another instance of an unknown, allegedly allelopathic tree becoming part of folklore originated from Brazil. The noted anthropologist Warren Dean (1995) recorded that, in the nineteenth century in Minas Gerais, there was a type of tree known as *solitaria*, and nothing would grow for twenty paces around it, although the botanical name was not given. It is also noteworthy that explorer and naturalist Auguste de Saint-Hilaire (1830) reported another phenomenon from Minas Gerais that may well be allelopathic:

> In this part of Brazil, when there has been a small number of harvests from an area of ground, one sees arise a very large fern from the genus *Pteris*. A grass, viscid, greyish and foetid-smelling, called *capim gordura*[39], or grease plant, succeeds soon after this cryptogam or possibly at the same time as it. Then, all of the other plants disappear rapidly. If any shrub grows above the level of the stems of the *capim gordura*, it is soon grazed by animals; the aggressive grass remains master of the terrain, and it can not be recommended as forage, for it is so fatty for beasts of burden and stock, that it perceptibly weakens their strength. The farmer has no hope of cultivating any new trees on this ground, such that it is said to be lost beyond point of return (*he uma terra acabada*).

It is to be hoped that traditional agriculture from various regions around the globe may give insight into possible practices in modern sustainable agriculture, in the same way that ethnobotany and traditional medicine are now regarded with much interest in the pharmaceutical industry. There are several regions apart from India and China where traditional agriculture has been practiced for centuries, but documentation is lacking, including parts of North America, South America and Africa. The wisdom of traditional agriculture, particularly in an allelopathic context has been presaged by von Uslar (1844), Coccannouer (1950) and undoubtedly others. More recently, Anaya and her coworkers have attempted to record the allelopathic implications of traditional agriculture in Mexico. For example, in the state of Tlaxcala, where often certain non-crop species are retained assumedly because their interaction with other species, whether in inhibiting weed growth or stimulating crop growth, is seen as ultimately beneficial to the farmer (Anaya *et al*. 1987). Recently Posey (2002) examined the culture of the Kayapó in Amazonia, Brazil, and speculated that the use of crop residues to kill weeds was within the cycle of their swidden agriculture. Denevan (2001) has also stated that native tribes of Amazonia and neighbouring

---

[38] In present-day Tanzania.
[39] *Melinis minutiflora*, is an invasive grass, originally native to Africa, currently being studied for its allelopathic properties.

regions, prior to colonisation, were thought to have used certain crops, such as tobacco, to control weeds amongst other crops.

Isolated comments such as those above highlight the fact we are largely ignorant of ethnobotanical knowledge, particularly from Latin America and Africa, and leads from traditional sources may shed new light on allelopathic interactions. For example, McKenna *et al.* (1995) noted in reference to the traditional botanical lore of eastern Peru, where the "spirit" of the plant is considered all-important in both its ecology and purpose, that the "compatibility and incompatibility of plants is often expressed in terms of friendship and enmity between the spirits of the plants". It would be fascinating to learn more about this.

It is not surprising that there are numerous bits of local folklore relating to plants that may have some real basis, whether allelopathic or otherwise. An example of this is given in *Old Wives Lore for Gardeners* (Boland and Boland 1976), where among other well known antipathies, it was stated that planting *Gladiolus* among peas, beans or strawberries could be injurious to these crops. As it turns out, the gladiolus is not allelopathic, but it is a host to several viral diseases that can severely affect the afore-mentioned crops.

REFERENCES

A.Z. 1845. *Die Mystères des grünen Tisches oder der europäische Bohon Upas bestehend in Beobachtungen und Bemerkungen gesammelt an den Spielbanken in Aachen, Alexisbad, Baden, Doberan, Ems, Homburg, Kissingen, Köthen, Pyrmont, Schwalbach, Schlangenbad, Wiesbaden und Wilhelmsbad während den letzten zwanzig Jahren*. G.F. Heyer, Giessen.

Anonymous. 1491. *Hortus Sanitatis*. Jacob Meydenbach, Mainz.

Agrippa von Nettesheim, H.C. 1651. *Three Books of Occult Philosophy*. Gregory Moule, London.

Alexander of Neckam. 1863. *De Naturis Reum. Libri Duo. With the poem of the same author, De Laudibus Divinae Sapientiae*. Edited by T. Wright. The Chronicles and Memorials of Great Britain and Ireland during the Middle Ages, no. 34. Longman, Green, Longman, Roberts and Green, London.

Anaya, A.L., L. Ramos, R. Cruz, J.G. Hernandez, and V. Nava. 1987. Perspectives on allelopathy in Mexican traditional agroecosystems: a case study in Tlaxcala. *Journal of Chemical Ecology* **13**: 2083-2101.

Austen, R. 1658. *Observations upon some part of Sr Francis Bacon's Natural History as it concernes, Fruit-trees, Fruits, and Flowers: especially the Fifth, Sixth, and Seventh Centuries, Improving the Experiments mentioned, to the best Advantage*. Thomas Robinson, Oxford.

Bacon, F. 1620. *Instauratio Magna*. [*Novum Organum*]. John Bill, London.

Bacon, F. 1631. *Sylva Sylvarum or A Natural History in ten Centuries*. W. Lee, London.

Bastin, J. 1985. New light on J.N. Foersch and the celebrated poison tree of Java. *Journal of the Malaysian Branch of the Royal Asiatic Society* **58**(2): 25-44.

Beekman, E.M. 1981. *The Poison Tree. Selected Writings of Rumphius on the Natural History of the Indies*. The University of Massachusetts Press, Amherst.

Belz, R.G., K. Hurle, and S.O. Duke. 2005. Dose-response – a challenge for allelopathy? *Nonlinearity in Biology, Toxicology, and Medicine* **3**: 173-211.

Bentley, R. 1692. *The folly and unreasonableness of atheism demonstrated from the advantage and pleasure of a religious life, the faculties of human souls, the structure of animate bodies, & the origin and frame of the world: in eight sermons preached at the lecture founded by the honourable Robert Boyle, Esquire; in the first year MDCXCII*. H. Mortlock, London.

Bock, H. 1560. *Neu Kreutterbuch*. Strasbourg.
Bodin, J. 1597. *Le Théâtre de la Nature Universelle*. François de Gougerolles, Lyon.
Boerhaave, H. 1743. *An Essay on the Virtue and Efficient Cause of Magnetical Cures: to which is added, a new method for curing wounds without pains, and without the application of remedies*. London.
Boland, B. and M. Boland. 1976. *Old Wives Lore for Gardeners*. The Bodley Head, London.
Bormann, E. 1895. The Shakespeare-Secret. Thomas Wohlleben, London.
Browne, C.A. 1944. *A Sourcebook of Agricultural Chemistry*. Chronica Botanica, Waltham.
Browne, T. 1658. *The Garden of Cyrus, or the Quincunciall, Lozenge, or Net-work Plantations of the Ancients, Artificially Naturally, Mystically Considered*. London
Burton, R. 1621. *Anatomy of Melancholy*. John Lichfield, Oxford.
Bush, G. 1854. *Illustrations of the Holy Scriptures*. Lippincott, Grambo & Co., Philadelphia.
Calabrese, E.J. and L.A. Baldwin. 2003. Hormesis: the dose-response revolution. *Annual Review of Pharmacology and Toxicology* 43: 175-197.
Cameron, V.L. 1877. *Across Africa*. Daldy, Isbister & Co., London.
Cardan, G. 1550. *De Subtilitate Libri XXI*. Nuremberg.
Cardilucius, J. H. 1686. *Das Buch von der Harmonie, Sympathie und Antipathie, der Kräuter und ihren vier ersten Materien*. Verlegung Johann Andreae Endters, Nürnberg.
Carpenter, W. 1832. *Scripture Natural History: or a Descriptive Account of the Zoology, Botany, and Geology of the Bible*. Third edition. Book Society for Promoting Religious Knowledge, London.
Cocannouer, J.A. 1950. *Weeds, Guardians of the Soil*. Devin-Adair Company, Old Greenwich, Connecticut.
Cogan, [T.]. 1584. *Haven of Health, chiefly made for the comfort of students, and consequently for all those that have a care for their health*. London.
Cowley, A. 1689. *The Third Part of the Works of Mr. Abraham Cowley Being his Six Books of Plants..* Charles Harper, London.
Culpeper, N. 1652. *English Physitian: or an Astrologo-Physical Discourse of the Vulgar Herbs of this Nation*. Peter Cole, London.
Curzon, H. 1712. *The Universal Library; or, Compleat Summary of Science. Containing Above Sixty Select Treatises*. George Sawbridge, London.
Darwin, E. 1789. *The Botanic Garden. A Poem in two parts. Part II. The Loves of the Plants*. J. Johnson, London.
Dean, W. 1995. *With Broadax and Firebrand: the Destruction of the Brazilian Atlantic Forest*. University of California Press, Berkeley.
Della Porta. G.B. 1588. *Phytognomonica octo libris contenta*. Orazio Salviani, Naples.
Della Porta. J.B. 1658. *Natural Magick*. Thomas Young and Samuel Speed, Lomdon.
Denevan, W.M. 2001. *Cultivated Landscapes of Native Amazonia and the Andes*. Oxford University Press, Oxford.
Ducourthial, G. 2003. *Flore Magique et Astrologique de l'Antiquité*. Belin, Paris.
Duret, C. 1605. *Histoire Admirable des Plantes et Herbes Esmerveillable & Miraculeuses en Nature: mesmes d'aucunes qui sont vrays Zoophytes, ou Plant-animales, Plantes & Animaux ensemble*. Nicolas Buon, Paris.
Erasmus, D. 1878. *The Colloquies of Erasmus concerning Men, Manners and Things (translated by N. Bailey)*. Reeves and Turner, London.
Estienne, C. and J. Liebault. 1689. *Maison Rustique*. Claude Carteron & Charles Amy, Lyon.
Evelyn, J. 1664. *Sylva, or, a Discourse of Forest Trees*. Royal Society, London.
[Foersch, J.N.]. 1783. Description of the poison tree, in the island of Java. *The London Magazine* **52**: 512-517. Also published as: Natural history of the Buhon-Upas, or Poison-Tree of the Island of Java. *The Universal Magazine* **1784** (January): 11-16.
Foucault, M. 1966. *Les Mots et les Choses*. Éditions Gallimard, Paris. Published in English as *The Order of Things*.
Fuchs, L. 1542. *De Historia Stirpium*. Michel Isingrin, Basel.
Gerard, J. 1597. *The Herball or General Historie of Plants*. John Norton, London.
Gottfried von Franken. 1530. *Pelzbuch*. Johann Weissenberger, Landshut.
Heresbach, K.1570. *Rei Rusticae Libri Quatuor*. I. Birckmann, Cologne.
Heresbach, C. 1577. *Foure Bookes of Husbandrie*. Richard Watkins, London.

Herrera, A. de. 1601-1615. *Historia General de los Lechos de los Castellanos en las Islas y Tierrafirma del Mar Oceano*. Imprenta Real, Madrid.
Herrera, G.A. 1513. *Obra de Agricultura*. Arnao Guillen de Brocar, Alcala de Henares.
Hill, T. 1594. *The Gardener's Labyrinth*. Adam Islip, London.
Johnson, T. 1595. *Cornucopiae, Or divers Secrets: Wherein is contained the rare Secrets in Man, Beasts, Foules, Fishes, Trees, Plantes, Stones and such like, most pleasant and profitable, and not before committed to be printed in English*. William Parley, London
Kircher, A. 1641. *Magnes sive de Arte magnetica*. Rome.
Köckemann, A. 1936. Albert der Grosse – der Entdecker der keimungshemmenden Wirkung des Fleisches saftiger Früchte. *Zeitschrift für die gesamte Naturwissenschaft* **2**: 266-367.
Konrad von Megenberg. 1861. *Das Buch der Natur*. Stuttgart.
Lauremberg, P. 1631. *Horticultura, Libris II.*. Matthaei Meriani, Frankfurt.
Lindley, J. and T. Moore. 1873. *The Treasury of Botany: a Popular Dictionary of the Vegetable Kingdom. Part I. - A to K.* Longmans, Green & Co., London.
Lippay, J. 1663. *De institionibuset seminatione*. Pozsony.
Marsden, W. 1811. *The History of Sumatra. Third Edition*. London.
McKenna, D.J., L.E. Luna, and G.N. Towers. 1995. Biodynamic constituents in Ayahuasca admixture plants: an uninvestigated folk pharmacopoeia. In: R.E. Schultes and S. von Reis (eds), *Ethnobotany: Evolution of a Discipline*, pp. 349-361. Dioscorides Press, Portland.
Melville, H. 1849. *Mardi: and a Voyage Thither*. Harper and Brothers, New York.
Mizauld, A. 1689. *Harmonia superioris naturæ mundi et inferioris; unà cum admirabili foedere et sympatheia rerum utriusque. Quibus annectuntur Paradoxa doctrinæ coelesti accommoda*.
Nieremberg, J.E. 1635. *Oculta filosofia de la sympatia y antipatia de las cosas, artificio de la naturaleza, y noticia natural del mundo, y Segunda partie de la Curiosa filosofia*. Imprenta del Reyno, Madrid.
Orlob, G.B. 1971. History of plant pathology in the Middle Ages. *Annual Review of Phytopathology* **9**: 7-20.
Orlob, G.B. 1973. Ancient and medieval plant pathology. *Pflanzenschutz Nachrichten* **26**(2): 65-294.
Paracelsus [Theophrastus Bombast von Hohenheim] 1930. *Von den naturlichen Dingen. Abteilung 1. Bande 2. 1525.* In K. Sudhoff (ed.), *Sämtliche Werke*. Johann Ambrosius Barth, Leipzig.
Paracelsus [Theophrastus Bombast von Hohenheim] 1915. *Sieben Defensiones. 1538* In K. Sudhoff (ed.), *Sämtliche Werke*. Johann Ambrosius Barth, Leipzig.
Parkinson, J. 1640. *Theatrum Botanicum*. Thomas Coates, London.
Posey, D.A. 2002. *Kayapo Ethnoecology and Culture*. Routledge, London.
Philips, J. 1708. *Cyder, a Poem in Two Books*. J. Tonson, London.
Rattray, S. 1658. *Aditus Novus as Occultas Sypathiae et Antipathiae Causas inveniendas: per Principia Philosophiae naturalis, ex Fermentorum artifiiosa Anatomia hausta, Patefactus*. Andreas Anderson, Glasgow.
Riotte, L. 1983. *Roses Love Garlic*. Garden Way Publishing Co., Pownal, Vermont.
Ruellius, J. 1536. *De Natura Stirpium Libri Tres*. Simon de Coline, Paris
Rumphius, G. E 1741-1750. *Herbarium Amboinense*. Franciscum Changuion, Joannem Catuffe, Hermannum Uytwerf, Amsterdam.
St. Francis de Sales. 1616. *Traité de l'Amour de Dieu*. Pierre Rigaud, Lyon. Published in English as *Treatise on the Love of God*.
Saint-Hilaire, A. de 1830. *Voyage dans les Provinces de Rio de Janeiro et de Minas Geraes*. Grimbert et Dorez, Paris.
Scheiden, J.M. 1848. *The Plant; a Biography*. Hippolyte Bailliere, London.
Sloane, H. 1698. A further account of the contents of the China cabinet mentioned last Transaction p. 390. *Transactions of the Royal Society* **20**: 461-462.
Swedenborg, E. 1753. *Sapientia Angelica de Divino Amore et de Divina Sapientia. Sapientia Angelica de Divina Providentia*. Amsterdam.
Sykes, W.H. 1837. Remarks on the origin of the popular belief in the upas, or poison tree of Java. *Journal of the Royal Asiatic Society of Great Britain and Ireland* **4**: 194-199.
Taylor, J. 1620. *The Praise of Hemp-Seed with the Voyage of Mr. Roger Bird and the writer hereof, in a boat of brown-paper, from London to Quinborough in Kent, etc.* H. Gosson, London
Thulesius, O. 1992. *Nicholas Culpeper, English Physician and Astrologer*. St. Martin's Press, London.

Turner, W. 1562. *The Seconde Parte of William Turners Herball.* Arnold Birchman, Cologne.
Uslar, J.L. von 1844. *Die Bodenvergiftung durch die Wurzel-Ausscheidungen der Pflanzen als vorzüglichster Grund für die Pflanzen-Wechsel-Wirthschaft.* Georg Blatt, Altona.
Vallemont, P. 1705. *Curiositéz de la Nature et de l'Art sur la Vegetation: ou l'Agriculture et le Jardinage.* Claude Cellier, Paris. Published in English in 1707 as *Curiosities of Nature and Art in Husbandry and Gardening.* D. Brown, A. Roper and F. Coggan, London..
Wagner, H. 2001. *Le Poireau Préfère les Fraises.* Terre Vivante, Mens.
Worlidge, J. 1677. *Systema Horti-Culturae or the art of gardening.* T. Burrell and W. Hensman, London.
Yule, H and A.C. Burnell. 1886. *Hobson-Jobson, The Anglo-Indian Dictionary.* John Murray, London.

CHAPTER 6

THE EIGHTEENTH CENTURY – ROOT EXCRETION

> Their sap thus like the blood of animals, has the need to be purified, it should supply particular secretions, that one may compare to the tangible and intangible transpirations of animals. Numerous experiments and a number of observations prove that plants are subject to these secretions, and that they seem to be even more essential to the plant economy than to the animal economy.
>
> *Physique des Arbres*
> Duhamel du Monceau (1758)

> What is absolutely singular, is that the ancients put all their effort into researching the properties of plants, and neglected the means of understanding with certainty the very plants that they used; whereas modern botanists, on the contrary, occupy themselves solely with the duty of distinguishing all the plants they can observe, without anything about them, that is to say, deign to apply themselves to indicate the use to which they can be put.
>
> *Histoire Naturelle, Génerale et Particulière des Plantes. Tome I*
> Lamarck and Brisseau-Mirbel (1803)

EARLY IDEAS CONCERNING ROOT EXCRETION

In the first half of the eighteenth century, there were scattered advancements in the understanding of how plants grow and function. In particular relevance to the discourse here, a topic that hitherto had received little attention, namely plant excretion, became a matter of progressive conjecture during the course of the eighteenth century.

The precept that plants are capable of excretion, notably via the roots, was to prove crucial to the development of the concept of allelopathy, particularly in the nineteenth century. This notion was in part a legacy of the Greek teachings which espoused that animals and plants have analogous functional systems, such as digestion, circulation, reproduction, etc. While many natural history writers freely accepted these ideas, remarkably little attention was paid to investigating in plants a process all too familiar in animals, and that was the elimination of waste, or excretion (e.g. Necker 1775, Home 1776, Smellie 1790). Part of the reason for this was retention of the Aristotelian teaching that the food of the plant was processed firstly in the soil, and thus in plants the excretion process was largely redundant. The first suggestion that plants may engage in excretion appears to be due to the important figure, Joachim Jung (1587-1657) in his posthumous work, *De Plantis Doxoscopiae Physicae Minores* (second fragment) of 1662. Jung (known also as Jungius) was reluctant to publish much of his scientific work during his lifetime for fear of reprisal, and consequently his works are both rare, and little known, as they have not been translated. He wrote that: "the openings in the root which take in liquid matter are so organised, that they do not allow every kind of juice to enter, and who can say that plants have the peculiarity of only absorbing what is useful to them, for like all other living creatures they have their excreta, which are exhaled through the leaves, flowers, and fruits." (see Sachs p. 454.) Another veiled suggestion of root excretion came from Malpighi (1671)[1], and according to Senebier (1800), Gauthier d'Agoty had favoured excretion in roots as well. There was debate as to which plant parts served in excretion; Hans Carl von Carlowitz (1713) in addressing the question of "do trees defecate", suggested that things such as resins, protuberant growths including galls and fungi, and even mossy growths were all manifestations of excretion, although these ideas had been expounded earlier (e.g. see Balduino 1694). Many botanists seemed to believe that root excretions existed even if they had not readily been observed. It has been suggested by Schroth and Hildebrand (1964), for example, that Micheli (1723) may have had the idea of root exudates in mind when he observed that seeds of the parasitic plant *Orobanche* only germinated when in the vicinity of the roots of host plants.

To understand the development of the concepts of root excretion and allelopathy during this period, one also should have an appreciation of concepts concerning plant nutrition. The following is simply an overview of the key ideas, and for those who are interested, a detailed account has been provided by Fussell (1971).

It is somewhat amazing that the fundamental importance of photosynthesis in the life of the plant has only been appreciated in last two hundred years or so. As noted earlier, the Greeks, Romans and indeed other early cultures, viewed the world as consisting of four or five fundamental elements, typically earth, air, water, fire, and sometimes ether. This view became only marginally more complex over the following centuries, and chemists in the Middle Ages added metal represented by mercury, sulphur, and salt. In the early eighteenth century, an imagined combustible component, phlogiston was added to the mix. It was largely during the eighteenth century that chemistry began to emerge as a rightful science, and knowledge of the modern elements grew, although their relationship to plant growth was slow to be appreciated. Regardless of the chemical elements progressively identified in plants, and discoveries regarding photosynthesis, the common view until at least 1800 was that plants acquired all of whatever it was they required for growth from the soil via their roots. Opinions varied with regard to the details of mechanism, for example whether the nutrients were preformed in the soil or whether materials were processed within the plant. In any case, the idea that the plant obtained its essential organic matter from the soil was commonly known as the "humus theory", and was championed well into the

---

[1] Malpighi linked excretion to flowering and fruiting, an idea which can be traced back to Theophrastus' *De Causis Plantarum* Book VI, Section 10.5.

nineteenth century, through notable agricultural writers such as Tull, Thaer and Dombasle. Leaves were recognised as essential to plant growth, but they were seen either as the power supply for root nutrition or some sort of processing site where sunlight was allowed to blend with and alter the organic matter obtained by the roots. Botanical writings of the seventeenth and eighteenth centuries were rich in analogies of plants with animals: roots were commonly compared with the animal alimentary system, leaves with the lungs, sap with blood, etc., and these ideas persisted well into the nineteenth century[2]. It is thus not surprising that the question of excretion in plants became topical.

The idea that plants absorb a raw mixture of substances, process (elaborate) them in the plant, and then eliminate the waste was clearly expressed for the first time by the Dutchman Herman Boerhaave (1668-1738; Figure 6.1). Boerhaave was a luminary figure at the University of Leiden, and achieved the remarkable feat of holding simultaneously the professorial chairs of medicine, botany and chemistry. At the time Leiden was one of the foremost academic centres in Europe, and Boerhaave's influence on subsequent botanists and chemists cannot be underestimated.

Boerhaave's lectures in chemistry were published firstly in book form in 1724 in a Latin edition entitled *Institutiones et Experimenta Chemiae*, and an English translation, *A New Method of Chemistry*, appeared in 1727. Several other editions of

Figure 6.1. Section of a rare engraving after a painting by J.W. van Borselen depicting Boerhaave in his role as a botanist (Lennep et al. 1868).

---

[2] For example, the following passage was written by Dadd in 1851, p. 319: "If you examine the potato, with its roots and stem, you will find the skin, including that of plant, stalk, leaf, and ball, is that to the potato what the skin and lungs are to animals; they, each of them, absorb atmospheric food, and throw off excrementitious matter; the roots and fibres are to the vegetable what the alimentary canal is to the animal. A large portion of the food of vegetables is found in the soil, and enters the vegetable system, through its capillary circulation, by the process of imperceptible elimination and absorption."

his chemistry lectures were published later, but this early version was claimed by Boerhaave to be an unauthorised "surreptitious edition", and he did not number it among his publications. In spite of the acknowledged piracy of the book, apparently by Boerhaave's students[3], the edition is still reckoned to be a fair summary of his chemistry lectures (Davis 1928). Oddly, it is this work alone which describes Boerhaave's views on excretion by the plant root, and the text below is extracted from the rare English edition (Boerhaave 1727):

> The root or part, whereby vegetables are connected to their matrix, and by which they receive their nutritive juice, consists of a number of *vasa absorbentia*, which being dispers'd thro' the interstices of the earth, attract and imbibe the juices of the same' consequently every thing in the earth, that is dissoluble in water, is liable to be imbibed; as air, salt, oil, fumes of minerals, metal, &c. and of these do plants really consist.
>
> These juices are drawn from the earth, very crude; but by the structure and fabric of the plant, and the various vessels they are strained thro', become changed, further elaborated, secreted and assimilated to the substance of the plant. (p. 144)
>
> The juice having thus gone its stage from the root to the remote branches and even the flower; and having, in every part of its progress, deposited something, both for aliment and defence; what is redundant, passes out into the bark, the vessels whereof are inosculated with those wherein the sap mounted: and thro' these it redescends to the root, and thence to the earth again. (p. 145)

The great English experimenter, the Reverend Stephen Hales (1677-1761) was a contemporary of Boerhaave, and is widely acknowledged as the founder of experimental plant physiology, with his ingenious experiments on fluids and gases in relation to plant function. His major botanical work, *Vegetable Staticks* first appeared in 1727. Despite being familiar with the teachings of Boerhaave, Hales did not believe that the structure of roots was suited to any excretory function. Hales did follow the mainstream in accepting, at least in principle, the analogous physiologies of animals and plants, and he credited the leaves solely with the role of excretion:

> I shall begin with an experiment upon roots, which nature has providently taken care to cover with a very fine thick strainer; that nothing shall be admitted into them, but what can readily be carried off by perspiration, vegetables having no other provision for discharging their recrement. (Chapter II)
>
> Thus the leaves, in which are the main excretory ducts in vegetables, separate and carry off the redundant watry fluid, which by being long detained, would turn rancid and prejudicious to the plant, leaving the more nutritive parts to coalesce. (Chapter VII)

This view was reinforced by experiments which showed that the collected exhalations of leaves were not pure water, and if allowed to stand for a few days, became putrescent (Experiment XVII). Hales' experiments were much later interpreted out of context: for example, in one experiment, the release of air bubbles from a severed pear tree root (Experiment XXI) was construed as root excretion (Clements 1921).

Duchartre (1868), perhaps guided by Gallic loyalty, stated that it was Duhamel du Monceau who discovered root excretions. Henri Louis Duhamel du Monceau

---

[3] Although the Latin edition of 1724 was stated as being published anonymously in Paris, this has become regarded as a deception, and the edition was likely printed in Leiden (Davis 1928).

(1700-1782) was a prolific and versatile scientist and writer, who authored many of the standard French works of his era in the disciplines of agriculture, botany and horticulture. Various authors (e.g. Clements 1921) have recorded that Duhamel du Monceau found the soil around old elm roots to darker and greasier than usual (ostensibly due to root excretions), but I have been unable to find the source of this reference[4]. Duhamel du Monceau favoured the idea that plants must rid themselves of waste products, and while it is in his *Physique des Arbres*, to which Duchartre alludes, that Duhamel du Monceau (1758) dabbled in the issue of whether roots are actually responsible for excretion, it was others who drew the conclusion. In growing tree seedlings in glass tubes, Duhamel du Monceau remarked:

> By means of the transparency of the glass, I saw that there formed on the roots soft tubercles which damaged them; nevertheless the tube filled with long filaments, and I managed to see a fox-tail, similar to those which clog the pipes of springs.
>
> Although I paid great attention to always keeping the tubes of glass full of very clear and very pure water, nonetheless there amassed around its roots a gelatinous material, which certainly had not been formed in the water I used without the cooperation of the roots. M. Bonnet is said to have seen at the ends of the roots which had formed in water, light earthy concretions; as for me, independently of the mucilage of which I have spoken. (p. 86)

Duhamel du Monceau also noted:

> For I have observed that elms, planted in a drive alongside fields of grain, exhaust the soil, principally in spots where their roots terminate; such that the grain does not come close to the young trees, while it is found much better at the foot of large trees, that have a distance of 4 to 5 toise[5]. (p. 89)

A lesser known commentary on ideas concerning plant excretion was provided by the Marquis de Saint-Simon (1720-1799) in his monograph on hyacinths (Saint-Simon 1768). In some ways, his ideas have presaged modern ideas[6], in which, for example, leaf abscission, is viewed as achieving the ridding the plant of wastes[7]:

> I do not regard the roots of the hyacinth, as aspirating pumps by which the sap is carried from the soil into the bulb, but on the contrary as excretory vessels which serve to discharge the bulb of too great an abundance of sap, which enters this solid and spongy body which is found in the region of the roots, and which one calls the eye of the root.
> (p. 16)
>
> If many productions ranked among the number of plants, such as the truffle, even the algae of which one makes the dikes in north Holland, do not present us with any suggestion of roots, leaves or flowers, does it offend the soundness of Natural Philosophy to admit a class of roots which would be only that of excretory vessels, and which would have no other functions than those which are common to all living creatures, from the moment of their conception until they leave the care of their mother

---

[4] This may originate from the German edition, *Natur-Geschichte der Bäume*, which was edited by Oelhafen von Schöllenbach in 1764-1765.

[5] A toise is an old French unit of length equal to approximately 1.95 m.

[6] The concept of plant excretion until a few years ago was distinctly unpopular. A relatively unremarkable note by Ford (1986) seems to have rekindled interest in the subject, which recently has enjoyed popularity due to its commercial possibilities in phytoremediation.

[7] This view was also taken by Théodore de Saussure according to Cuvier (1834).

to come into the world? Their excretory vessels do not discharge to the outside, and are in the placenta just as the roots of the hyacinth are in the soil. The root withers and drops off each year, carrying with it these excretions, which are found to be no less real in plants as in animals: a circumstance which one would not have perhaps thought allowed to admit in the parallel so often repeated in the animal and plant realm. (p. 27)

Root excretion was cautiously endorsed by Jean Senebier (1742-1809), an influential figure in plant physiology. Senebier was born in Geneva, and trained as a pastor, but maintained a strong interest in natural history. He was influenced by Bonnet, and Senebier is best remembered for his consequent discoveries concerning photosynthesis. Senebier was a central figure in the natural history circle of Geneva, and Augustin Pyramus de Candolle was among his protégés. Senebier (1791), seemingly ignorant at this point of the writings of Brugmans (*q.v.*), wrote:

> Is it not possible that roots are excretory organs, and that they operate on the sap already drawn by the roots? One is disposed to think so, when one sees the first sap in Spring already elaborated, although the vines do not have any leaves. Consequently it is clear that this elaboration can only be made by a particular excretion; and it is possible to imagine these excretory organs, for the descending sap or inherent juices, as I have already remarked, it is no less probable to think of the existence of similar organs for the ascending sap or lymph. These are the excretions which fertilise the soil near large roots; it is often seen that the soil which surrounds a large root, is more dark than that at a distance. (p. 244)

ANTIPATHY AND SYMPATHY

The concept of antipathy and sympathy among plants was still current, notably in rural areas. Hans Carl von Carlowitz (1645-1714), noted previously, wrote an important treatise, *Sylvicultura Oeconomica*, which was the first forestry book from continental Europe, and he is best known today for his seminal statements on forest sustainability[8]. Von Carlowitz recorded the familiar antipathies between the olive and the oak, and the walnut and the oak, and, in the case of the latter, provided a local saying:

> Der Nuss-Baum und die Eichen
> Sich nicht können vergleichen.[9]

Von Carlowitz also stated that there was antipathy between the hawthorn (*Crataegus* sp.) and the blackthorn or sloe (*Prunus spinosa*). He also raised the issue of soil sickness, as he indicated that when certain former forest soils were ploughed and layers of so-called "dead soil" were brought to the surface, it required a period of fallowing for two years to rectify the infertility.

Just as other authors had previously queried the origins of the French word *noyer* and Spanish word *nogal* for walnut (see Chapter 5), von Carlowitz suggested that the Latin name *Juglans*, rather than being derived from *Jovis glans*, as usually accepted,

---

[8] Von Carlowitz was the Director of Metallurgy for the Electorate of Saxony. In Saxony, forestry was closely allied to mining, as both mine construction and smelting required huge quantities of wood.

[9] The walnut and the oak tree
Are not able to agree.

was in fact derived from *jugulet glandes*[10]. He elaborated further in stating that the local German name for the walnut tree is *Eichelmörder*, or oak-killer.

An Englishman, identified only as S.J. (1727) wrote an account of vineyards; however, in considering the topic of antipathy and sympathy of plants, he concluded that any negative or positive effects were simply due to either overlapping or differential use of "juices" from the soil. He explained that the sympathetic effect of garlic in making the rose more fragrant was due to garlic claiming the foetid substances from the soil, and the like effect between a yew tree and a fig was due to the yew claiming the bitter substances and the fig drawing sweet substances.

It has been largely forgotten that Scotland was a major agricultural center, and the earliest agricultural society in Europe (including the British Isles) was formed in Scotland in 1723, and eventually became the Highland and Agricultural Society of Scotland, which was a major forum for debate on the subject of root excretion in the nineteenth century (see Chapter 7). Francis Home (1719-1813) was a Scot who studied medicine, and who developed an interest in chemistry, likely through his opportunity to study occasionally at the University of Leiden when he served as an army surgeon during the Seven Years War. Home returned to Edinburgh, and completed his medical degree in 1750. In 1756 the Edinburgh Society of Scotland offered a gold medal for the best dissertation on "Vegetation[11] and the Principles of Agriculture", and Home (1757) replied with his publication, aptly titled, *Principles of Agriculture and Vegetation*, in which one finds:

> Among the class of external accidents we may place the effects which arise from the contiguity of certain plants. There are some plants which do not thrive in the neighbouhood of others. This is observed of the cabbage and cyclamen, of hemlock and rue, of reeds and fern. We have many examples of such like antipathies amongst animals. These effects seem to be produced by the effluvia which are emitted by all organised bodies. (pp. 171-172)

> Farmers think it [paring and burning] acts by dispelling a sour juice which land has contracted from lying long untilled.[12]

A contemporary of Home was John Randall (1764) who wrote a rather prolix work *The Semi-Virgilian Husbandry*, which largely espoused the virtues of ploughing to comminute clods of soil and increase soil porosity. Randall also tackled the issues of sympathy and antipathy among plants, and adopted more or less the arguments of Bacon. Randall believed that roots were totally unselective in absorbing materials from the soil, and acted simply as capillary tubes:

> It is affirmed, that there is a sympathy, or mutual friendship, between rue and the fig-tree, the rose and the garlick[13], the wild poppy and wheat; all which it is said, are

---

[10] This means literally "to cut the throat (or kill) of acorns (oaks)".

[11] The term vegetation formerly meant the process of plant growth.

[12] This passage does not appear in the edition of 1757, and likely stems from a later edition, e.g. third edition of 1762. This quote is cited by Rennie (1834).

[13] The lore of the sympathy of the rose and alliaceous plants has been remarkable durable, and persists to the present. Note also Kerner (1811): "Each plant can, when it is nearly already wilting, become refreshed anew through a particular other plant, which is planted beside it. A wilting rose bush is

observable to delight and flourish most in the neighbourhood of each other: but this is a mistaking of the cause and the effect, with regard to the supposed friendship, for instance, between the wild poppy and wheat; and imputing that to sympathy, which proceeds from a voracious disposition in the poppy; which, in truth, consumes much of the nutritive ingredients, and thereby robs part of the wheat of its due nourishment, and, at length, destroys all within its reach, which does not look like friendship, but rather shews, that many of the fibres of the poppy attract the nutritive fluids powerfully in the same direction with the horizontal fibres of the wheat, and intermix with them.

In like manner we can conceive of the friendship, or sympathy, between the rue and the fig-tree; or, rather, there is more of the appearance of kindness between them, as they do grow well together, instead of one starving the other: but their being able to live in the neighbourhood of each other, can arise from no other cause, than a manifest dissimilitude in the ranges of their fibres, and the disproportions of their wants at certain depths of the soil. Those who affirm, that the rank and bitter nature of the rue arises from the rank and bitter aliment it imbibes from the heterogenous moisture in the earth, mistake it for the effect of configuration and motion in vegetation; and when they say, that the rue leaves the milder and sweeter vegetative particles, for the nourishment of the fig tree, it is ascribing a power to the fibres to attract and repel, at one and the same instant of time, which cannot be.

Thus, also, it is said, that garlick, set near a rose tree, will consume the foetid juice which descends from the atmosphere in rain drops, and leaves the odoriferous ingredients for the fibres of the rose tree to imbibe, in order to increase the sweetness of its flowers; but the different motions and dispositions of the parts, whereof each species consists, give those different sensations, when applied to the senses, and not the different particles of which the nutritive principles are formed, which are, indeed heterogenous, but, as a fluid, are all imbibed by the mouths of those tubes which stretch out to them. From what has been hinted concerning the sympathy, we may judge of the antipathy, which is said to be between some plants, and presumed to be so odious to each other, that if any two of them are set together, one, or both, will die: but the truth is, as was mentioned before, all plants, that are greater depredators of the nutritive moisture, than those near them, they only defraud their neighbours of their requisite nourishment, and, in that case, may be called voracious, without paying any attention to a secret antipathy: and thus hemlock is a dangerous neighbour to rue, because being, by much, the more succulent plant of the two, it deceives and starves the latter, by depriving it of sufficient sustenance, and makes it pine away for want. (pp. 293-295)

Another Scottish agricultural writer of note was Adam Dickson (1721-1776). Dickson authored the *Treatise of Agriculture* which appeared firstly in 1762, and in an expanded two-volume second edition in 1765. Dickson's *Treatise* was perspicacious and was rich in observations, especially concerning plants growing under Scottish conditions. Dickson, like Randall, subscribed to the idea that plants were unselective in absorbing food from the soil, but differed in their requirements and use of foodstuffs. His discussion of antipathy and sympathy was concerned primarily with the interactions of soil properties with environmental factors. Agricultural writers of this period generally lacked knowledge of chemistry, but Dickson (and others such as Hale) described instances of crop failure after repeated cultivation, in particular flax, that much later became tied to allelopathic interpretations:

> There is not farmer that has tried the culture of this plant, that will ever be prevailed upon to raise even three successive crops of it on the same field; for he is convinced that he would have bad crops, and would destroy his land. (Volume II, p. 261)

---

brought to life again, when a leek is planted nearby. Thus each plant seeks one friendly to it, separation from it or never finding it is fatal to it."

The Scottish philosopher Henry Home, or Lord Kames (1696-1782), in later life devoted his attention to agriculture. In *The Gentleman Farmer* (Home 1776), he alluded to root excretions, poisoning of the soil by weeds, staling of the plant environment with repeated cultivation, and soil sickness in red clover.

The topic of the antipathy of trees was addressed briefly by another Scotsman, Walter Nicol (17??-1811) who pursued a gardening and horticultural career in England. In *The Practical Planter*, first published in 1799, Nicol indicated that the antipathy of trees was often touted as a reason for the maintenance of tree diversity, and that the supposed harmful effects of tree leaf leachates were well known, although he subscribed to competitive factors:

> Some are advocates for planting in *groupes*, from the idea that there is an antipathy between trees, or that the shade of one kind of tree is hurtful to another. That the shade of any one tree is hurtful to another, cannot be doubted; but that there is an antipathy between the kinds, seem a doctrine founded in chimera.
>
> That the *drop* of one kind is hurtful to another, is also advanced in support of this kind of planting, and the Ash is generally held out as an example. If one Ash tree overhang another, or if an Elm overhang an Ash, is the consequence different? Does not every tree, who lords it over his neighbour, not only *over-drop* him, exclude him from sun and air, but also out of his food, by greedily extending his roots, and devouring *his portion*? Hence, the Ash has generally been quoted for the support of their argument, from the circumstance of his being a quick frower, and great impoverisher of the soil, to the detriment of his fellows in all *mixt* and *neglected* plantations.
>
> (Second edition 1803, pp. 111-112)

It was also likely he who, perhaps in recollecting Worlidge (see Chapter 5), admitted that the artichoke seemed inimcal to other plants growing underneath, although he suspected shading was the cause (N.W. 1803)

The exudations and excretions of plants was still regarded with great suspicion, and claims of their pernicious nature were sometimes exaggerated, because of insufficient knowledge of phytopathology. Good observation, but accompanied by inadequate understanding of the complexities of plant diseases such as common rust[14], led to the common acceptance among farmers that plant exudations were largely responsible for crop failure, as suggested in the following description by Davies (1810):

> In July 1808, a season nearly as blighty, in some districts, as the former one of 1804, an exhalation was observed in Shropshire, about ten in the evening, after several hot days, skimming the surface of the plains; and visibly attaching itself to the leaves of the wheat and other vegetables, so as to be rubbed off, in a whitish film, by the hand. A few days afterwards, a hue and cry of blighted wheat crops became general through the country. Whether this vapour was the efficient cause of the mildew, or only an unconnected phenomenon, happening at the time, may not be easily decided.
>
> The barberry bush, like the witches of old, has been frequently condemned, and even executed upon a supposition of its causing the mildew; but apparently without a fair examination into the nature of causes and their effects. The venom of the fabulous Bohon-upas tree of Java, could scarcely be equal to the effects attributed to this apparently harmless shrub. But farmers positively assert that their wheat crops were successively blighted in its vicinity; and upon hearing its deleterious quality, and

---

[14] While wheat rust had been thought for centuries in some way due to *Berberis*, the heteroecious nature of a causative fungus, *Puccinia triticina* (formerly known as *P. recondita*), was not shown until 1865 by Anton de Bary.

stocking it up, their crops were as sound as those of their neighbours. They even trace the extent of its influence, between converging lines, to a certain distance. (p. 90)

The topic of antipathy in plants in relation to its occult bases was critiqued briefly by Thomas Cooper (1759-1841), a radical Englishman who attained prominence in law, chemistry, and medicine, and who eventually settled in the United States. Cooper wrote (1791):

> The theosophers however, in this as in the doctrine of signatures extended their notion of the subject far beyond the mere medicinal application of it, including under the denomination of sympathy the consent or connection between celestial and terrestrial objects. Indeed, the term itself has been variously applied; as 1. to the cure of wounds by the application of some medicinal salve or powder (called weapon-salve or sympathetic powder) to the instrument which inflicted the wounds: 2. to the supposed cure of diseases by means of the magnetic effluvia: .3. to the influence beneficial or otherwise of certain plants over others growing within their reach; to the poisonous effect of this kind attributed to aconite for instance, and the antipathy of oak and the olive: 4. to the unknown (but supposed real) connection between certain plants and artificial preparations of their produce, as the fermentation of wines when the vine flowers: 5. to the indirect affection of one part of the body when another distant part is immediately affected, without any apparent direct connection; as the stomach and the uterus: this is the modern medical doctrine of sympathy, and was not unknown at the period in question. (pp. 454-455)

The Frenchman Maupin, of whom few personal details are recorded, was in the employ of Queen Marie Leszcinksa, and became an expert vigneron and wine-maker. The results of his experiments at estates at Sèvres and Belleville were published in various works, and a likely posthumous volume (Maupin 1799), edited by Buc'hoz recorded that the vine was antipathetic to certain plants.

FRESH OBSERVATIONS

During the latter part of the eighteenth century there were scattered comments amongst the literature alluding to allelopathy. It is often forgotten that Sweden was an important cultural center, and a number of comments relating to allelopathy emanate from Swedish authors. The great Linnaeus (1745) during his travels in Gottland observed the following:

> The farmers say that when this plant [ramsons, *Allium ursinus* L.] grows, it drives away other herbs and weeds; we had proof of this before our very eyes, since under those bushes where the ramsoms grew there were no other plants. The farmers also told us that they plant it among the hops, to keep wild chervil [*Anthriscus sylvestris* (L.) Hoffm.] and other weeds away.

According to Aamisepp and Osvald (1961), the Swedish Royal Academy sponsored a competion in the middle of the eighteenth century which sought solutions to the problem of combating the notorious weed, wild oats (*Avena fatua*). In response, both Siosteen (1749) and Johan Brauner (1751) recognised that rye (*Lolium* spp.) could be injurious to wild oats – an early proposal of biological control with a possible allelopathic basis.

An important early figure in agricultural chemistry was Johann Gottschalk Wallerius (1709-1785), professor at the University of Uppsala. His views were espoused in

1761 through a thesis by his student Gustavus Adolphus Gyllenborg[15], which was also published in English in 1770. Many of the views in this work seem to coincide with those of Francis Home. Thus we can extrapolate that it was widely accepted among farmers that certain plants had a chemical effect on fields, and "soured" the soil. On this point, Gyllenborg (1770) wrote:

> Every tree or shrub in a field should be rooted up, in order to turn the ground they stand on to better purpose, and prevent the inconveniences of their shade and leaves, which smother other plants, as well as sour the ground. (p. 195)

The inherent appeal of allelopathic explanations of inhibition of plant growth is evident in accounts of vegetation from visitors to parts of the New World. Felix de Azara (1746-1811) was a Spanish military officer, who later in his career spent time in the Rio de la Plata region, as he had been authorised to investigate border disputes concerning the Spanish and Portuguese territories (Paraguay and Brazil respectively). During the course of his travels from 1781 to 1801, he recorded his observations, in particular on the natural history of the region (de Azara 1809). He noted that there is little plant growth within groves of orange trees, and wrote:

> As the shade of these trees [orange trees] or the juice of the rotting oranges does not allow any other tree or plant to grow, when some of these, that were previous to the orange trees, gets to die of old age, or by accident, those are alone without even suffering agarics (fungi) or other parasitic plants, and are as the old vegetation perishes little by little, without being replaced. I presume that these forests of orange trees postdate the conquest, because they are ordinarily near places formerly populated or which are so at the moment. They are very dense and the ground is almost completely lacking of plants. One does not see anything other than a great number of young orange trees growing, and every so often trees which were in the region before the orange trees.

A remarkably similar, ingenuous comment, free of dogma of its day, originated from another part of the New World, a hemisphere away, in the infant colony of New South Wales, in eastern Australia. David Burton, having both surveying and gardening experience, had been recommended by Sir Joseph Banks, to serve in New South Wales. Banks' provided the young man with an additional stipend to collect seeds, live plants, and other botanical specimens exclusively for export to Banks. While surveying the Parramatta district, west of Port Jackson (Sydney), Burton reported to Governor Arthur Phillip on 24 February 1792 (Britton 1892):

> I beg leave to observe here that where different species of red gum-trees grow, the earth has a great portion of oils mixed with it, and unless the ground is properly worked and turned over to meliorate and disolve those oils, the first crop will come to little account.

Burton died a few weeks later, having accidentally shot himself.

THE BLACK WALNUT (*JUGLANS NIGRA*)

Most modern works concerning allelopathy inform us that the earliest observation of the celebrated toxicity of black walnut (*Juglans nigra*), a native of eastern America,

---

[15] Among European dissertations of this period, it is debatable as to who actually wrote the document. – the professor or the student.

Figure 6.2. Title page of Plappart's thesis on Juglans nigra.

was published in 1881 by Stickney and Hoy (see e.g. Rice 1984), and recently it has been found that similar observations were recorded earlier in the American literature by Galusha and others as early as 1870 (Willis 2000; see Chapter 8). Oddly, the first report of the harmful effects of black walnut actually emanates from an eighteenth century European report (Plappart 1777). Joachim Friedrich Plappart von Frauenberg (1753-1845), an Austrian minor noble who became a veterinarian, wrote his doctoral dissertation (Figure 6.2), under the direction of Nicholas Joseph von Jacquin, largely on the medical properties of black walnut. His thesis was published in a small edition as was generally required, but was also included amongst *Miscellanea Austriaca as Botanicum, Chemiam et Historiam Naturalem* by Jacquin (1781), where it commonly has acquired attribution to Jacquin. Jacquin, was a Dutchman (1727-1817) who found patronage in Austria and travelled extensively in the Caribbean and South America, and later was appointed as Professor of Botany at Vienna.

In his dissertation Plappart reported:

> As it will soon be shown the [black walnut] tree is, in fact, very useful on account of its spreading crown; nevertheless it presents these obstacles, that it is believed to kill meadows, pastures, flower-filled and scented gardens more effectively than any other tree. For the tree destroys any neighbouring plants, such as apple trees, cherry trees, corn, flax and vegetables, as if it is killing them off completely.
>
> Seedlings and cuttings cannot grow well nearby but gradually die. Many Swiss farmers have told me that the only Swiss explanation for why I destroyed my apple trees is because black walnut trees were growing nearby. In fact one of these men asserted that he himself lost forty or more apple trees because he left the black walnut trees which I gave him in the orchard; for this reason it is impossible to cultivate healthy apple trees near black walnut trees; but after their removal the apple trees grow favourably.

Eighteenth Century – Root Excretion

Truly the real cause of this phenomenon is still unknown. Yet many people will attribute this to the vapours of the black walnut tree carry across to neighbouring trees.

(pp. 12-13)

Subsequently, the idea that black walnut was harmful to neighbouring plants became relatively widely known in England and Europe, as it was reiterated in major botanical compendia, such as in the later editions of the monumental *Gardener's Dictionary* by the Englishman Philip Miller (later editions were edited by Thomas Martyn (e.g. 1807), as Miller died in 1771), where it was stated that:

> The growth of the tree is remarkably quick; it spreads out roots horizontally to a considerable distance and will not suffer any thing to grow under its shade. When planted in an orchard, it destroys all the apple trees that are planted near it.

A very similar passage appeared in *The Universal Herbal* by Thomas Green (1820).

SEBALD JUSTINUS BRUGMANS AND JULIUS VITRINGRA COULON

The Dutchman Sebald Justinus Brugmans (1763-1819; Figure 6.3) was a leading figure at the University of Leiden, and at one point surpassed Boerhaave's achievement, as he was the holder of four professorial chairs: botany, natural history,

Figure 6.3. A lithograph c. *1850 of S.J. Brugmans by L. Springer, Leiden from* Galerij van Hoogleraaren aan de Hoogeschool te Leijden, naar de oorspronkelijke afbeeldsels op de Senaatzaal aldaar.

medicine and chemistry. However, it was as a young man that he submitted an essay in 1785 to the Royal Berlin Academy of Science which was to set the scene in allelopathy for the next sixty years, as it offered what, at the time, was perceived to be the first experimental demonstration of root excretion. Until this time, a common view among botanists was that the various excrescences, on the surfaces of plant parts, were the plant excretions, and consequently excretion in plants was a function primarily of leaves, stems, flowers and fruits (see e.g., Anonymous 1773).

The title of the Brugmans' essay was *De Lolio Ejusdemque Varie Specie, Noxa et Usus*, and it was awarded a prize of fifty silver ducats by the Royal Berlin Academy of Science. According to Treviranus (1838), it had been translated into German by Gleditsch in Leipzig; however, no copy of this work, either printed or in manuscript, in Latin or German, is now known to exist[16], and other sources described it as unpublished (Anonymous 1789) The only record of text from this essay comes from a dissertation by one of Brugmans' students, Julius Vitringa Coulon (1767-1848), who went on to a career in medicine. The title of Coulon's thesis was *De mutato humorum in regno organico indole a vi vasorum vitale derivanda*, and it was published in 1789 (Figure 6.4).

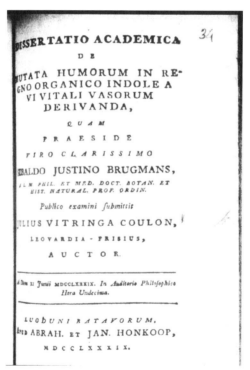

Figure 6.4. The title page of Coulon's thesis, De mutata humorum in regno organico indole a vi vitale vasorum derivanda, *1789.*

---

[16] Rafn (1796), likely in error, cited the essay of 1785 as being published by Lugd. Batavorum (p. 123).

Coulon's thesis had a strong vitalistic bent, reminiscent of the views of Theophrastus (see Chapter 2), and subscribed to comparing plant systems to those of animals. Coulon asserted that the nutrition of a plant had much to do with the ability of vessels to convert juices from the soil into the nutrients required by the plant. However, it was in a note on pages 77-79, that Coulon quoted from Brugmans' work:

> *Lolium* still seems to harm a cultivated plant through the root, in a way not yet well understood. It is established that certainly *Lolium* grows next to a cultivated plant its roots may thus be affected, as if it is consumed by insects: from this cause it grows weak and often even dies. When I first observed this I persuaded myself that this damage arose, in truth from insects, judging from the apparent damage to the roots. Nevertheless he [Brugmans] considered this conclusion doubtful after carrying out many trials, since he never happened to observe insects either with the naked eye or with magnification. It happens that, with a seed of *Lolium* sown next to cultivated plants, which is accustomed to attack, only that which is nearest to the emerging *Lolium* seems poisoned through the root, the roots of the remaining plants being evidently intact. I have always seen the same event happening, despite differing numbers of seeds. If this effect is to be attributed to insects, why then are only those plants eaten away at the roots which have *Lolium* nearby? To establish this more certainly, I devised the following experiment.
>
> I planted one of those cultivated, which I will at once enumerate, in a glass vessel, not so spacious that there would be so much continuous growth of roots but that I might daily observe the appearance of those extending themselves among the inner surface of the glass, in case by chance some of them were injured. The roots pushed themselves out as described but when Lolium we have designated by the title of "harmful" grew happily enough next to the cultivated plant in that vessel when the growth of that was clearly weakened and, what is the heart of the trouble, I saw those tender fibrils of the root extending next to the inner walls of the vessel soon attacked in the way I have described. I am not convinced so much that the enfeeblement of those plants, which we cultivated, invaded in this way through the roots, arose from insects but from a special and harmful relationship particularly of *Lolium* to certain plants, but it is not possible to determine anything beyond the name of this relationship. Evidently all plants in the first instance emit droplets at night, through the ends of their roots, which have likewise been observed by others since they do not escape the observation of good observers with the naked eye or at least with magnification. Thus, they are recognised clearly by the rather moist sand in which they terminate. These droplets so distilling, although they seem similar, nevertheless it is probable that they flourish with properties, often even harmful to another plant, when then by a similar reasoning a poison is produced which the tender roots of cultivated plants receive. This can be attributed to the liquid which is exuded through the ends of the fibres from the roots of *Lolium* so that, that part of the roots of a cultivated plant ought to be first injured which was closest to the root of *Lolium*. In this manner are injured:
>
> *Avena* by *Serratula arvensis*,
> *Linum* by *Euphorbia peplus* and *Scabiosa arvensis*[17],
> *Triticum* by *Erigeron acris*,
> *Polygonum fagopyrum* by *Spergula arvensis*,
> *Daucus carota* by *Inula helenium*, etc.

These ideas on the possible effects of root excretions in crop rotation appeared in the popular press only a few months later. It is not generally known that Coulon's thesis was reviewed in the English periodical, *The Monthly Review* in late 1789, and the reviewer, in addition to supplying the Latin text concerning Brugman's work, wrote the following (Anonymous 1789):

---

[17] = *Knautia arvensis*

> It being observed that the same species of plants, or of grain, will not continue to flourish with equal vigour in the soil where they were first planted, the ancients generally imagined that this phenomenon proceeded from their having exhausted those nutricious juices in the soils which were adapted to the peculiar nature of the vegetable; while the other juices which were capable of furnishing nutrition to the plants of a different species, remained un-absorbed. Many of the moderns entertain the same opinion; and attribute the necessity of sowing different grains in succession, to this principle. M. Coulon, and Professor Brugmans, his patron, ascribe this necessity to a very different cause. They assert that the grain, instead of being deprived of nutrition, by continuing long in the same soil, ejects, from its roots, a fluid which is pernicious to its own growth, and to the growth of some other vegetables; while it is highly beneficial to those of another class. This curious hypothesis was suggested to the Professor, by observing that all plants, though they absorb juices from the earth during the day, emit from the extremities of their roots, during the night, a fluid in the form of a drop, which is very different in different plants; and which being applied to the root of a neighbouring plant, sometimes proves pestiferous. (p. 685)

Brugmans and Coulon's ideas on plant excretions steadily gained wide exposure both in Europe and England. Their ideas were restated and/or amplified by several influential botanists. The earliest was Alexander von Humboldt (1769-1859) who was born into a wealthy German family, and from an early age had the opportunity to fraternise with many individuals who were or would become leading figures in science and politics. On the botanical side, at the age of eighteen he befriended Karl Ludwig Willdenow, who first whetted Humboldt's interest in botany, and a couple of years later he became greatly influenced by Georg Forster, at the University of Göttingen. Von Humboldt's restless nature and his love of geology eventually led him to pursue study at the Mining Academy at Freiberg in Saxony. His practical experiences in the mine pits fired his botanical curiosity, as he observed a surprisingly rich flora of bryophytes and lichens growing in the dim light of the mine pits. This led to publication in 1793 of von Humboldt's first book, *Flora Fribergensis* to which was appended von Humboldt's collected thoughts on plant physiology, *Aphorismi ex Doctrina Physiologiae Chemicae Plantarum*. This appendix was translated into German in 1794 and published as *Aphorismen aus der chemischen Physiologie der Pflanzen*[18]. It was von Humboldt who first suggested that the ideas of Brugmans could explain processes such as fallowing:

> Through these phenomena, it perhaps remains clear why the field is left to rest and why there is harmony of the plants, which have troubled man since the oldest times.

Similarly Joseph Jacob Plenk cited Brugmans' ideas in the various editions of his *Physiologia et Pathologia Plantarum*, firstly in Latin (1794), then German (1795), French (1802) and Italian (1804). An earlier citation in Italian of Brugmans' ideas was due to Carradori (1803). A little known author to have cited Brugmans and Coulon was Julius Johann von Uslar[19] (1752-1829) in *Fragmente neuerer Pflanzenkunde* (von Uslar 1794), and a translation of this work provided further exposition of these ideas in English, in an obscure book, *Chemico-physiological Observations on*

---

[18] This also appeared in 1798 in a rare work on plant nutrition by Ingenhousz, as Gotthelf Fischer was the translator for this as well as von Humboldt's *Aphorismen*.

[19] Julius Johann von Uslar was the father of Justus Ludewig von Uslar (see Chapter 8).

Plants, commonly attributed to Schmeisser (1795), who was basically the translator. Von Uslar's book is interesting also for other reasons: he provided an early statement that harm due to root excretions (allelopathy) could occur along side of competition, and he also suggested that root excretions can play a role in forestry:

> Though plants receive their food through various canals or mouths, and do not collect them in one reservoir like animals; yet, the mode and effect of their digestion resemble much that of animals. They part like animals with the superfluous and useless matter; and this separation they effectuate, not only by their respiratory organs, the leaves and stems, but also by other secretions similar to what we find obvious in animals.
>
> Mr Humbold, a gentleman of great reputation among the learned in Germany, has observed, that plants really secrete impurities through the extremities of their roots during the night, which excrements may, like those of animals, prove sometimes useful, sometimes hurtful, to other neighbouring plants. "*sic laevitur,*" says he, "*avena a serratula arvensis, triticum ab erigeron acri, linum ab euphorbia puplo et scabiosa arvensis, polygonum fagopyrum, a spergula arvensis, etc.*" From this he derives the effect of fallowing, and the harmony among plants.
>
> It is a well known fact, that some trees will not grow well near others, or that the one is hurt or suppressed by another of a different kind. The cause of this was thus explained, that the one deprived the other of food; but Mr Uslar supposes, with much plausibility, that the secreted matter of one kind of plant or trees, may likewise add to the cause of the destruction or injury of others. There are plants which do not allow others to grow near them, and which seem to prefer a solitary life. This circumstance has given rise to a division of plants, into Sociatae and Solitariae. Similar antipathy we observe likewise among animals; as certain genera of animals will not live together in harmony.
>
> Mr Uslar is of opinion, that this partly arises from physical causes; as he observed, that certain animals cannot bear the effluvia of others.
>
> Though certain plants show a great antipathy to one another, yet there are some which assist the growth of others. So, for instance, we see the birch often nourishing the oak and the beech.

The appearance of Brugmans' ideas in English is more usually attributed to the celebrated Erasmus Darwin (1731-1802), grandfather of Charles Darwin, in his *Phytologia* (Darwin 1800). Nonetheless Darwin's remarks are very interesting and astonishingly modern, as with so many of his observations. Firstly he wrote regarding plant acidic secretion:

> I suppose it is secreted both for the defence of those plants from the depredation of insects and larger animals; and also for the purpose of its being converted into a saccharine juice by the digestion of the young bud in the bosom of the leaf.

With regard to the excretion observed by Brugmans and Coulon, he noted that:

> But this I suspect to have been produced by the death and consequent decomposition of the extremities of the roots in their unnatural situation.

Senebier (1800), having become familiar with Brugmans ideas through Plenk's work, wrote:

> What happens eventually to these inherent juices[20] which continually descend in the roots; one can only imagine that they are completely combined, because one hardly sees

---

[20] In the original, the term used was *sucs propres*, which is essentially the fluid other than that found in the sap.

> a secretion without an excretion. What then will be the means of ridding the roots of this excess? I made some experiments to resolve this question during a very dry summer. I enclosed some roots of trees and shrubs that I had laid bare in some very dry bottles; I replaced them then in the place where they were, I covered them with soil, and I visited them often. When the roots were very small, I noticed nothing new in the bottle; but when they were large, I saw the roots covered with moisture, droplets formed, and their number increased as the experiment went on, so long as the root did not suffer too much from its confinement; it is true that one could attribute this effect to evaporation; but Brugman in a dissertation *De lolio ejusque varia specie, noxa & usu*, assures that the roots of these plants gives off during the night droplets harmful to other plants. Plenck teaches in his *Physiologia plantarum* that oats suffers for this reason in the vicinity of corn thistle; flax in that of *Euphorbia peplus* and *Scabiosa arvensis*, wheat in that of *Erigeron acris*; and *Daucus carotus* in that of *Inula helenium*.
>
> Although I have never been able to assure myself of this root excretion, there are several reasons why it is likely. The soil which surrounds the roots is greasier and moister than the other[21]; it can only be produced by the moisture which the outer part of roots are able to release. Malpighi and Gautier, like Brugman, have seen this excretion of roots. But one suspects that it is indispensable for grafted trees, since the leaves of the peach so different in all regards from those of the plum, would only know to elaborate the same as its bark the juices appropriate to the food of the roots of the latter, if these roots were not processing these juices for them, and one can hardly imagine this elaboration without a prior excretion. (pp. 315-317)

Also, Senebier in conjunction with his colleague François Huber (Huber and Senebier 1801) provided some of the earliest data on the effects of various organic substances on germination. In particular, were described experiments on the effect of camphor of seed germination, in which it was recorded that the camphor largely inhibited germination, but that the degree of inhibition depended to some extent on the species tested.

A similar account to that of Senebier was offered by the French anatomist and prolific botanical writer Charles François Brisseau-Mirbel (1776-1854). In his compendial *Histoire Naturelle, Génerale et Particulière des Plantes*, Brisseau-Mirbel (1802) wrote:

> The root also performs the functions of the excretory organ. The soil which surrounds it becomes greasy and takes on a darker colour, proving unequivocally that it imbibes the juices that the plant excretes. One sees all the time roots penetrating into canals full of water, becoming thinner, and dividing into a multitude of small threads which are fringed to their extremity, and are covered in a gelatinous material, which without doubt the soil would absorb if they were living buried there. It is to root excretions that one may perhaps attribute often the sort of antipathy that one observes between certain plants, which are never found together. Sympathies seem due to the same causes; it is of plants which seem to look for another and to follow on another; this phenomenon is so well known to botanists that the encountering of such a plant is sometimes for them a certain indicator of the presence of another that they have seen yet. This part of the history of plants has not been studied well enough, that concerns in some way their deaths and their sociability, and moreover it is probable that agriculture would draw there great illumination. (pp. 147-148)

---

[21] The notion that soil in the vicinity of older roots is darker and greasier than soil surrounding young roots has been alleged to originate with Duhamel du Monceau (Clements 1921), but Clements' citation seems in error (see his note 4).

Brisseau-Mirbel in *Elémens de Botanique*, (1815)[22] wrote:

> Some species of plants let juices flow out of their roots, which are, according to the opinion of Plenk and Brugmans, mortal poisons for other plants. But is it not more likely, that if certain plants of different species cannot live together on the same soil, that it shows that one of them removes from the soil nutritive elements necessary for the vigorous development of the others? This hypothesis explains in a plausible enough manner what one calls the antipathy of plants. (Volume 1, pp. 366-367)

The leading German botanists of the early nineteenth century cautiously endorsed the concept of root excretion (e.g. Link 1807), although Hedwig (1794) did not believe that the excretions observed by Brugmans were natural. Another authority to cite Brugmans with a degree of uncertainty was Willdenow, whose text on botany was translated into English in 1811.

> Brugmanns observed a particular kind of aqueous transpiration in the roots of some luxuriant plants; he had put some plants of this kind into a glass filled with earth, and observed at night a drop of fluid in the top of the radicles; he remarked as soon as such a drop touched the roots of other plants, they dried immediately. If this happened frequently, the plant decayed.
>
> Thus Oats, (*Avena sativa*), were destroyed in this manner by *Serratula arvensis*.
> Flax, (*Linum usitatissimum*), by the *Scabiosa arrvensis* and *Euphorbia Peplus*.
> Wheat, (*Triticum aestivum*), *Erigeron acre*.
> Buck-Wheat, (*Polygonum Fagopyrum*), by *Spergula arvensis*.
> Carrots, (*Daucus Carota*) by the *Inula Helenium*.
>
> Hence he concludes, that weeds with the fluid dropping from their radicles, suppress the growth of the contiguous plants. But might not the weed destroy the cultivated plant, owing to its absorbing the alimentary matter with greater rapidity, and expanding sooner, and thus prevent the further growth of the adjacent plant? (p. 325)

Modern Russian and Ukrainian plant ecologists (e.g. Gortinskii 1966, Grodzinskii 1973) have previously tried to claim that the eighteenth century Russian, Nestor Maksimovich-Ambodik, offered amongst the earliest statements concerning the chemical interactions of plants. Maksimovich-Ambodik (1796) authored a botanical text, in which he wrote: "such excretions by plants, often both in excreting and in growing in the vicinity, sometimes are of use, and sometimes inflict harm. Hence it is clear why often one plant suppresses another." However, it must be stated that Maksimovich-Ambodik was likely familiar with the writings of others, at least Plenk, for whom he had translated medical texts.

A curious synopsis of the idea of plant excretion was provided by Pitt (1810):

> The perspiration of trees, especially when they put out their leaves, is a fact which has been known for a long time. The effluvia, or attenuated substances, which they exact, and which excites in us the sensation of smell, show that these particles, which may be considered with regard to the tree as feculent matter, sometimes extend their influence to an astonishing distance. Nature, by rejecting them, plainly proves that they are baneful to vegetation. On this account, trees which have been planted by themselves in a favourable situation, have almost in every case, a finer appearance, and their timber is

---

[22] This excerpt was reprinted in *Dictionnaire des Sciences Naturelles*, vol. 28, p. 433.

more perfect than that of such as grow in forests. Whenever any of the latter acquire strength, they cause the contiguous ones to perish, that they may at the same time find room to extend their roots and get rid of their effluvia which incommode them.

(pp. 357-358)

The idea of root excretion became fairly standard fare in botany textbooks of the early nineteenth century. There is little need to cite all of these, but a typical view was expressed by the Frenchman Achille Richard (1794-1852), who authored a number of popular botany textbooks, that appeared in numerous editions, as well as several languages. Richard (1828) wrote in regard to root excretions:

> It is to this material, which as we have spoken, is different in each species, that are attributed the sympathies and antipathies that certain plants have one for another. On knows, in effect, that certain plants look for each in some way, and live constantly one beside the other; these are the social plants; however, on the contrary, other plants seem not to be able to grow in the same place. (p. 41)

REFERENCES

Anonymous. 1773. *Démonstrations Élémentaires de Botanique. Tome Premier.* Jean-Marie Bruyset, Lyon.

Anonymous. 1789. *Dissertation Academica*, &c. i.e. Inaugural Dissertation, attempting to prove that the changes observable in the fluids of all organised bodies, proceed from the vital influence seated in their vessels. By Julius V. Coulon. 8vo. Pp. 97. Leyden. 1789. *The Monthly Review* **81**: 682-687.

Anonymous. 1798. Notice sur les excréments des végétaux. *Journal de Physique, Chimie et d'Histoire Naturelle* **4**: 388.

Aamisepp, A. and H. Osvald. 1961. Influence of higher plants upon each other – allelopathy. *Nova Acta Regiae Societatis Scientiarum Upsaliensis, series 4,* **18**: 1-19.

Azara, F. de, 1809. *Voyages dans l'Amérique méridionale depuis 1781 jusqu'en 1801. Tome I.* Dentu, Paris.

Balduino, P.F. 1694. *Dissertatio Physica de Excrescentiis Plantarum Animatis, quam Inclytae Facultatis Philosophicae Indultu Publicae Eruditorum disquisitioni proponit M. Simon Weiss / Thor. Pruss.* Christopher Fleischer, Vienna.

Boerhaave, 1724. *Institutiones et Experimenta Chemiae.* Paris.

Boerhaave, 1727. *A New Method of Chemistry including the Theory and Practice of that Art.* J. Osborn and T. Longman, London.

Brauner, J. 1751. *Tankar om akerns rätta anlaggning, skötsel och såning, samlade gensm försök, samt noga beskriftning på de der til nödwändige redskaper.* Stockholm.

Brisseau-Mirbel, C.F. 1802. *Histoire Naturelle, Génerale et Particulière des Plantes. Tome I.* F. Dufart, Paris.

Brisseau-Mirbel, C.F. 1815. *Élémens de Physiologie Végétale et de Botanique. Première Partie.* Magimel, Paris.

Britton, A. 1892. Letter to Gov. A. Phillip, dated 24 February 1792 from D. Burton. *Historical Records of New South Wales* **1**(2): 599-600.

Carlowitz, H.C. 1713. *Sylvicultura Oeconomica.* Johann Friedrich Braun, Leipzig.

Carradori, G. 1803. *Della Fertilita della Terra.* Tipografia della Società Letteri, Pisa.

Clements, F.E. 1921. *Aeration and Air-content: the Rôle of Oxygen in Root Activity.* Publication No. 315 Carnegie Institution, Washington, D.C.

Cooper, T. 1791. Observations respecting the history of physiognomy. *Memoirs of the Literary and Philosophical Society of Manchester* **3**: 408-462.

Coulon, J.V. 1789. *De Mutato Humorum in Regno Organico Indole a Vi Vasorum Vitale Derivanda.* A. and J. Honkoop, Batavia.

Cuvier, G. 1834. *Histoire des Progrès des Sciences Naturelles depuis 1789 jusqu'à ce jour. Tome Premier.* Librairie Encyclopédique de Roret, Paris

Dadd, G.H. 1851. *The American Reformed Cattle Doctor.* Phillips, Sampson, and Company, Boston.

Darwin, E. 1800. *Phytologia; or the Philosophy of Agriculture and Gardening.* J. Johnson, London.

Davies, W. 1810. *General View of the Agriculture and Domestic Economy of North Wales*. Richard Phillips, London.
Davis, T.L. 1928. The vicissitudes of Boerhaave's textbook of chemistry. *Isis* **10**: 33-46.
Dickson, A. 1765. *A Treatise on Agriculture*. A.Kincaid and J.Bell, Edinburgh.
Duchartre, M. 1868. *Rapport sur le Progrès de la Botanique Physiologique. Recueil de Rapports sur les Progrès des Lettres et des Sciences en France*. L'Imprimerie Impériale, Paris.
Duhamel du Monceau, M. 1758. *Physiques des Arbres*. Paris.
Ford, B.J. 1986. Even plants excrete. *Nature (London)* **323**: 767.
Fussell, G.E. 1971. *Crop Nutrition: Science and Practice before Liebig*. Coronado Press, Lawrence, Kansas.
Gortinskii, G.B. 1966. 'Allelopathy and experiments of Soviet studies from the early twentieth century'. *Byulletin Moskovski Obshchestva Ispytarelei Prirodi, Otdeleniye Biologicheskii* **71** (5): 128-133.
Green, T. 1820. *The Universal Herbal; or Botanical, Medical, and Agricultural Dictionary*. Volume 1. Caxton Press, Liverpool.
Grodzinskii, A.M. 1973. *Osnovy Khimichnoi Vzaemodii Roslin*. Naukova Dumka, Kiev.
Gyllenborg, G.A. 1761. *Agriculturae Fundamenta Chemica*. Uppsala.
Gyllenborg, G.A. 1770. *The Natural and Chemical Elements of Agriculture*. Translated by J. Mills. John Bell, London.
Hale, T. 1756. *A Compleat Body of Husbandry*. T. Osborn, London.
Hales, S. 1727. *Vegetable Staticks: or, an account of some statical experiments on the sap in vegetables: being an essay towards a natural history vegetation. Also, a specimen of an attempt to analyse the air, by a great variety of chymio-statical experiments; which were read at several meetings before the Royal Society*. W. and J. Innys, London.
Hedwig, J. 1794. Appendix to: F.A. von Humboldt, *Aphorismen aus der chemischen Physiologie der Pflanzen*, pp.158-164. Voss, Leipzig.
Home, F. 1757. *Principles of Agriculture and Vegetation*. G. Hamilton and G. Balfour, Edinburgh.
Home, H. 1776. *The Gentleman Farmer, Being an Attempt to Improve Agriculture by Subjecting It to the Test of Rational Principles*. W. Creech, Edinburgh.
Huber, F. and Senebier, J. 1801. *Mémoires sur l'Influence de l'Air et de Diverses Substances Gazeuses dans la Germination de Differentes Graines*. J.J. Paschoud, Genève.
Humboldt, F.A. von 1793. *Flora Fribergensis specimen plantas cryptogamicas praesertim subterraneas*. Augustus Rottmann, Berlin.
Humboldt, F.A. von 1794. *Aphorismen aus der chemischen Physiologie der Pflanzen*. Voss, Leipzig.
Ingenhousz, J. 1798. *Über Ernährung der Pflanzen und Fruchtbarkeit des Bodens. Aus dem Englischen übersetzt und mit Anmerkungen versehen von Gotthelf Fischer. Nebst einer Einleitung über einige Gegenstände der Pflanzenphysiologie von F. A. von Humboldt*. Leipzig.
J., S. 1727. *The Vineyard: Being a Treatise, etc.* W. Mears, London
Jacquin, N.J. von 1781. *Miscellanea Austriaca ad Botanicum, Chemiam et Historiam Naturalem*. Officina Kraussiana, Vienna.
Jung, J. 1662. *De Plantis Doxoscopiae Physicae Minores*. Johann Naumann, Hamburg. (Also published as part of *Opusculae Botanico-physica* in 1747.)
Kerner, J.A.C. 1811. *Reiseschatten*. G. Braun, Heidelberg.
Lefebure, E.-A. 1801. *Les Expériences sur la Germination des Plantes*. Levrault Frères, Strasbourg.
Lennep, J, Moll, W. and ter Gouw, J. 1868. *Nederlands Geschiedenis en Volksleven in Schetsen*. A.W. Sijthoff, Leiden.
Link, H.F. 1807. *Grundlehren der Anatomie und Physiologie der Pflanzen*. Göttingen.
Linnaeus, C. 1745. *Öländska och Gothländska Resa på Riksens Högloflige Ständers Befallning förratad Ahr 1741 med Anmärkinger uti Oeconomien, Natural-Historien, Antiquitater, &c.* Stockholm. This has been translated into English in: Åsberg, M. and W.T. Stearn. (1973). Linnaeus's Ökland and Gotland Journey. *Biological Journal of the Linnean Society* **5**: 1-220.
Maksimovich-Ambodik, N. 1796. *Pervonachalniya Osnovaniya Botaniki*. St. Petersburg.
Maupin. 1799. *La Méthode de Maupin, sur la Manière de Cultiver la Vigne et l'Art de Faire le Vin*. Delaplace, Paris.
Micheli, P.A. 1723. *Relazione dell'erba da'Botanici Orobanche e volgarmente Succionale, Fiamma, Emald'Occio, etc.* Florence.
Miller, P. and T. Martyn. 1807. *The Gardener's and Botanist's Dictionary, corrected and newly arranged*. F.C. and J. Rivington, London.

Necker, N.J. de. 1775. *Physiologie des Corps Organisés*. A. Bouillon, Paris.
Nicol, W. 1803. *The Practical Planter, or, a Treatise on Forest Planting: comprehending the culture and management of planted and natural timbers, in every stage of its growth: also on the culture and management of hedge fences, and the construction of stone walls, &c.* J. Scatchard and H.D. Symonds, London.
N.W. [Nicol, W.] 1803. Answers to queries on vegetation. *The Farmer's Magazine* **4**: 44-53.
Pitt, W. 1810. *A General View of the Agriculture of the County of Worcester, with Observations on Means of its Improvement.* Sherwood, Neely, and Jones, London.
Plappart, J.F. 1777. *De Juglande Nigra, Dissertatio Inauguralis Medicina*. Gerold, Vienna.
Plenk, J.J. 1794. *Physiologia et Pathologia Plantarum.* A, Blumauer, Vienna.
Plenk, J.J. 1795. *Physiologie und Pathologie der Pflanzen*. C.F. Wappler, Vienna.
Plenk, J.J. 1802. *Physiologie et Pathologie des Plantes*. J.-F. Barrau, Paris.
Plenk, J.J. 1804. *Fisiologia e Patologia delle Piante*. G. Pezzana, Venice.
Rafn, C.G. 1796. *Udkast tif en Plantephysiologie, grundet paa de unere Begreben I Physik og Chemie.* Sebastian Popp, Copenhagen.
Randall, J. 1764. *The Semi-Virgilian Husbandry, Deduced from Various Experiments, or, an Essay towards a New Course of National Farming.* B. Law, London.
Rennie, J. 1834. The practice of fallowing, of paring and burning, of irrigation, and of draining, explained on new scientific principles. *Quarterly Journal of Agriculture* **5**: 1-32.
Richard, A. 1828. *Nouveaux Éléments de Botanique et Physiologie Végétale*. Quatrième édition. Bechet Jeune, Paris.
Sachs, J. von 1889. *History of Botany (1530-1860).* Clarendon Press, Oxford.
Saint-Simon, M.H. Marquis de. 1768. *Des Jacintes, de leur Anatomie, Reproduction et Culture*. Amsterdam.
Schmeisser, J. 1795. *Chemico-physiological Observations on Plants.* William Creech, Edinburgh.
Schroth, M.N. and D.C. Hildebrand. 1964. Influence of plant exudates on root-infecting fungi. *Annual Review of Phytopathology* **2**: 101-132.
Senebier, J. 1791. *Physiologie Végétale. Encyclopédie Méthodique*. Panckoucke, Paris.
Senebier, J. 1800. *Physiologie Végétale. Tome I*. J.J. Paschoud, Geneva.
Siösteen, K. 1749. Landtmann Försök, at Fördrifva Land-eller Flyg-Hafra utur äkerjorden. *Kongliga Svenska Vetenskaps Academiens Handlingar* **10**: 187-190.
Smellie, W. 1790. *The Philosophy of Natural History*. Heirs of Charles Elliot, Edinburgh.
Treviranus, L.C. 1838. *Physiologie der Gewächse. Zweyter Band*. Adolph Marcus, Bonn.
von Uslar, J. 1794. *Fragmente neuerer Pflanzenkunde*. Schulbuchhandlung, Braunschweig.
Willdenow, C. 1811. *The Principles of Botany, and of Vegetable Physiology*. William Blackwood, Edinburgh.

CHAPTER 7

AUGUSTIN PYRAMUS DE CANDOLLE, AND HIS ERA

> New theories must struggle only with difficulty,
> and from them a kernel rarely drops.
>
> *Die Bodenvergiftung durch die Wurzel-Ausscheidungen der Pflanzen*
> *als vorzüglichster Grund für die Pflanzen-Wechsel-Wirthschaft*
> Justus Ludewig von Uslar (1844)

AUGUSTIN PYRAMUS DE CANDOLLE – EARLY YEARS

Interest in allelopathy in the first half of the nineteenth century has been linked primarily to one man, Augustin Pyramus de Candolle (Willis 1996, 2002). A.P. de Candolle (1778-1841; Figure 7.1) was born in Geneva into a moderately affluent Protestant family. At the age of seven, he was stricken with hydrocephalus, but survived seemingly with no ill affects. He became fluent in Latin while at school, and seemed destined for a literary career. The revolutionary fervour in France spread to the republic of Geneva, and de Candolle's family, being both Protestant and of privileged position, was forced to seek refuge during 1792-4 in Vaud on the shores of Lac Neuchâtel. This period was undoubtedly important for de Candolle's health, and for the development of his lifelong love of botany.

De Candolle was able to return to Geneva in 1794 where he studied botany under Jean Pierre Vaucher, who became his friend and mentor, as did the plant physiologist Jean Senebier, whose prolix style likely implanted in de Candolle the seed to produce his own logical account of plant physiology. In 1796 de Candolle visited Paris for the first time, and at the young age of 18, he met several of the leading French scientists in natural history including Cuvier, Desfontaines and Lamarck. In 1798 the political situation again became unstable in Geneva, as the district was annexed by France, and de Candolle decided that it was an opportune time for him to pursue medical studies in Paris.

De Candolle's principal interest remained botany; however, the key to obtaining a botanical position was in completing a medical degree. De Candolle did not enjoy the practical side of medicine, and he spent as much time as possible in botanical pursuits. He received various commissions, among which writing the text for *Plantarum Historia Succulentarum*, a project initiated by the ill-fated L'Héritier de Brutelle, established his reputation and led to his being invited by Lamarck to prepare the third edition of the *Flore Française*, of which publication began in 1805. He also had begun giving botany lectures at the College de France in 1804. In the same year he completed his dissertation for his degree in medicine, and this was published successfully as *Essai sur les Propriétés Médicales des Plantes*.

Figure 7.1. *A lithograph of Augustin Pyramus de Candolle by Hébert after a drawing by Alexandre Calame (from Choisy 1843).*

The first volume of the new *Flore Française* was almost entirely the work of de Candolle, although Lamarck appears as first author. The volume is actually a primer on botany, which de Candolle probably intended as a textbook for his botany course in Paris. De Candolle was sufficiently pleased with the work that he had a major proportion of the volume published privately and anonymously under the title *Principes Élémentaires de Botanique et de Physiologie Végétale* (1805; Figure 7.2). It is here (and in *Flore Française*, volume 1) that we find de Candolle's first mention of root excretions and the chemical interaction of plants; however, there is nothing original in it; indeed, it is very reminiscent of comments by Senebier (1791, 1800), and there is no mention of crop rotation:

Finally, roots themselves present, in some plants, particular secretions; it is observed in *Carduus arvensis, Inula helenium, Scabiosa arvensis*, several euphorbs, and several members of the Chicoraceae. In these last plants, these secretions are very visible, because they have latex, as in the ordinary sap: it seems that these root secretions are nothing other than part of the inherent juice, which having not served in the nutrition are rejected outwards where they arrive at the lower part of the vessels. Perhaps this phenomenon, sufficiently difficult to see, is common in a great number of plants. Ms. Plenck and Humboldt had the ingenious idea to seek in this fact the cause of certain habits of plants. Thus one knows that the corn thistle is harmful to oats; euphorbia and scabiosa to flax; elecampane to carrot; bitter fleabane and cockle to wheat, etc. Perhaps the roots of these plants release substances harmful to the growth of others. On the contrary, if the loosestrife grows neighbouring to the willow, orobanche near the hornbeam, etc., is it not that the root secretions of these plants are advantageous to the growth of others?

In the following years, de Candolle had little to add regarding the concept of root excretion, and it must be remembered that the vast majority of his botanical contributions were in the fields of taxonomy and systematics. In 1808, after experiencing disappointment in not securing an academic appointment in Paris, he accepted the position of professor within the Medical School, later professor of Natural History, at Montpellier, a quiet regional centre in southern France. Whilst at Montpellier he authored *Theorie Élémentaire de Botanique* (1813), which eventually went through three editions. Plant excretion was addressed fleetingly in this work, but perhaps what is more important in this work is that de Candolle introduced the concept of the spongiole, a structure with a spongy, absorptive texture allegedly found in pistils, seeds, and notably, root tips; although again the concept of the root spongiole was reminiscent of the words of Boerhaave (1727). The indiscriminant absorptive qualities ascribed to the root spongioles were later to play a central role in de Candolle's flawed theory of plant nutrition.

In 1816 de Candolle returned to Geneva to assume the chair of Natural History at the Université de Genève, a position that had been created for him. The root spongiole is again described in 1827 in *Organographie Végétale*, and de Candolle also briefly foreshadowed his theory of crop rotation based on root excretions:

> Many roots exude, it is said, by their extremities excremental juices, of which the origin and history are still little known, but they seem to be the cause of several important phenomena. These excretions of roots were seen especially by Brugmans, and will deserve special attention on the part of physiologists. It is probable that if one will study them with care, one will then discover the true theory of the affinities and repulsions of certain species, and what is more important, the true theory of crop rotation.[1]

Similarly in 1820, he had touched briefly on the issue of substances in the soil in his famous essay on "Géographie Botanique", which appeared as an entry in the mammoth *Dictionnaire des Sciences Naturelles*:

> Thus the diverse nature of substances dissolved in waters [of the soil] is evidently one of the numerous causes which determines the status of plant species. (vol. 18, p. 374)

---

[1] The last sentence in this excerpt does not appear in the English translation, *Vegetable Organography*, published in 1841, as assumedly by this time, publication of *Physiologie Végétale* had rendered the statement superfluous. Furthermore this statement is perhaps rather coy, as by 1826 de Candolle had actually elaborated his theory of crop rotation sufficiently to include it in his lectures.

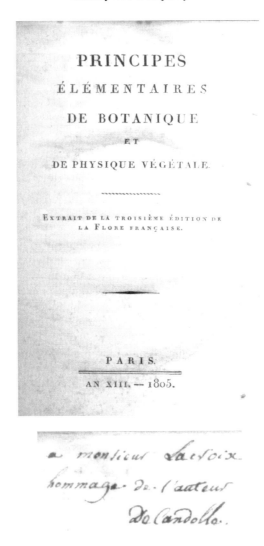

Figure 7.2. The title page of Principes Élémentaires de Botanique et de Physique Végétale *(de Candolle 1805), above; and details of an inscription, from the author to M. Lacroix, (assumedly the mathematician Silvestre-François Lacroix, a fellow member of the Société Philomathique de Paris) on the half-title, below.*

The specific concept that crop rotation (*théorie des assolemens*[2]) could be explained through the differing effects of root excretions is generally ascribed to de Candolle. The core idea had certainly been given earlier in the 1789 dissertation by

---

[2] The modern spelling provides *assolements*.

Julius Vitringa Coulon (see Chapter 6), and amongst French publications, it was fairly clearly stated in an anonymous note that appeared in 1798 in the *Journal de Physique, Chimie et d'Histoire Naturelle*. The most probable author was the journal's editor, J.C. Delamétherie[3]. The note discussed root excretions as observed by Brugmans in his essay of 1785:

> These phenomena can explain why farmers are obliged to leave their fields to rest for a year; because in this interval, this humour has time to decompose.
>
> This will explain, as well, why a soil tired of a plant will be vegetated by others with force: the faeces of the first are harmful to the plants of the same species, and serve as fertiliser to the others. A field, for example, fatigued from producing clover, and planted with wheat, will give an abundant yield, because doubtless the faeces of clover are fertiliser for the wheat.

It is tempting to speculate that de Candolle was the author of the note, and indeed, de Candolle had contributed an article on lichen nutrition to the same journal a few months earlier. However, there is really no further evidence to support any link[4], and de Candolle, age 21, did not meet Brugmans, in Leiden, until several months later in the Spring of 1799, during a tour of Holland, although de Candolle had access to Brugmans and Coulon's ideas through numerous written sources. Certainly, it is well established that in the domain of plant physiology, de Candolle was primarily a synthesiser of information, not an innovator.

While de Candolle is frequently given much of the credit for the later botanical writings which appeared under Lamarck's name, notably the third edition of the *Flore Française*, it should not be forgotten that Lamarck made his own important contributions, particularly in the earlier parts of his career. An often ignored publication is *Histoire Naturelle des Végétaux*, authored by Lamarck and Brisseau-Mirbel, and later A.L. de Jussieu, which is often catalogued among the works of Buffon, as it forms part of his series, *Histoire Naturelle*. In the first volume (1803), apparently authored mainly by Lamarck, the writer raises, seemingly for the first time, an important ecological issue, which would not have been lost on de Candolle: what happens to the varied constituents of plants after the death of the plant or the plant parts?

> All these substances (herbaceous or woody matter, mucilages, gums, resins, oily substances, salts, etc.) that could have been produced only through the action of life, having ceased to be maintained by this action, come successively to increase, or at least keep up with the mass of *corps bruts* that exists. But, as with time these same substances undergo changes in the combination and proportion of their constituents, changes which alter their state, even their nature, and consequently all their qualities, they contribute to the maintenance of the diverse mineral matters which we observe all around, where living beings are in abundance (plants and animals), there leaving their detritus. (pp. 303-304)

---

[3] The General Index to the journal for the years 1787-1802, published *c*. 1803, gives the author as Brugmann (*sic*), but this is unlikely, and not so stated in the individual index for volume IV.

[4] That de Candolle's theory of crop rotation was not entirely original is also inferred in Jane Marcet's version (see excerpt below).

The concept that one crop could be poisonous to a successive crop, especially of the same or similar species, was more widely known than generally has been credited. In an early contribution, Braconnot (1807) cited the findings of Brugmans and Coulon, and added that the success of paring and burning may be attributable to the combustion of the harmful excrementitious matters ejected by roots. Similarly, Féburier (1812) invoked the Lamarckian concept of elemental "fire" as the universal transforming agent in explaining the noxious effects of one plant on another of the same species:

> The excretions of leaves and roots of certain plants are also favourable or adverse to the growth of other plants. Thus root excretions, just as with those of the leaves of the plant, are harmful to the growth of the same species, when placed there after the death of the first one, because these substances, already rejected as useless as food of the first plant, cannot serve as food for the second until after new combinations. This reason seems all the more likely, for if one has mixed the ashes of an oak, for example, with the soil where on has planted another oak, the growth of this plant will increase perceptibly, and more so if one places there the ashes of another species, because these parts, separated by fire, and reduced to the appropriate state to enter again into the sap vessels and to supply food to the plant, are thus found to be in the necessary proportions. (p. 352)

Yvart (1821), in a thorough account, also cites von Humboldt and Brugmans, and in particular, addresses the phenomenon of the poor growth of replanted orchards, and concludes:

> It seems to us then well proven, that by these facts and by many others which are similar, that either by their excretions or by their rotting debris, plants are harmful more or less to the same species that follows immediately in the soil.
>
> This very pronounced repugnance that is manifest in plants that immediately replace those of the same species without a preliminary preparation of the ground, seems also to extend more or less to all the species of the same genus, even to all those of the same natural family. (Volume 2, p. 138)

Furthermore, in considering the failure of replanted elms, he quotes Thouin, a close friend of de Candolle's, as having written:

> That roots which decay in the ground impart to those which belong to the same type of plant a principle of death, while they furnish a fertiliser to others. (Volume 2, p. 138)

The issue of "plant antipathy", which had been so topical in late eighteenth century, was still very much a subject of controversy. For example, the *Hollandsche Maatschappij der Wetenschappen* (Holland Society of Sciences) at Haarlem in 1823 (Anonymous 1823) posed a prize essay question:

> Does evidence adequately show that there are species of trees or plants, especially those that are useful, which can grow well when they are near other plants? And in this case, which are the data one could cite about it? This antipathy between species, could it in some way be explained in knowing something of the nature of these plants? What useful instructions could be extracted for the cultivation of trees and useful plants? (Question III)

A related question was issued by the Society in 1830 (Anonymous 1830):

> Given that ryegrass (*Lolium temulentum*) is the only plant, which of all the grasses, because of its noxious quality, seems to be the exception to the uniformity and to the

general analogy of the properties, for which the class of grasses is characterised, one asks: What causes the harmful quality of ryegrass? Is it constant and inseparable from the nature of this plant, or rather is it only by accident or produced by some particular cirncumstance? Perhaps, in the latter case, can this noxious property be prevented? (Question V)

While debate centred chiefly on the role and effects of root excretion, it was still widely regarded that trees, in particular, could exert a harmful effect on the vegetation underneath due to the injurious qualities of the drip from the leaves. The walnut tree maintained its reputation has being harmful to crop plants. The English botanist James Edward Smith (1809) provided a logical explanation for certain "noxious" plants:

> So the bad effects, observed by Jacquin, of Lobelia longiflora on the air of a hot-house, the danger incurred by those who sleep under the Manchineel-tree, Hippomane Mancinella, or, as it is commonly believed under a Walnut-tree, are probably to be attributed as much to poisonous secretions as to the air those plants evolve. (p. 204)

In *Pomarium Britannicum*, Henry Phillips (1820) related that a well-known strawberry-grower, Mr. Keen, had found that "the walnut is so injurious to strawberry beds, that they seldom bear fruit in the neighbourhood of that tree." This was repeated by Loudon (1838), who added that the poor growth of the oak near the walnut may be due to "the interference of their roots in the subsoil", and that the harmful effects of the walnut on grasses and other ground plants was due to the effects of the decaying leaves. The French author Marie-Henri Beyle (better known by his pseudonym Stendahl) in his 1830 novel *Le Rouge et Le Noir* wrote:

> Each of these cursed walnuts, said M. de Rênal when his wife admired them, costs me the harvest of a demi-arpent[5], wheat cannot grow under its shade. (Chapter VIII)

The issue of the alternation of species, which was discussed by de Candolle (1830) in relation to trees, and which was at the core of his *théorie des assolemens*, was also widely topical at the time in Europe. In particular, Dureau de Lamalle (1825) considered the apparent antagonism of certain plants in relation to their dominance over one another. He noted that grasses tended to suppress saintfoin and lucerne when grown together, although the grasses rarely totally eliminated the legumes. Indeed, he had observed in certain isolated plateaus, that had never been manured, a regular alternation of dominance by grasses and legumes.

It is also seldom acknowledged that root excretion theory was advocated simultaneously with de Candolle's major work *Physiologie Végétale*, if not earlier, in Germany, and the Hannoverian agricultural chemist Carl Sprengel[6] (1787-1859) described the fundamental points in 1832 in his *Chemie für Landwirthe, Forstmänner und Cameralisten*:

---

[5] A demi-arpent is about 344 m² or roughly one twelfth of an acre.
[6] During roughly the same period, there were three different German botanical writers known as "C. Sprengel", and many have confused them. Carl Sprengel (1787-1859) was an agricultural chemist. Christian Konrad Sprengel (1750-1816) was not closely related, and was known for his work on floral biology. He was the uncle of Kurt (sometimes given as Curt) Polycarp Joachim Sprengel (1766-1833), a botanist known especially for his important history of botany, *Historia Rei Herbariae*.

> That certain plant substances, which among others those also belong, that occur in the roots of the dandelion and coltsfoot, affect several plants thus as poisons, although they are often not at all harmful to the animal organism, is indeed a phenomenon most important for us to notice because we received thereby the conviction that still much research must be undertaken, in order to determine what plants with the greatest use after certain plowed-in green plants may be cultivated; also it thereby probably occurs to us from it some rules that may be derived regarding the rotation of crops that can be cultivated. We may perhaps see then, as potato roots contain a certain substance, the cause of why rye grows badly after potatoes and so forth. (Volume 2, p. 313)
>
> Many [acids] are eliminated by the roots, and in conjunction with associated acids in the ground remain most probably for a while undecomposed in the soil, and serve succeeding plants then for either beneficial or unfavorable effect; from this it may be possible to explain some of the phenomena due to crop rotation. (Volume 2, p. 340)

Also, there are scattered early references to root excretion in an ecological context. Cotta (1806) described various experiments with the roots of trees, and claimed that the finding of condensation in his observation vessels supported the idea of root excretion. He concluded that:

> Surely a general knowledge of plant physiology must leave us to suspect, that plants release certain fluids as well through their roots, just as these occurs via their upper parts through the exhalation, particularly through the leaves. However, Brugmans was the first investigator, who called attention to this, and he conveyed the observation that the dripping of fluid occurs from the tips of its rootlets, to which are ascribed the term plant excrement. (p. 47)

K.P. Sprengel (1812)[7] stated that the coastal grasses *Arundo arenarius* and *Elymus arenarius* excreted material from their roots, and that this may increase the fertility of dune sands. Meyer (1830), like Duhamel du Monceau (see Chapter 6) had noted that the occurrence of root mucilage in several coastal plants, and suggested that the release of this material may aid in the growth of neighbouring plants.

JANE MARCET

According to de Candolle's memoirs (de Candolle 1862, 2004), his theory of crop rotation, or *théorie des assolemens*, was first presented as a coherent entity, as part of his *Cours de Botanique* at the Université de Genève in 1826[8]. His students over the years included many individuals who became respected in the field of botany, including Nicolas Seringe, Charles Daubeny, etc. Another of these was Jane Marcet (1769-1858; Figure 7.3) who became the first to publish de Candolle's ideas on root excretions in relation to crop rotation. Jane Marcet (née Haldimand) was an Englishwoman, who was born into a wealthy banking family with Swiss connections, and

---

[7] This statement appears on p. 405 of the 1812 edition of *Bau und der Natur der Gewächse*, but the 1817 edition of similar title, which forms the first volume of *Anleitung zur Kenntniss der Gewächse* bears little resemblance.

[8] The content of de Candolle's lectures was informally published at the time, possibly using lithography from handwriting. A set of such notes from 80 lectures of the *Cours de Botanique* was acquired by the Huntington Library, California, and is believed to date from about 1830.

who had married Alexandre Marcet, a Geneva-born doctor working in London from 1805 to 1819. De Candolle had visited England in 1816, and there he relied heavily on the Marcets for social contacts, as his English was not very good[9]. The Marcets provided de Candolle with many opportunities to meet important English figures, as they moved amidst an affluential London society that included academic luminaries

Figure 7.3. Painting of Jane Haldimand Marcet (courtesy of Mr. Chris Pasteur).

---

[9] In the early part of the nineteenth century, there was still considerable enmity between the English and the French, even in the sciences. A double edged insult was hurled in the *Edinburgh Review* **34**: 375, "It is because Botany is one of the sciences which demands the smallest range of intellect, that the French have made themselves more conspicuous in it than in most others – and may absolutely claim a superiority over England!"

such as Thomas Malthus, Michael Faraday, Humphrey Davy and Peter Mark Roget (Polkinghorn 1993). In 1819 the Marcets returned to Geneva, where they socialised amongst a circle of academics that included Pictet, de la Rive and de Candolle. While Jane Marcet was living in Geneva, she attended de Candolle's lectures, and with his agreement she wrote an instructive book on plant botany (Marcet 1829), based largely on his teachings, which subsequently had a strong emphasis on plant physiology. Jane Marcet had already achieved considerable acclaim for her textbooks in a wide range of fields including economics and chemistry, and generally these quaint works were written as conversations between a teacher and two students (Figure 7.4).

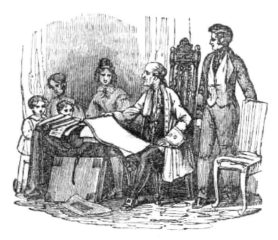

Figure 7.4. A woodcut from the title page to Lessons on Animals, Vegetables, and Minerals, *by Mrs Marcet (1844).*

In *Conversations on Vegetable Physiology*, we find the following:

Mrs B.

The theory which M. De Candolle is most inclined to favour, if indeed he is not its author, is the following. A plant, being under the necessity of absorbing whatever presents itself to its roots, necessarily sucks up some particles which are not adapted to its nourishment, and in consequence – after having elaborated the sap in its leaves, and re-conducted it downwards through all its organs, each of which takes in the nourishment it requires; after having extracted from it the various peculiar juices, and, in a word, turned it in every possible way to account, - finds itself encumbered with a certain residue, consisting of the particles it had unavoidably absorbed, and which were not adapted to its nourishment: these particles, having passed through the system without alteration, are exuded by the roots which had absorbed them, and thus return into the soil, which they deteriorate for a following crop of the same species of plant, but improve and fructify for one of another family; thus affording an admirable proof of the wise economy of Nature, in multiplying her vegetable produce by feeding different plants with different substances, and enabling beings, incapable of distinguishing their food, to obtain that which is appropriate to them.

Emily.

It is, indeed, admirable! Then, though the roots of plants can make no choice, their organs are in some measure capable of selecting, since they reject, and will not elaborate, substances which are not adapted to the nourishment of the plant.

Mrs. B.

If we cannot exactly allow them the nice discrimination of the chemist, we must at least suppose their laboratory to be so arranged as to act only on bodies congenial to the plant.

Caroline.

And the rejected substances, which would be poison to one family, when transferred into the soil, is greedily devoured by a succeeding crop of a different family.

Emily.

Yet, Mrs. B., there is land in the Vale of Glastonbury, in Somersetshire, which is celebrated for growing wheat for many years together without any manure; and I have heard that in the neighbourhood of the Carron iron-works, in Scotland, wheat has been raised above thirty years, without injury either to the crops or the soil.

Mrs. B.

Those soils must not only abound with vegetable nourishment, but the land be particularly well adapted to growing wheat; consequently, the roots would have little or nothing to exude, and successive crops of wheat might be raised so long as the land was not exhausted. This explanation would reconcile your difficulty to the theory of exudations; but interesting and plausible as this theory is, it requires the confirmation of facts to rest on a solid foundation: few experiments have yet been made relative to it. Mr. Brookman[10] has raised some plants in sand, and ascertained that they exuded by the roots small drops during the night, which there is reason to suppose was the object in research; but experiment has not yet been pushed far enough fully to verify it.

(Volume 1, pp. 261-263)

Thus, ironically, de Candolle's theory of crop rotation first appeared in English, a language he did not well understand. Jane Marcet's book achieved wide circulation as it went through three editions, and numerous plagiarised editions appeared in the United States, ostensibly authored and edited for the American market by the Rev. S.F. Blake. In 1840 the titled was changed to *Conversations on Botany* to better reflect the content. The French edition appeared in 1830 with the title *Conversations sur la Physiologie Végétale*, but curiously the author's name is not directly mentioned, whereas that of the translator, Macaire-Prinsep[11], appears boldly on the title page. Macaire-Prinsep was eventually to provide the experimental cornerstone that de Candolle had decreed lacking to his theory.

---

[10] The author has anglicised the name of Brugmans.

[11] Jean Francois Macaire (1796-1869) was a pharmacist who was a member of the Société de Physique et d'Histoire Naturelle de Genève, and a colleague of de Candolle. His name appears variously as he also early adopted the name Isaac-Francois Macaire, and after marriage to Caroline Prinsep in 1824, he frequently used the hyphenated surname Macaire-Prinsep, which he also latinised sometimes as Macaire-Princeps. Curiously, his second wife also bore the name Prinsep, although she was English, as Jean Francois married the widow of his brother-in-law, Agnes Catherine Prinsep (née Blake). Although he is often cited as Macaire in references, I have uniformly given his name here in the text as Macaire-Prinsep, as the bulk of his relevant work was post-1825.

DE CANDOLLE'S *THÉORIE DES ASSOLEMENS*

According to de Candolle himself (1832a), he did not present his theory to the public until 1826, within his lecture series on botany (*Cours de Botanique*[12]) at the Université de Genève. Until this time, the subject of crop rotation had largely been dominated by de Candolle's compatriot Charles Pictet de Rochement, who authored *Traité des Assolemens* (1801), the Frenchmen Victor Yvart (1821), and de Candolle's friend, André Thouin, who was head of the Jardin du Roi in Paris, and an influential agronomist who had edited the definitive *Nouveau Cours Complete d'Agriculture*. However, one may conjecture that the deaths of Thouin in 1824, Pictet in 1825, and Yvart in 1831 cleared the way for de Candolle to express his ideas with greater confidence.

The *Théorie des Assolemens* is generally cited as having been published in de Candolle's *Physiologie Végétale*, which indeed is true. However, the *Théorie des Assolemens*, authored by de Candolle, was published on two separate occasions immediately prior to this. The first of these was in February 1832 as an essay within the *Bulletin de la Classe de la Société des Arts de Genève* (de Candolle 1832a), and, curiously, a second version also appeared in June 1832, before *Physiologie Végétale*, since his former student, Nicolas Seringe (1832), published a lengthy extract of de Candolle's essay in his journal, based at Lyons. The explanation for de Candolle's action is likely twofold. *Physiologie Végétale*, which was to form the second printed part of de Candolle's *Cours de Botanique*, had been in press since August 1831, and publication was delayed. Firstly, one suspects that on account of these delays, de Candolle was worried that his crop rotation theory might be usurped by someone else, and indeed, as discussed following, subsequent claims of priority did arise in England. However, equally importantly, while *Physiologie Végétale* was in press, de Candolle's colleague Macaire-Prinsep, had announced his experimental findings at the December 1831 meeting of the Société de Physique et d'Histoire Naturelle de Genève. These were able to be included in de Candolle's 1832 separate and now very rare essay on *les assolemens*, although he did accordingly amend the manuscript of *Physiologie Végétale*, but in the section on root excretions. The essay on *Théorie des Assolemens* (de Candolle 1832a) began with a gentle introduction to de Candolle's thinking on soil infertility in crop rotation:

> The whole theory of crop rotation rests on the fundamental fact that plants grow poorly on land which has just supported plants of the same species, genus, or even the same family. Now this important fact rests on the distinction it is necessary to establish between *exhaustion* and *sickness* of the soil. Soil exhaustion occurs when various plants have drawn from a given soil all the extractive material, and sickness occurs when a certain plant causes the sterility of the soil, either for individuals of the same species or those of the same genus or family, but often leaves it fertile for other plants. Exhaustion occurs for all plants whatever, it acts in impoverishing the soil, in removing from it the

---

[12] De Candolle as professor of natural history was required to give lectures in both botany and zoology. He recorded that from 1816 until his resignation from professorial duties in 1835 he gave nineteen courses, ten in botany and nine in zoology, each with 108 lectures (perhaps three per week for 36 weeks?). One can assume that botany and zoology courses were given in alternate years. The botany course was evidently given in the even-numbered years.

nutritive material; soil sickness is something more specific which approximates the nature of poisons, and which seems to act as if the ground were corrupted by the addition of some material. (p. 2)

Later in 1832, *Physiologie Végétale* was finally published in three volumes (de Candolle 1832b), not two as originally planned, and the *Théorie des Assolemens* appeared in Volume 3, but without reference to Macaire-Prinsep's new work.

Physiologie Végétale received wide acclaim, and in 1833, in recognition of his contributions to plant physiology, de Candolle was awarded the prestigious Royal Medal from the Royal Society of London (Sussex 1833). *Physiologie Végétale* was never published in English; however, some extracts relating to the discussion here have been published in Browne (1944) and Willis (1996), and extracts relevant to the present discussion are provided in the following paragraphs:

> All plant substances, when they are no longer protected by the vital force, yield more or less quickly to the action of external bodies and are decomposed. If this decomposition occurs in open air, their elements, mixed in the vast mass of the atmosphere, have no perceptible effect on the vegetation; but if it occurs in the soil, the soil becomes enriched with all the directly or indirectly soluble substances which can be found in these decomposed plants.
>
> The effect of this type of enrichment may be useful or injurious to the plants destined to be fertilised, depending on the chemical nature of the plants incorporated. Thus, the effect will be useful to the vast majority of plants, if the plants incorporated contain much gummy, starchy, sugary, or woody substances, or in general, substances which are not bitter: the general effect will be on the contrary harmful if the plants incorporated contain much acrid, astringent, alkaline bitter, etc. substances. Thus, farmers well know that they ameliorate the soil in incorporating cereals or legumes, whereas they damage the soil in incorporating poppies or spurges. They know that barks which contain little tannin and gallic acid can improve the soil, whereas the bark of oak damages it. When one considers that neutral or insipid substances form the most considerable proportion of the bulk of plants, it should be concluded that incorporation tends generally to favour the vegetation and to improve the soil. (pp. 1490-1491)

On crop rotation, he wrote:

> The whole theory of crop rotation rests on this fundamental fact that plants do not grow well on land which has just supported plants of the same species, genus, or even the same family as themselves. Thus, cereals do not succeed on soil which has grown cereals the year before. Fruit trees become poorly in nurseries in places where the previous year there were the same sort. Street trees which have just died are difficult to replace with trees of the same species, etc., etc. This law is so general that it is certain that even mushrooms (*Agaricus albellus*) do not come up two years in a row in the same place.
>
> This remarkable fact is based on the distinction it is necessary to admit between exhaustion and specific sickness of the soil.
>
> Soil exhaustion occurs when a large number of plants consume from a given soil all the extractive material, and specific sickness occurs when a certain plant causes the sterility of the soil, either for individuals of the same species as itself or those of the same genus or same family, but leaves the soil fertile for other plants.
>
> Exhaustion occurs for all plants whatever; it acts in impoverishing the soil, in removing from it the nutritive material. Soil sickness is something more specific; it acts in corrupting the soil, and as we have indicated in discussing root excretions (Bk. II, chap. IX, §. 12), in incorporating a dangerous substance. Thus a peach tree injures the soil for itself, in such a way that, if without changing the soil, one replants a peach tree in soil where there had already lived another before, the second languishes and dies,

whereas any other tree can grow there. If the same tree does not produce this result, it is because its own roots, in always elongating continually meet veins of earth where they have not yet deposited their excretions. It is conceivable that its own excretions should be self-injurious almost as if an animal were forced to feed on its own excrement. This effect, in one or another example, is not confined to individuals of the same species, by virtue of their organization, should suffer when they take up by their roots a substance excreted by an analogous species, just as a mammalian animal is generally loathe to touch the excrement of other mammals. Therefore one may conceive easily enough how each plant tends to corrupt the soil for its congeners: how certain plants with bitter juice, such as poppies and spurges, damage the soil for most plants." (pp. 1495-1496)

While the root excretion theory is commonly tied to de Candolle, often in a pejorative way, one needs to recognise that his view of allelopathy was more balanced than is usually credited. He envisaged that root excretions played a prominent role primarily among agricultural species, particularly where plants, especially annuals, were grown at high density and in an artificial assemblages, and he acknowledged that other factors were more likely to be important in the wild:

> In the state of nature, this reciprocal influence of plants, one on another does not seem very important, or at very least is masked by the co-occurrence of several phenomena. We see certain plants which seem to be favoured and to be harmed by their proximity; but on the one hand this effect is produced by other causes, such as shading, the intertwining of stems, etc.; on the other hand, the dispersion of plants on the soil offers phenomena so varied and so complex, that it is difficult to appreciate the exact influence of each of them; and this dispersion reveals more weakly the individual action of each plant. The difficulty is reckoned even when it concerns recognising the consecutive effect; in essence, we do not have generally enough of an interest to examine this succession for us to be concerned with it, and the length of the life of certain plants makes it difficult to observe. (p. 1501)

De Candolle's theory of crop rotation[13] was summarised:

> After all these agricultural considerations, which are modified one and the other within certain limits, remain the fundamental and physiological principles to know:
>
> 1° One should not make two crops of the same species follow one another: thus one does not sow wheat after wheat, or clover after clover; for the soil impregnated with the excretions or debris of a plant, is not exactly suitable to this plant, as one cannot feed an animal with the excrement of another being of the same species. The truth of this principle was well known as a fact before one had reflected on its cause. Farmers perform this nearly always for annual plants. Gardeners know well to alternate their legumes and to not replace a fruit tree with a similar tree. If, from place to place, one sees exceptions to this law, one obtains them by changing the soil where on wishes (as in uniform plantations) to replace a dead tree by another of the same species. One sometimes sows wheat for several years in certain soils which hint at renewing themselves, or in certain soils so fertile by themselves that they can resist this method; but these cases are so rare, and it is so doubtful that there would be any advantage to follow this way, or that in the general thesis one should recommend it.
> 2° Not only should one not replace a crop by the same species, but one should not even replace it by a plant of the same natural family. The excretions and the debris of a plant are harmful a those which have the same organization, a bit like a mammal or a bird cannot be fed any excrements from animals analogous to itself. Thus agriculturalists

---

[13] De Candolle (1832a) noted (p. 1504) that there is no satisfactory word in English for *assolemens*, although it is commonly translated as "crop rotation". The meaning of *assolemens* is broader than this, and encompasses the whole realm of cropping systems.

alternate between the legumes and the grasses, the nursery growers between rosaceous trees and amentaceous trees, etc. The exception the most remarkable in fact that I know to this principle is the biennial cropping system of the Garonne, where one alternates wheat and maize. The ground is so extraordinarily fertile, that it supports this methode; but when one has extended it indiscreetly to the less fertile soils of the neighbouring provinces, Périgord, etc., one obtains results of which the meagerness confirm the rule instead of destroying it.

3° All plants with a bitter and milky juice are evidently harmful to the quality of the soil, and;

4° Plants with an insipid and mucilaginous juice improve the soil for the plants of other families., both by their excretions and by their debris or their cast off material. The legumes occupy in this regard the highest rank, and their culture is the usual basis of improvements obtained in cropping systems. This effect is detectable even for those which lose very little of their leaves, such as gorse or broom, or for those which do not leave any of their own stubble to be buried, such as beans, vetches, etc.; but it is more pronounced in those which involve all these conditions, such as clover, lucerne, etc., and in general the perennial and leafy legumes. The grasses seem to occupy the second rank in this series. With regard to other families, the number of species that are in cultivation is too limited in Europe for one to be able to appreciate their general effects.

(pp. 1508-1510)

De Candolle similarly elaborated a theory of forest succession:

It seems, however, that in some cases, one observes in wild nature, a spontaneous renewal or alteration of forests: but the extreme slowness of these phenomena makes this subtle observation difficult to confirm. It is no less certain to my eyes that one of the causes of decline of forests which are regularly cut, is that the soil, impregnated for many centuries with the excretions and debris of the species, can no longer adequately nourish the tree of this species. I have shown moreover that that a statute from Louis XIV, which in defending against the deterioration of forests, forced property owners in France to conserve all the forests in places that they had already occupied for a long time, produced a troubling effect on this culture, and that the real way of remedying it is to gradually clear the forests, in a manner to encourage the laying waste of the most mediocre, and the plantation of new forest in land little useful for other things.[14]

(pp. 1502-1503)

The noxiousness of certain weeds was easily explained:

Certain plants are dreaded by informed farmers because they damage the soil in a distinct manner: such are the various species of spurges or *Tithymalus*, corn poppy and the other species of poppy: it is that these plants with bitter juice and latex exude from their roots substances which alter the quality of the soil. I am led to believe cockscomb (*Rhinanthus crista-galli*) injures neighbouring plants by some excretion of its roots.

(pp. 1479-1480)

It has rarely been appreciated that de Candolle was aware not only of the effects of root excretions, but also of leaf leachates:

Trees sometimes seem to be harmful to delicate plants underneath them, because the rainwater, in falling from their foliage, is charged with soluble substances which could have been excreted by the leaves, and that this water, according to the nature of these excretions, can be harmful to the plants on which it falls: it is probable that part of these

---

[14] The title to the reference to this, provided by de Candolle, "Considérations générales sur les forêts de la France", is wrong. See de Candolle (1830).

effects which the trees of heaven, the walnuts, the manchineels[15] exert on plants that they encompass, is due to these special leachates. (p. 1470)

Despite the acclaim of *Physiologie Végétale*, the only language into which it was translated was German. The German botanist Johannes Röper (1801-1885) had visited Geneva and worked with de Candolle during 1825-6. Commencing in 1833 he began producing a German edition of de Candolle's new text. However, only volumes one and two were subsequently published, and for no accountable reason[16], volume three containing the theory of crop rotation was never published in German. The relationship between de Candolle and Röper, whilst initially very cordial, seems to have become strained. Röper's edition of *Pflanzen-Physiologie* was essentially unauthorised, and included numerous footnotes, often containing critical remarks by Röper. De Candolle chose not to include it amongst the lists of his published work included in his memoirs (de Candolle 1862, 2004), nor was it mentioned by his biographer De la Rive (1845, 1851); moreover, de Candolle claimed to have not actually ever seen a copy.

It is important to bear in mind why de Candolle attached such importance to his theory of crop rotation. The theory tied in with his personal view of plant nutrition, which had already been outlined by Marcet (1829). The real nature of carbon assimilation through photosynthesis, had been largely unraveled through the experiments of numerous important investigators, including Bonnet, Ingenhousz, Senebier, and de Saussure, of which the latter two were well known personally to de Candolle. However, despite the advances made in realising the fundamental importance of photosynthesis and the role of the assimilation of carbon dioxide in the growth of the plant, de Candolle clung to some old-fashioned ideas. While he accepted that carbon dioxide, green matter (chlorophyll), and light were all of great importance, he maintained that carbon dioxide, as well as other organic substances, mainly entered the plant already dissolved in water, largely through the roots.

> We have just seen that the plant presents a series of formations and decompositions of carbonic acid, of which it is difficult to yield an exact story. Here is the manner which to me seems the most natural to understand the process.
>
> 1°. Water which enters plants by its roots is charged with carbonic acid, which is transported by the sap into the green parts, and is there decomposed by the action of sunlight; carbon is fixed there and oxygen escapes in the form of a gas.
> 2°. Carbonic acid, which the coloured parts of plants have formed with oxygen from the air, is in part dispersed in the atmosphere, in part dissolved in the water of the plant and transported with it, such as that absorbed by the roots, towards the leafy parts, where it is decomposed.

---

[15] The tree *Hippomane mancenilla*, of the Euphorbiaceae, a native of the coastal tropical Americas, acquired a somewhat fabulous reputation, rivaling that of the upas, *Antiaris toxicaria*. It was alleged that loggers had to burn around the base of the tree before felling, to get rid of otherwise noxious fumes and residues. The tree contained substantial amounts of a highly irritating latex, and rain falling from the leaves, was regarded as poisonous, not only to animals but also to plants (Lindley and Moore 1884).

[16] De Candolle recorded that from about 1834 onwards, Röper became overly critical of methodology, and furthermore, became part of the German romantic naturalism movement, which the former saw as distinctly detrimental. Communication apparently ceased in 1835 (de Candolle 2004).

3°. Water absorbed by the roots contains a certain quantity of soluble plant or animal material, which contains carbon: this carbon is carried by the sap into the green parts; it is combined during the night with oxygen absorbed by the leaves, the following day, this carbonic acid, formed in the leaves, is decomposed by sunlight, such that the carbon could be deposited usefully only when it comes from the decomposition of the carbonic acid gas.

4°. The green parts of plants which are in contact with their surrounds (air or water) charged with a small quantity of carbonic acid, seize it, decompose it, and expel oxygen; if the quantity is too great (more than an one twelfth), it acts as a sort of poison on the leaf, and alters it or kills it.

It appears that de Candolle was also substantially influenced by rapidly expanding research in the nineteenth century on the effects of diverse substances, ranging from gases to metallic salts to natural substances such as tannin, on plant processes. With the advent of analytical and preparatory chemistry, this was a research area that grew rapidly, and Macaire-Prinsep (1825) was among the early contributors, and others included Senebier and Huber (1801), de Saussure (1804), Francois Marcet (1825) who was Jane Marcet's son and a colleague of Macaire-Prinsep, Schübler and Zeller (1826, 1827), Göppert (1828), and Leuchs (1829). A detailed review of research of the era on substances poisonous to plants may be found in Wolff (1847). It was not simply a matter that that certain substances were harmful to plants; there was a rekindling of the idea that plants were organic beings more similar to animals than supposed, as expressed in the following review of Francois Marcet's work (Marcet 1825), from the *Quarterly Journal of Agriculture* (Anonymous 1828):

> M. Marcet's experiments upon vegetable poisons are no less interesting, and still more wonderful, as indicating a degree of irritability in plants somewhat similar to that which depends on the nervous system in animals. After having ascertained that the bean plants could exist in a healthy state for five or six days, if immersed in the same quantity of spring water, he tried them with five or six grains of opium dissolved in an ounce of water, the consequence of which was, that in the evening the leaves had dropped, and, by the middle of next day, they were dead beyond recovery. Other vegetable poisons of the narcotic class produced a similar effect. Hemlock was equally fatal, and six grains of dry powdered foxglove, in an ounce of water, began to operate, by wrinkling some of the leaves of the bean in a few moments, which it completely killed in twenty-four hours. Oxalic acid or salt of sorrel, though found in common and wood sorrel, and a great many plants, proved a very fatal poison to others. The absorption of one-tenth of a grain, killed a rose branch and flower in forty–eight hours.

Another area of burgeoning interest that briefly lent support to the theory of root excretions was that of the water relations of cells and tissues. In particular, R.J.H. Dutrochet (1776-1847), beginning in 1820, published numerous papers and books that explained many physiological phenomena in plants (and animals) on the basis of osmosis, or the movement of water across membranes. Dutrochet called the movement of water into cells (which had solutes) endosmosis, but he erred in believing that there was a substantial and compensatory outflow or exosmosis of material from the cell. This alleged exosmosis was viewed by some as a reasonable explanation of root excretion, particularly in view of the absorptive function of the root (e.g. Jackson 1840). Dutrochet (1826) also claimed to have found microscopic structures in the cell wall, reactive to acids and alkalis, which he believed were analagous to a diffuse nervous system, and thus involved in the irritability of plants. This offered further credibility at the time to the animal-like responses of plant structures.

In summary, de Candolle considered that the plant was required to absorb large quantities of water, that the soil water contained all manner of organic and inorganic substances - beneficial, harmful and neutral, and that the absorption of matter by the roots was essentially a passive process governed by the so-called "spongioles". Consequently the plant could become loaded with potentially harmful substances, arising from both the metabolism of the leaves and the passive absorption of the roots. The roots then played a vital role in excreting these materials into the soil, a process which led possibly to soil becoming less favourable to the growth of similar plants.

> The spongioles of the root, by their vital contraction, assisted by capillarity and the hygroscopic force inherent in its tissue, pumps the surrounding water along with saline, organic or gaseous substances, with which it is impregnated.
>
> This water is raised by the woody mass, and in particular by the intercellular canals, up to the leafy parts, by a vital effect manifested principally by the contractility of the cells and perhaps that of the vessels, assisted by the hygroscopicity and the capillarity of the tissue, by the vacuum caused by exhalation, etc.
>
> It arrives at the leafy organs, calling vertically by the leaves and laterally in all seasons, but above all in spring, by the cellular envelope; a large part is exhaled outside during the day by the stomates in the form of pure water, it leaves behind in the organs where it has made this exhalation, all the saline parts, and notably all the mineral parts which it contained.
>
> The sap which arrives in the leafy parts is hit by sunlight, and by means of this agent the gas carbonic acid dissolves in the sap (from which it was provided, or from the water pumped by the roots, or from the carbonic acid of the air, or from that which the oxygen of the air has formed with the superabundant carbon of the plant), is decomposed during the night; the carbon is fixed in the plant, and the oxygen is expelled outside in the form of gas.
>
> The immediate result of this operation seems to be the formation of the gummy matter which is composed of an atom of water and an atom of carbon, and which is susceptible by very rapid modifications, to be changed into starch, sugar, lignin, materials of which the composition is nearly similar.
>
> The nutritive juice supplied by these elaborations, and of which the gummy material seems to be the simplest and most ordinary state, redescends from the leaves towards the roots during the night, the length of the bark and the sap-wood, in the exogens [17], and the length of the woody mass in the endogens.
>
> It encounters along its way, especially in the bark and near its origin, glands or glandular cells which imbibe it and form in their cavities special substances, most of which are incapable of serving as food, destined to be expelled to the outside or transported in the tissue.
>
> It deposits along the way nutritive substances which, more or less mixed in the woody mass with the ascendant sap or absorbed with the water which the cellular envelope respires transversely by the medullary rays, are absorbed and elaborated by the cells, and above all by the rounded or slightly elongated cells.
>
> This deposition of nutritive substances, mainly composed of gum, starch, sugar, perhaps lignin and sometimes fixed oil, operates often in organs prepared in advance, where the substances repumped later serve next to feed other organs. (pp. 420-422)

---

[17] The term exogen refers to the manner of deposition of the vascular tissue in the stem, through an outer ring of vascular cambium, and is equivalent now to dicotyledonous plants. The term endogen is equivalent to monocotyledonous plants.

From the foregoing, it is important to realise that de Candolle's theories of root excretion and crop rotation, while highly attractive through their careful logic, rested upon some serious misconceptions about the nature of the functioning plant. The root spongiole or its equivalent, in most plants, was largely a myth; plants are essentially selective in their uptake of soil solutes; but most importantly, plants do not take up significant amounts of organic material from the soil.

MACAIRE-PRINSEP

De Candolle was clearly aware that his theory of crop rotation was nothing more than that, and he beseeched fellow scientists to conduct experiments that would shed light on the matter. Macaire-Prinsep, who had already published a paper on the effects of various chemicals on plant growth in 1825, conducted a number of experiments, which were summaried in a paper first published in 1832. This paper (Macaire 1832, 1833a) attracted even greater interest than de Candolle's theory, and it was soon translated into English (Macaire 1833b) and German (Macaire 1833c), and appeared in numerous different publications in Europe and America.

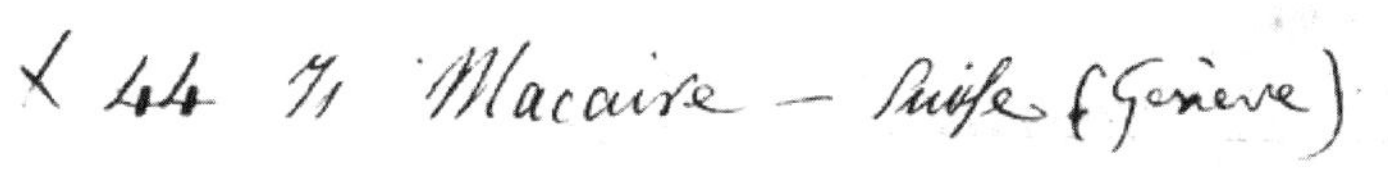

Figure 7.5. Macaire's signature (1816), taken from the attendance roll at a course given by Lamarck at the Muséum d'Histoire Naturelle, Paris (courtesy of Pietro Corsi, Oxford University).

As indicated previously, Macaire-Prinsep was a Genevois colleague of de Candolle's and a young man with expertise in pharmacy and, consequently, analytical chemistry. No portrait of him is known, but his signature is given in Figure 7.5. De Candolle thought highly enough of him to name the genus *Macairea* (Melastomataceae) in his honour. Macaire-Prinsep's publications are often difficult to track because of the numerous variations to both his given names and surname. In any case, Macaire-Prinsep, as we shall call him here, wrote two articles which were relevant to de Candolle's theory of crop rotation. The first of these is fairly mundane, and describes some simple experiments in which various organic substances were administered to a variety of plants (Macaire 1825). His most important paper is the one that seemingly answered de Candolle's plea in *Physiologie Végétale* for experimental work supporting his theory of crop rotation. The paper was written while *Physiologie Végétale* was in press, and de Candolle's text was revised at the last minute to include some of Macaire-Prinsep's new information. However, it is difficult to believe that de Candolle was unaware of Macaire-Prinsep's experiments in finalising the manuscript for *Physiologie Végétale,* and one has a suspicion that de Candolle's plaintive call and Macaire-Prinsep's almost immediate response were orchestrated to achieve

144   History of Allelopathy

maximum effect in the scientific community[18]. In any case, this paper became published widely, sometimes in edited or summarised form, in at least three languages: French, English and German.

The following then is a full translation of Macaire-Prinsep's key article, "Mémoire pour servir à l'Histoire des Assolemens", as published in 1832 in *Mémoires de la Société de Physique et d'Histoire Naturelle de Genève* **5**: 287-302.

> Of all the numerous advancements that have been promulgated in agriculture at the beginning of the century, one of the most important, without doubt, has been the diffusion of the theory and practice of crop rotation. One knows that one thus invokes a certain rotation of crops determined in advance, in which one avoids the too close repetition of the same plants in the same ground. Moreover, while the theory about it is new, the practice is as old as agriculture itself. Indeed, for a long time, it has been noticed that the great object of agriculture, the production of wheat, was very considerable, despite the time lost, when instead of sowing the field each year, one left the soil to rest, as it is said, for a year of fallow. But as tireless as was the work of the labourer during this rest interval, he could not prevent the soil being covered with herbs of every sort, it happened that after all it is only a rotation of wheat and some adventitious herbs. The progress of the science has been then to substitute crops for these plants, without other usage apart from that sometimes as a lean pasture, and to show that it is the variety of cultures and not the repose, which is impossible, that maintains the fertility of the soil. But how does this effect, so remarkable, operate? The ideas on this question are not yet entirely established. Some farmers, stirred by the need to clean their fields of weeds, have noticed this beneficial effect, notably produced by the large and numerous leaves of leguminous plants, usually called forage plants, such as clover, lucerne, and have seen in this cleaning of the soil the total effect of crop rotation. But, as M. De Candolle noticed, they have forgotten that which the gardener knows very well, that is, a fruit tree, if it happens to die, cannot be replaced by another of the same species, without changing the ground; and it is the oversight of this necessity to vary the culture, which covers the walls of our gardens so much, with trees that are weak and without yield. This is surely not because of the effect of weeds which the gardener is always careful to dig out in hoeing his trees. Others have imagined that the plants absorb different juices in the same soil, and that one soil exhausted by a culture can be fertile still for another class of plants. But this supposition is contrary to the fact well known to physiologists, that plants absorb by their roots all the soluble matter which is presented to them in the soil, without having the ability of eliminating that which could be harmful to them, and one sees them gorge themselves with poisonous substances, as long as these are soluble, that are totally contrary to their make-up. It is said that the beneficial effects of crop rotation are related to the difference in the length of roots of the different plants which follow, which allows them to exhaust by twists and turns the different layers of the same ground; but it must be remembered that following germination of the seeds, all the roots are found in the same layers of soil, and consequently, according to this opinion, would always be firstly in the exhausted layers. Furthermore, the operation even of cultivation, tilling turns over and mixes the various layers of the soil, and one knows also that plants of the same family, such as clover and lucerne do not succeed at all after one another, although their roots are very different in length. Without my looking at another hypothesis of which the success of a new culture is dependent on the plant residues left by the preceding, which should render the alteration of plants more harmful than useful, since these residues are always there, those which would be of the same nature as the plant which they are

---

[18] It is possible that Macaire-Prinsep's article actually preceded publication of *Physiologie Végétale*. The paper was read before the *Société de Physique et d'Histoire Naturelle* de Genève in December 1831, and the month of publication in 1832 is unknown.

supposed to nourish, and should be more freely assimilated, I cross over to the theory of crop rotation which is due to M. De Candolle. Some facts, already given in the *Flore Francaise** by this learned naturalist, seem to have furnished him with the first opportunity of turning his thoughts to this important subject; he thus expresses himself, p. 167. "M. Brugmans, having placed some plants in dry sand, saw some small drops of water exude from the extremity of the radicles." And further on, in p. 191; "Finally, the roots themselves in some plants present particular secretions, this may be observed in the *Carduus arvensis, Inula Helenium, Scabiosa arvensis,* several *Euphorbias* and several of the chicories.... It seems that these secretions of the roots are only parts of the juices, which having not served for nourishment, are rejected when they arrive at the inferior parts of the vessels. Perhaps this phenomenon, which is not easily seen, is common to a great number of plants. MM Plenck and Humboldt conceived the ingenious idea of seeking from this fact the cause of certain habits of plants. Thus, we know that the thistle is injurious to oats, the *Euphorbia,* and *Scabiosa* to flax, the *Inula betulina*[19] to the carrot, the *Erigeron acre* and cockle to wheat, &c. Perhaps the roots of these plants give out a matter which is harmful to the vegetation of others. On the contrary, if the *Lythrum salicaria* grows freely near the willow, and the branching *Orobanche* near the hemp, is it not because the secretions from the roots of these plants are beneficial to the vegetation of the others?

Extending these ideas still further, and applying them to the theory of the rotation of crops, both in his public lectures and in a book not yet published, his *Physiologie Végétale*[20], M. De Candolle admits, that every plant, in ejecting all the moisture that extend to the roots, cannot fail to eject also such particles as do not contribute to nourishment. Thus when the sap has been spread by circulation throughout the vegetable, elaborated and deprived of a great quantity of water by the leaves, and then redescending has furnished to the organs all the nourishment it contained, there must be a residue of particles which cannot assimilate with the vegetable, being improper for its nourishment. M. De Candolle asserts that these particles, after having traversed the whole system without alteration, return to the earth by the roots, and thus render it less proper to sustain a second crop of the same family of vegetables, by accumulating soluble substances that cannot assimilate with it; in like manner, he observes, that no animal whatever can be sustained by its own excrement.[21] Besides, it may also follow that the action even of the organs of a vegetable converts the mixed particles into substances deleterious to the plant which produces it, or to others, and that a portion of this poison is also rejected by the roots. Some experiments which I had formerly the honour of communicating to the Society, have shown that, in fact, vegetables may suffer from the absorption of the poisons which they themselves furnish. The continual elongation of the roots renders the effect hurtful not to the same generation of plants; it is the following of the same species which suffers from it, while it is possible to imagine that, on the contrary, these same excrements will furnish wholesome and abundant nourishment to another order of vegetables. The examples drawn from vegetables here offer themselves again with the force of analogy which is very remarkable. It was still, perhaps, necessary to this very ingenious theory, which accounted so reasonably for most of the facts obtained, to be more clearly confirmed by the results of direct experiments; and by the invitation of M, De Candolle I endeavoured to obtain them.

---

[19] = *Inula helenium.*

[20] In the English translation of this article, a publication date of 1827 was provided for *Physiologie Végétale*. While *Physiologie Végétale* was published in 1832, the earlier date has occasionally been cited, as it was the date for *Organographie Végétale*, the first part of de Candolle's planned series, titled *Cours de Botanique*. *Physiologie Végétale* was published as the second part, and the planned third part, *Géographie Botanique*, and fourth part, *Agronomie*, were never completed.

[21] In the English translation, the editor noted that the ostrich and cassowary always devour their own excrement.

The thing was, however, not very easy, and my first attempts were unavailing. I first strove to obtain the supposed exudation directly from plants plucked up by the roots; but, with the exception of some very doubtful cases, it, was impossible ever to obtain any sufficient quantity, and the rapidity with which the plants perished in this state destroyed all chance of succeeding by this means. I, afterwards attempted to sow the seeds in substances purely mineral, such as pure siliceous sand, pounded glass, &c. Also on clean sponges, white linen, &c, but although they germinated well, the existence of the plants was always short and precarious, and when I endeavoured to collect their exudation by the use of earths, I found that the decomposition of the refuse from the seeds gave the same character to the whole of them, and that a sort of vegeto-animal substance was always obtained, of which it was impossible to mistake the source, and which entirely concealed the results of the real exudation, if any were present in plants so imperfectly developed. As a last resource, with the use of rainwater, the purity of which I had ascertained by the usual reagents, and which left no residue after evaporation, I endeavoured to preserve plants that were entirely developed. Their roots being taken from the ground with the greatest care; I washed them minutely in rainwater to remove all the mould, and when they were entirely cleansed from all impurity, they were dried and placed in phials with a certain quantity of water. I soon observed that they flourished in it, developing their leaves, blossoming, and after some time, giving by the evaporation of water in which the roots were plunged, and by the reagents, evident marks of exudation by the latter. Much time is required for studying a great number of families, and at present I am able to present to the Society only a kind of preface to a more complete work. I have, however, seen the phenomenon repeated with a sufficient number of vegetables, and whose theory of the rotation of crops is the basis of my observations, in considering it nearly: general, at least[1] among all the phanerogamous vegetables.

Vigorous plants of *Chondrilla muralis*[22] when placed in rain water filtered, having their roots first cleansed: as I above described, grow and bloom freely. These were; thrown away when in full bloom, and replaced by fresh ones every two days, to allow no time, for a change of regimen. After eight days, the water acquired a yellow tint, and a strong odour very similar to that of opium, and a bitter and rather a pungent taste; it precipitated in small brown flakes the solution of subacetate and neutral acetate of lead, rendered, turbid, a solution of gelatine, &c., and by slow evaporation deposited a residue of a brown-reddish colour, which I shall examine hereafter, and which leaves no doubt that the water was perfectly free from any observable substance whatever. In order to ascertain, whether this substance was produced or not from the vegetation of roots, I steeped, during the same time, the roots only of the *Chondrilla* and in another phial, the stalks only, cut from the same plant. They continued fresh and in flower, but the water was not charged with any remarkable colour, had no taste, nor smell resembling opium, did not precipitate the acetate of lead, and contained scarcely any thing in solution. It was now clear to me that the produce obtained from the entire plant was the result of exudation from the roots, which took place only while the vegetable followed its natural course. The same experiments repeated on several other plants produced similar results, as will be seen when I speak of the produce of a small number of families which I have had time to examine. When once assured that plants rejected by their roots the parts improper for their nourishment, it remained for me to ascertain at what time of the day the phenomenon took place. For that purpose I steeped a vigorous plant of the kidney bean *(Phaseolus vulgaris)* with the root in rain water during the day; at night the plant was taken out, washed carefully, dried, and replaced in another bottle full of rain water: the experiment continued eight days, the plant continuing to grow with great vigour. On examining .the two liquids, I found in both evident marks of the excretion, from the roots; but the water in which the plant had grown during the night contained a considerably greater quantity. Both were clear and transparent; the

---

[22] = *Mycelis muralis*

experiment being repeated many times on plants of different natures, always produced similar results. I am convinced that by causing artificial night for the plants during the day, the excretion of the roots would be instantly much increased; but in all the plants that I have tried, I always found that it continued slightly during the day. As it is well known that by day the action of the light causes the roots of the plants to absorb the liquid which contains their nourishment, it is natural to suppose that the absorption would cease during the night when the excretion takes place.

It appeared probable that by means of the roots the plants might throw off the substances which they had imbibed, which were injurious to vegetation. To satisfy myself on this point, and at the same time, as the result was another means of verifying the existence of the excretion of roots, I tried tire following experiments: some plants of annual mercury (*Mercurialis annua*), carefully taken up, and washed with great precaution in distilled water, were so placed that a portion of their roots was plunged in a slight solution of acetate of lead, and the other portion in pure water. They continued to live very well during several days; after which the pure water evidently precipitated the black hydrosulphate of ammonia, and consequently had received a certain quantity of salt of lead, rejected by the roots which were soaked in it. Groundsel (*Senecio vulgaris*), cabbages, and other plants, placed in the same manner, produced the same results. Some plants, which were placed in a slight solution of acetate of lead, lived very well during two days, after which they were taken out. Their roots were washed in a large quantity of distilled water, carefully dried, again washed in distilled water, which precipitated no hydrosulphate, after which they were left to grow in rain water: in two days the reagents demonstrated in the water a small quantity of acetate of lead.

The experiments were made in limewater, which being less harmful to vegetation than acetate of lead, was preferable for the object sought after. When part of the roots were steeped in lime water, and part in pure water, the plants lived very well, and the water considerably whitened the oxalate of ammonia which demonstrated the presence of lime. Also a plant that had been kept in limewater, and washed precipitated the oxalate of ammonia, then transferred into pure water, after some time discharged a great quantity of lime, which was demonstrated by the reagents.

I repeated the same trials with a slight solution of sea salt, and the nitrate of silver also demonstrated that-the salt, which the plant had imbibed by absorption, was, partly ejected by the same roots which had imprudently admitted it.

When speaking to M. De Candolle of these results, he related to me a curious, fact which he had himself observed. The plants that are cultivated, near the sea for the produce of-soda, sometimes thrive very well at a great distance from the ocean provided they are placed within the influence of the sea air, which, it is well, known, transports the particles of salt with which it is charged to a great distance. M. De Candolle was persuaded that the land where the kali thus placed had grown, contained more salt than, the land adjoining; so that, instead of extracting it from the earth, these plants appeared to have furnished it by the exudation of their roots. Reflecting on this experiment, I imagined that I could perform it myself on a small scale with common plants, and I placed the roots with the plants of the groundsel, swine thistle *(Sonchus oleraceus,)* mercury, &c. in rainwater, and proceeded to bathe the leaves with a solution of sea salt. My solution being too concentrated acted forcibly on the leaves, I diluted it with water, and with a pencil touched the lower part of the leaves and stalks; I even moistened all the green part of the plant, but the reagents never, indicated any trace of salt rejected by the root, although, .the plants had flourished. Hence it appears, that either solutions of salt cannot imitate the proceedings of nature, or that perhaps the soda vegetables alone have the power of absorbing the marine salt, and of rejecting a portion of it by their roots. I should like very much to be able to repeat my experiment on a *Mesembryanthemum* or a *Salsola*. There is, then, no doubt that the plants have the power of rejecting by their roots those soluble salts injurious to vegetation, which are found in the water which they absorb; though but a small portion of these salts appeared in the residue which I obtained in my own-experiments, because the plants, imbibing only pure water and carbonic acid, could reject by their roots only the small quantity of salt which, they contained at the time they were taken out of the earth. I could gather little more than the result of the action of their organs on the aliment, not of foreign bodies, which only spread through the vegetable system without being decomposed. I shall now enter into some details on the small number of families which I have examined; each of them has produced results nearly similar in the divers individuals or kinds under experiment, but unhappily the number is very small.

Leguminosae.

The only plants examined of this family were kidney beans, peas, and beans of the species generally cultivated in this country. These plants exist and develop extremely well in rain water. After they have grown in it some time, the liquid, when examined, has but little taste, and the smell is slightly herbaceous; it is clear, and scarcely coloured by the kidney bean, but turns more yellow with the pea and common bean; it precipitates the acetate of lead, and nitric acid re-dissolves the precipitated gum without effervescence; nitrate of silver gives a slight precipitate soluble in acid, (carbonic acid); oxalate of ammonia renders it turbid; the other reagents cause no change. By slow evaporation a yellowish or brownish residue is obtained, more or less abundant, according to the plant under experiment, increasing in this order: kidney beans, peas, beans. In all other respects these residua are similar to each other. Ether separates an oily substance; alcohol nothing, and a substance remains analogous to gum and a little carbonate of lime.

In the course of the experiments on these plants, I perceived that when the water in which they had been kept was charged with much excrementitious matter, the fresh flowers of the same species that were put into it faded quickly, and did not live well in it. To ascertain if this resulted from the want of carbonic acid, although they might draw it from the air, or from the effect of the matter excreted, which these plants refused to absorb, I replaced the leguminous plants by those of another family, especially that of corn[23]. The latter lived in it, and the yellow colour of the liquid diminished in intensity; the residue was less considerable, and it was evident that the new plants absorbed a part of the matter excreted by the former. It was a kind of rotation of crops in a bottle, and the result tends to confirm the theory of M. De Candolle, of which trying this experiment on a great number of plants, we may arrive at some results which may be applicable to the practice of agriculture: for example, by supposing, as I feel disposed to believe by my trial, that the exudation from the roots of cultivated legumes contributes to the nourishment of corn[24], I should be disposed to conjecture, according to the relative quantity of these exudations, that the bean will produce the finest wheat, then the pea, next to that the kidney-bean. I am not sufficiently a practical agriculturist myself to know if experience has confirmed this view of the fact.

Gramineae.

The plants examined were wheat, rye, and barley.

These plants do not thrive so well in rainwater as the *Leguminosae,* and I suppose that this difference arises from the great quantity of mineral substances, especially silica, which they contain, and which they do not imbibe from pure water. The water in which they have grown is very clear, transparent, without colour, smell, or taste. The reagents demonstrate the presence of salts, muriates, and carbonates, alkaline and earthy; and the residue from evaporation is scanty and but slightly coloured, containing but a very small proportion of the gummy matter, no oily matter, and the aforesaid salts. I should be led to believe that the exudation from the roots of these plants tends to do little more than to reject the saline matter which is foreign to vegetation.

---

[23] Corn was an English term for wheat.

[24] While this appears superficially to accord with modern practice, this experiment was interpreted as justifying excretion theory in relation to planting wheat after a legume crop, not the particular nitrogenous value of legumes (e.g. see Johnson 1848, p. 93).

Chicoraceae.

The plants examined were the *Chondrilla muralis* and the *Sonchus oleraceus*. They live very well in rain-water; the latter acquires a clear yellow colour, a strong odour, and tastes bitter and somewhat noxious. It precipitates abundantly brown flakes of neutral acetate of lead, and renders turbid a solution of gelatine. Evaporated slowly, the liquor, when concentrated, has a very strong and persistent taste. The residue of a reddish brown, by boiling absolute alcohol, partly dissolves; the alcohol evaporating leaves a yellow, slightly brown, substance, of a very bitter taste, soluble in water, alcohol, and nitric acid, precipitated in brown flakes from its solutions by nitrate of silver, and appears to be very analogous to the bitter principle of the English chemists. The residue, re-dissolved in water, has a very strong noxious taste, similar to that of opium; it contains tannin, a brown gummy extractive substance, and some salts.

Papaveracece.

Plants of the corn poppy *(Papaver Rhoeas)* were not able to live in rain-water; they faded in it immediately.

The white poppy *(Papaver somniferum)* will exist in it; the roots impart to the water a yellow colour; it acquires a noxious odour, a bitter taste, and the brownish residue might be taken for opium. This plant is one of those where I put the roots and the stalks to soak separately, such that neither imparted to the water any of the properties which it acquired from the entire living plant.

Euphorbiaceae.

The plants tried were the *Euphorbia Cyparissias* and *E. Peplus*. These are the euphorbias with which Brugmans says he had observed the phenomenon of small drops oozing from the roots during the night. Possibly I did not adopt the right method, as I could not verify the fact by my own observations. The euphorbias grow extremely well in rainwater; the liquor becomes slightly coloured, but acquires a strong and persistent taste, especially after it is concentrated by evaporation. Boiling alcohol dissolves almost all the residue, which has but little colour, and by evaporation deposits a granular substance, gummy, resinous, yellowish, white, very acrid, and unpleasant to the throat.

Solaneae.

The only plant of this family that I had time to grow for a few days is the potato. It lived well in rainwater, and developed its leaves. The water was scarcely coloured, leaves very little residue, and the taste is very slight; which makes me think that the plant is one of those of which the excretions are of little abundance, and have no pronounced characteristics. But this conclusion is drawn from a single and very short experiment made on a plant scarcely developed.

In concluding this memoir, which should have contained the examination of more families and individuals had the time permitted, I shall recount that the results deduced are: First, That most vegetables exude by their roots substances useless to vegetation; second, That the nature of these substances varies according to the families of the vegetables that produce them; third, That some being pungent and resinous may hurt, and others being sweet and gummy may contribute to, the nourishment of other vegetables; fourth, That these facts tend to confirm the theory of the rotation of crops suggested by M. De Candolle.

REACTION TO THE ROOT EXCRETION THEORY

The publication of de Candolle's theory of crop rotation, and subsequently complemented by Macaire-Prinsep's experimental work, created great interest outside of France, especially in the Great Britain, likely because of de Candolle's celebrity.

Two journals, in particular, became focal points for debate on the matter: the *Quarterly Journal of Agriculture (Edinburgh)* and the *Gardeners' Magazine (London)*. In 1833 unknown correspondent to the former gave an optimistic account of the matter (Anonymous 1833). James Rennie, in an essay on fallowing, endorsed de Candolle's ideas, and went so far as to suggest that fallowing was beneficial as it allowed sunlight to destroy any toxins that had accumulated during the previous cropping (Rennie 1834)[25]. However, this latter theory was criticised at length by Main (1834) who noted that shielded areas in fallowed fields yielded better growth. An correspondent known only as S.W., in considering the claim by Macaire-Prinsep that plants excrete substances abundantly during flowering, speculated that annual crops should be those which are most injurious to the soil because of the yearly deposition of excreted matter exacerbated by flowering (S.W. 1834, 1835), an opinion that was subsequently challenged (Anonymous 1835a). The lively discussion and generally positive reception of de Candolle's theories became known to de Candolle, and in gratitude he became a contributor to the *Quarterly Journal of Agriculture*. His article on the diseases of larch (*Larix* spp.) in Great Britain was followed by a contribution from Andrew Gorrie who cited correspondence from a Mr Young that suggested that the decline of larch on soils which had previously grown Scot's pine (*Pinus sylvestris*) may have been due to poisoning of the soil through the decomposition of the pine roots (Gorrie 1833, 1835).

Another correspondent had a vested interest in the debate, and that was George Towers, a self-described "horticultural chemist". Towers (1833) presented a very favourable report on the issue of plant excretions in crop rotation, for the simple reason that he claimed to have pronounced the theory himself in 1830, in his own publication, *The Domestic Gardener's Manual*, which to his chagrin he had published anonymously. In any case, it is worth examining what Towers did write:

> Whenever raspberry plants are removed to another situation, the old ground ought to be well manured, deeply digged, and turned, and then it should be placed under some vegetable crop. By this mode of treatment, it will be brought into a condition to support raspberries again in two or three years. This is a curious and interesting fact, one which proves that it is not solely by *exhausting the soil* that certain plants deteriorate, if planted on the same ground year after year; for were this the case, manuring would renovate the ground: but it fails to do so, and thus, if peas or wheat, for example, be grown repeatedly on a piece of land, the farmer may manure to whatever extent he choose, his crops will dwindle, and become poorer and poorer. This is remarkably the case in the Isle of Thanet[26], where, to use the local term, if the land be "*over-peaed*", it becomes, as it were poisoned; and, if peas be again planted, though they rise from the soil, they soon turn yellow, are "*foxed*", and produce nothing of a crop. To account for this specific poisoning of the soil, we must suppose, that *particular plants convey into the soil, through the channels of their reducent vessels, certain specific fluids, which in process of time saturate it*, and thus render it incapable of furnishing those plants any longer with wholesome aliment; in fact, the soil becomes replete with fecal or excrementitious matter, and on such, the individual plant which has yielded it, cannot feed; but it is not *exhausted*; so far from that, it is to all intents and purposes manured

---

[25] This idea has gained currency in recent times via a process named "soil solarization" (Chou *et al.* 2000)

[26] The Isle of Thanet is a region of coastal Kent, and lies approximately 100 km east of London.

for a crop of a different nature; and thus by the theory of interchange between the fluids of the plant and those of the soil, we are enabled, philosophically, to account for the benefit which is derived from a change of crops. (pp. 397-398)

However, a little investigation reveals that at least the harmful qualities of peas were well known, as described earlier by Vancouver (1810):

On the contrary, when the pease do not take, there is nothing tends more to the fouling of the soil, than the great burthen of rubbish they uniformly give rise to.

A subsequent explanation of the physiology of root excretion (Towers 1833), however, also offered very little different from what Boerhaave had uttered a century earlier:

I have throughout my work maintained that the vital powers of every living vegetable are stimulated by the electrifying principle of light; by the agency of which, the nutritive substances being about the radicals, are decomposed, and then attracted and propelled into the recipient vessels of the roots. The element so prepared, I consider and designate the *"fluids of the soil"*, which by the same exciting energy, are, I conceive, carried upwards through the cellular system, till at length they are deposited in the leaves, wherein they are elaborated, and become the vital, nutritive, *proper* juices of the plant. These juices are the, I argue, carried back from the leaves, and distributed in due specific proportions, into cells or vessels appropriated to every required function of the plant; but certain portions are carried to and through the roots, and propelled into the soil; not, however, in the simple bland state of those taken up by the vessels of supply, but imbued with peculiar compound, odorous, and sapid qualities, the effects of the process of elaboration within the vessels of the leaves of the bark. These *exuded juices* I style the *"fluids of the plant"*; and as the processes of supply and ascent, and those of return and exusion, are unintermitting and coincident, - the results of the same mighty electrifying principle, - I view and describe them as acting *interchangeably*, as by the law of electric *induction*, whatever is excited positively induces a *negative* condition in a body immediately within the range of its energy. It is evident that the vital principle stimulates the decomposition of the previously inert matters of the soil about the roots of plants, and that to a considerable distance, otherwise no food could be introduced into the inconceivable fine vessels of the fibrils.

Towers (1836) persevered with his claims for English priority, despite mounting controversy surrounding the root excretion theory. There was some substance to his argument, although it must be remembered that de Candolle's theory was actually published as early as 1829, in English, through the efforts of Jane Marcet. This may have been pointed out to Towers, as in yet another article, he claimed that he actually wrote these ideas in 1829 (Towers 1834). Towers sought to further undermine de Candolle's celebrity by citing that the prolific botanical author, John Lindley, Professor of Botany at the University of London, had also published on the matter in 1832. Lindley (1832) had published a small précis of a much larger work that did not appear until 1840. Lindley himself made no claims as to the originality of the ideas presented, and they were likely borrowed from de Candolle and/or Jane Marcet, particularly in view of the terminology used. Lindley wrote:

52. Spongioles excrete excrementitious matter, which is unsuitable to the same species afterwards as food: for poisonous substances are as fatal to the species that secrete them, as to any other species.
53. But to other species the excrementitious matter is either not unsuitable, or not deleterious.

54. Hence, soil may be rendered impure (or, as we inaccurately say, worn out) for one species, which will not be impure for others.

55. This is the true key of the theory of the rotation of crops. (p. 19)

Further evidence for the widespread acceptance of excretion theory prior to 1832 comes from publications such as *Observations on the Rural Affairs of Ireland* by an Irish farmer, Joseph Lambert (1829). A review of Lambert's book indicated that he had presaged de Candolle's ideas on forest succession (Anonymous 1829):

> Thus, a wood of oak, ash, and hazel, may succeed to one of beech; but after a long course of years, the old stumps and roots of the primary beeches, favoured by the exudations of a different race of trees may shoot out, and in the end supplant the new forest, and a second forest of beech will be re-established.

John Loudon (1838), in his monumental *Arboretum et Fruticetum Britannicum*, paid scant attention to the lore of root excretions; however, he did regard that the raspberry was "a good example of the doctrine of the excretions of plants." The ephemeral nature of wild raspberry plants, and need to continuously replant the cultivated raspberry were cited in evidence. Loudon also mentioned that besides Towers, a Mr Sheriff (*sic*)[27], a Scottish farmer of Mungos Wells also presaged the root excretion theory of crop rotation.

Another cause of interest in de Candolle's work was its endorsement by P. M. Roget (1834). Roget had been requested by the Royal Society to prepare a text on physiology, as part of the Bridgewater Treatises. The eighth Earl of Bridgewater, the Rev. Francis Henry Egerton, died in 1829 without an heir, and bequeathed £8000 in trust for the publication of works in natural history, in celebration of the Creator. Various other contemporary works freely adopted the concept of root excretion; for example the English writer Clement Hoare (1835), in a work that later became widely available in the United States stated that:

> The excrementitious matter discharged from the roots of a vine is very great, and if this be given out in a soil that is close and adhesive, and through which, the action of the solar rays is feeble, the air in the neighbourhood of the roots quickly becomes deleterious, and a languid and diseased vegetation immediately follows. (p. 47)

The excretion theory was also widely disseminated through popular and cheap encyclopaedias such as the *Penny Cyclopaedia* (Anonymous 1841).

The work of de Candolle and Macaire-Prinsep was also widely discussed and debated in the United States, and new theories accorded well with the American revolutionary spirit. Macaire-Prinsep's memoir on crop rotation had been published

---

[27] The actual name was Shirreff, although it is unclear whether it was John Shirreff, or Patrick Shirreff. The issue is confused, as Patrick Shirreff (1791-1876), noted for a book about his travels to the United States and Canada, and his work on plant selection, was the third son of a John Shirreff (1746-1830) of Mungoswells. Both were buried at Prestonkirk, Scotland. The *Oxford Dictionary of National Biography* records a John Shirreff (1759-1818), agricultural writer, also buried at Prestonkirk. As the surname is rare, these two individuals may be the same (John Martin, pers. comm.). John Shirreff was the author of works such as *A General View of the Agriculture of the West Riding of Yorkshire* (1794), in which were discussed issues such as crop rotation and fouling of the soil by crops. Loudon's reference may be to a footnote in an article by Patrick Shirreff (1831): "The influence of grasses as affecting the succeeding grain crops, is an important and perhaps neglected branch of inquiry. Farmers generally agree in thinking the effects of a rye-grass crop injurious."

in the United States in 1833, and been presented by others (e.g. Collamore 1834). The influential agriculturalist, Judge Jesse Buel, wrote to journals in both England and the United States concerning root excretion (Buel 1835b, 1836), and he used his own publications, notably to dispense his ideas (Buel 1835a, 1839; Anonymous 1835b). Buel had substantial influence, as in 1832 he had founded a lay agricultural journal, *The Cultivator*, which became one of the most popular journals among the farming community in both the United States and Canada. Buel was happy to accept the idea that plants were capable of excreting waste material, but he opposed the notion that this played a dominant role in the rotation of crops. In Rhode Island, Jackson (1840) also accepted the root excretion theory.

Despite the criticism, and indeed praise, that the root excretion theory attracted both in Europe and the United States, neither de Candolle nor Macaire-Prinsep participated substantially further in the debate[28]. De Candolle seemed content to let others do his bidding, and Macaire-Prinsep concentrated on his career in chemistry. In France itself, there appears to have been little controversy, and indeed, initially, there was little commentary, perhaps out of polite deference to de Candolle, who was greatly respected as a botanist. Francois Desire Roulin, who had spent several years in South America, reported on Macaire-Prinsep's work as means of reconciling the ever-changing vegetation that followed successive slashing and burning of forest[29] widely practised in agriculture in countries such as Brazil and Colombia. Generally, the topic of cropping, at least in France, was considered to lie within the domain of agriculture, and French agricultural experts such as Leclerc-Thouin[30] (1835), and even Yvart (1843[31]), after his death, were held in higher esteem on the practical matters of agronomy. It is recorded that a discussion of de Candolle's theory of crop rotation was held at the February, 1837 meeting of the Académie Royale du Metz, and the key speakers were invited to contribute papers. Lapointe (1837a[32], 1837b) noted that de Candolle's theory was generally supported by observed crop performance, and exceptions were commonly associated with root crops, were the roots themselves were harvested, or with plants growing on unusual soils, such as heaths and beat-bogs. Piobert (1837a, 1837b, 1837c) adopted an opposing view that soil nutrients, especially nitrogen, better explained the succession of various crops. Similarly, it was recognised that there was little evidence to support the existence of root excretions, at least in any appreciable quantity, and it was suggested that crop rotation might be the result of residues, rather than excretions, being unfavourable for the same succeeding crop (Anonymous 1837)[33]. The fact that de Candolle's root

---

[28] In his chemistry text, Macaire-Prinsep (1836) maintained the importance of root excretions in the func- functioning of the plant.

[29] The brushland that followed exhausted land was known as *capoeiras* in Brazil.

[30] Oscar Leclerc-Thouin (1798-1845) was the nephew of de Candolle's friend, André Thouin.

[31] An edition of Yvart's writings was published posthumously and annotated by his grandson Victor Rendu in 1843.

[32] A slightly abridged version of this was published separately (Piobert 1837b).

[33] This note apparently appeared firstly in the rare journal *La Flandre Agricole et Manufacture*, but was reproduced in various other journals. It was even translated into Italian in the *Giornale Agrario*

excretion theory made surprisingly little impact in the French agricultural world is perhaps indicated by the writings of Valcourt (1841). In considering the role of plant excretion in soil sickness, Valcourt mentions that this was brought to his attention by having read an 1834 article by the Scotsman, James Rennie.

De Candolle retired from his academic duties at the Université de Genève in 1835, which were taken over by his son, Alphonse de Candolle. The elder de Candolle found that his health was failing, and he spent his declining years attempting to complete his taxonomic work, until his death in 1841. In summary, it may be stated that the temporary success of the root excretion theory, especially in the English-speaking world, was due not so much to its originality, as most of the ideas had been stated earlier by others, but its cohesive and concise statement by the celebrated botanist.

REFERENCES

Anonymous 1798. Note sur les excrémens des végétaux. *Journal de Physique, Chimie et d'Histoire Naturelle* **4**: 388.

Anonymous 1824. Programme de la Société Hollandoise des Sciences à Harlem, pour l'année 1823. *Jahrbuch der Chemie und Physik* **11**: 242-256.

Anonymous. 1828. On the poisoning of vegetables. *Quarterly Journal of Agriculture* **1**: 78-81.

Anonymous. 1829. Art. IV. *Observations of the Rural Affairs of Ireland; or a Practical Treatise on Farming, Planting and Gardening, Adapted to the Circumstances, Resources, Soil and Climate of the Country; Including Some Remarks on the Reclamation of Bogs and Wastes, a Few Hints on Ornamental Gardening. The Monthly Review* **1829** (51): 359-374.

Anonymous. 1830. Extrait du Programme de la Société Hollandoise des Sciences à Harlem, pour l'année 1830. *Annalen der Physik* **95**: 156-160.

Anonymous. 1833. De Candolle's theory of the rotation of crops. *Quarterly Journal of Agriculture* **4**: 320-327.

Anonymous. 1835a. Remarks on the excretory theory of plants. *Quarterly Journal of Agriculture* **5**: 586-590

Anonymous. 1835b. The new theory. *The Cultivator* (Second Edition) **2**: 8.

Anonymous. 1837. Théorie des assolemens. *Journal d'Agriculture, Sciences, Lettres et Arts, rédigé par des members de la Société Royale d'émulation de l'Ain* **1837** (6): 190-192.

Anonymous. 1841. Root. In: *The Penny Cyclopaedia of the Society for the Diffusion of Useful Knowledge. Volume XX. Richardson-Scander-beg*, pp. 152-153. Charles Knight, London.

Braconnot, H. 1807. Sur la force assimilatrice dans les végétaux. *Annales de Chemie* **61**: 187-246. An English précis appeared in 1813 as "Nutrition of vegetables" in *The Emporium of Arts and Sciences*, new series **1**: 295-303; 463-466.

Brisseau-Mirbel, C.F. 1802. *Histoire Naturelle, Générale et Particulière des Plantes*. Volume 1. F. Dufart, Paris.

Browne, C.A. 1944. A source book of agricultural chemistry. *Chronica Botanica* **8**(1): 1-290.

Buel, J. 1835a Excretory powers of plants. *The Cultivator*. Second Edition **1**: 135-136

Buel, J. 1835b. On the excrementitious matter thrown off by plants. *American Journal of Science and Arts* **28**: 267-8.

Buel, [J.] 1836. On the excretory functions of plants. *Gardeners' Magazine* **12**: 229-30.

Buel, J. 1839. *The Farmer's Companion; or, Essays on the Principles and Practice of American Husbandry*. Marsh, Cape, Lyon and Webb, Boston.

Candolle, A.P. de. 1805. *Principes Élémentaires de Botanique et de Physiologie Végétale*. Paris.

Lombardo-Veneto e continuazione degli annali universali di agricoltura di industria e d'arti economiche, serie 2 **8**(11/12): 193-195.

Candolle, A.P. de. 1813. *Théorie Élémentaire de la Botanique, ou Exposition des Principes de la Classification Naturelle et de l'Art de Décrire et d'Etudier les Végétaux*. Déterville, Paris
Candolle, A.P. de. 1820. Géographie botanique. *Dictionnaire des Sciences Naturelles* **18**: 359-422
Candolle, A.P. de. 1827. *Organographie Végétale*. Déterville, Paris.
Candolle, A.P. de. 1830. Considérations sur la legislation forestière, ou examen des recherches statistiques sur les forêts de France, par Faisseau Lavanne à Paris, 1829. *Revue Française* **16**: 1-22.
Candolle, A.P. de. 1832a. Essai sur la théorie des assolemens. *Bulletin de la Classe d'Agriculture de la Société des Arts de Genève* **73**: 1-22.
Candolle, A.P. de. 1832b. *Physiologie Végétale, ou Exposition des Forces et des Fonctions Vitales des Végétaux, pour servir de suite à l'Organographie Végétale, et d'introduction à la Botanique Géographique et Agricole*. Volumes 1-3. Béchet Jeune, Paris.
Candolle, A.P. de. 1833-1835. *Pflanzen-Physiologie, oder Darstellung der Lebenskräfte und Lebensverrichtungen der Gewächse*. Volumes 1-2. J.G. Cotta, Stuttgart
Candolle, A.P. de. 1862. *Mémoires et Souvenirs de Augustin-Pyramus de Candolle écrits par lui-même et publiés par son fils*. Joël Cherbuliez, Geneva.
Candolle, A.P. de. 2004. *Mémoires et Souvenirs (1778-1841)*. Edited by J.-D. Candaux and J.M. Drouin. Georg Editeur, Geneva.
Choisy, J. 1843. Augustin-Pyramus de Candolle, né à Genève, le 4 Février, mort le 9 Septembre 1841. In: *Album de la Suisse Romande. Premier Volume*, pp. 15-16. Ch. Gruaz, Geneva.
Collamore, A. 1834. A Dissertation: On the course of tillage most suitable for soils in Plymouth County, and the rotation of crops, most conducive to the interests of the inhabitants. Read before the Plymouth County Agricultural Society, at their annual meeting at Bridgewater, October 15th, 1834. *New England Farmer* **13**(28): 217-219; **13**(29): 225-226.
Cotta, H. 1806. *Naturbeobachtungen über die Bewegung unf Funktion des Saftes in den Gewächsen, mit vorzüglicher Hinsicht auf Holzpflanzen*. Hoffman, Weimar.
Daubeny, C. 1834. On excretions from the roots of vegetables. *Report of the British Association for the Advancement of Science* **3**: 598.
De la Rive, A. 1845. *Notice sur la Vie et les Ouvrages de A.-P. de Candolle*. Fredinand Ramboz, Geneva.
De la Rive, A. 1851. *A.-P. de Candolle: sa Vie et ses Travaux*. Cherbuliez, Geneva.
Dureau de Lamalle, A.J.B.C. 1825. Mémoire sur l'alternance, ou sur ce problème: la succession alternative dans la reproduction des espèces végétales, vivantes en société, est elle une loi générale de la nature? *Annales des Sciences Naturelles*, 1ère Série **5**: 353-381.
Dutrochet, R.J.H. 1826. *L'Agent Immédiat du Mouvement Vital dévoilé dans sa Nature et dans son mode d'Action chez les Végétaux et chez les Animaux*. Baillière, Paris
Féburier. 1812. Essai sur les phénomènes de la végétation, expliqués par les mouvements des sèves ascendente et descendente. *Journal de Physique, de Chimie, d'Histoire Naturelle et des Arts* **74**: 343-359.
Gorrie, A. 1833. Soil and site for the larch, rot of the larch, etc. *Gardener's Magazine* **7**: 374.
Gorrie, A. 1835. On the rot in larch. *Quarterly Journal of Agriculture* **5**: 537-545.
Hoare, C. 1835. *A Practical Treatise on the Cultivation of the Grape Vine on Open Walls*. William Mason and Son, Chichester.
Home, F. 1757. *Principles of Agriculture and Vegetation*. G. Hamilton and J. Balfour, Edinburgh.
Huber, F. and J. Senebier. 1801. *Mémoires sur l'Influence de l'Air et de Diverses Substances Gazeuses dans la Germination de Différentes Graines*. J.J. Paschoud, Geneva.
Jackson, C.T. 1840. *Report on the Geological and Agricultural Survey of the State of Rhode-Island*. B. Cranston & Co., Providence.
Johnson, C.W. 1848. *British Husbandry; Exhibiting the Farming Practice in Various Parts of the United Kingdom*. Volume 2. Robert Baldwin, London.
Lamarck, J.B.P.A. and B. Mirbel. 1803. *Histoire Naturelle des Végétaux, classés par Familles*. Volume 1. Deterville, Paris
Lamarck, J.B.P.A. and A.P. de Candolle. 1805. *Flore Française, ou Descriptions Succinctes det Toutes les Plantes qui Croissent Naturellement en France, disposés selon une nouvelle méthode d'analyse, et précédés par un exposé des principes élémentaires de la Botanique. Troisième Édition. Tome Premier*. Desray, Paris.
Lambert, J. 1829. *Observations of the Rural Affairs of Ireland; or a Practical Treatise on Farming, Planting and Gardening, Adapted to the Circumstances, Resources, Soil and Climate of the Country;*

Including Some Remarks on the Reclamation of Bogs and Wastes, a Few Hints on Ornamental Gardening. Curry & Co., Dublin.

Lapointe, L.E. 1837a. Note sur la physiologie des assolemens. *Mémoires de l'Académie Royale du Metz, Mémoires et Rapports publiés par les Membres* **1836-1837**: 41-50.

Lapointe, L.E. 1837b. Deuxième article sur les considérations généales qui peuvent servir de base à la théorie des assolemens. *Mémoires de l'Académie Royale du Metz, Mémoires et Rapports publiés par les Membres* **1836-1837**: 67-73.

Leclerc-Thouin, O. 1835. Des assolemens. In: *La Maison Rustique du XIXème Siècle*, vol. 1, (ed. Bailly, Bixio, A. and Malpeyre) pp. 256-285. Librairie Agricole de la Maison Rustique, Paris.

Leuchs, E.F. 1829. Ueber die Wirkung einiger Stoffe auf die Pflanzen. *Annalen der Physik* **91**: 153-157.

Lindley, J. 1832. *An Outline of the First Principles of Horticulture*. Longman, Rees, Orme, Brown, Green, & Longman, London.

Lindley, J. and T. Moore. 1884. *The Treasury of Botany: A Popular Dictionary of the Vegetable Kingdom; with which is incorporated a Glossary of Botanical Terms*. New and Revised Edition, with Supplement. Longmans, Green, and Co., London.

Loudon, J.C. 1838. *Arboretum et Fruticetum Britannicum*. Longman, Orme, Brown, Green, and Longmans, London.

Macaire, J. 1825. Note sur l'empoisonnement des végétaux par les substances vénéneuses qu'ils fournissent eux-mêmes. *Mémoires de la Société de Physique et d'Histoire Naturelle* **4**: 91-93. Also in: *Annales de Chimie et de Physique* **39**: 95-97 (1828)

Macaire, J. 1832. Mémoire pour servir à l'histoire des assolemens. *Mémoires de la Société de Physique et d'Histoire Naturelle* **5**: 287-302. Also in: *Bibliothèque Universelle* **50**: 33-49 (1832); *Annales de Chimie* **52**: 225-240 (1833).

Macaire, J. 1833a. Expériences sur les excrétions racinaires, extraites d'un mémoire pour servir à l'histoire des assolemens. *Annales des Sciences Naturelles*, sér. 1 **28**: 402-416.

Macaire, J. 1833b. Vegetable physiology in relation to the rotation of crops. *American Journal of Science and Arts* **23**: 138-43. Also in: *Edinburgh New Philosophical Journal* **14**: 215-220 (1833); *Field Naturalist* no. **9** (1833); *Quarterly Journal of Agriculture* **4**: 882-890 (1834); Baxter, J. (1836), *The Agricultural and Horticultural Gleaner, Containing Important Discoveries and Improvements in Farming, Gardening, and Floriculture*, pp. 125-131. Sussex Agricultural Press, Lewes.

Macaire, J. 1833c. Denkschrift zur Geschichte der Wechselwirtschaft. *Journal für technische und ökonomische Chemie* **15**: 43-56. Also in: *Annalen der Chemie und Pharmazie* **8**: 78-92 (1833).

Macaire-Prinsep. 1836. *Précis de Cours de Chimie Expérimentale donné à l'École Industrielle de la Classe d'Industrie*. P.-A. Bonnant, Geneva.

Main, J. 1834. Remarks on a new theory of fallowing. *Quarterly Journal of Agriculture* **5**: 241-249.

Marcet, F. 1825. De l'action des poisons sur le regne végétale. *Mémoires de la Société de Physique et d'Histoire Naturelle* **3**: 37-65.

Marcet, J. 1829. *Conversations on Vegetable Physiology comprehending the elements of botany with their application to agriculture*. Two Volumes. Longman, Rees, Orme, Brown, and Green, London.

[Marcet, J.] 1830. *Conversations sur la Physiologie Végétale Comprenant les Élémens de la Botanique et leurs Applications à l'Agriculture*. Volumes 1-2. Translated by Macaire-Prinsep. Ab. Cherbuliez, Paris.

Marcet, [J.] 1844. *Lessons on Animals, Vegetables, and Minerals*. Lindsay and Blakiston, Philadelphia.

Meyer, E.H.F. 1830. *De plantis labradoricis libri tres*. Leipzig.

Piobert. 1837a. Note sur la théorie des assolemens. *Mémoires de l'Académie Royale du Metz, Mémoires et Rapports publiés par les Membres* **1836-1837**: 31-66.

Piobert. 1837b. *Note sur la théorie des assolemens*. S. Lamort, Metz

Piobert, 1837c. Deuxième note sur la théorie des assolemens *Mémoires de l'Académie Royale du Metz, Mémoires et Rapports publiés par les Membres* **1836-1837**: 74-80.

Polkinghorn, B. 1993. *Jane Marcet: an Uncommon Woman*. Forestwood Publications, Aldermaston.

Rennie, J. 1834. The practice of fallowing, of paring and burning, of irrigation, and of draining, explained on new scientific principles. *Quarterly Journal of Agriculture* **5**: 1-32.

Roget, J. 1834. *Animal and Vegetable Physiology Considered with Reference to Natural Theology*. Volumes 1-2. William Pickering, London.

Roulin, F.D. 1833. Les jachères de France et les capoeiras du Brasil. *Revue des Deux Mondes, 2ème sér.* **4**.

Schübler, G. and E.A. Zeller. 1826. *Untersuchungen über die Einwirkung verschiedener Stoffe des organischen und unorganischen Reichs auf das Leben der Pflanzen.* Schönhardt, Tübingen.
Schübler, G. and E.A. Zeller. 1827. Untersuchungen über die Einwirkung verschiedener Stoffe auf die Pflanze. *Flora (Jena)* **5**: 753-768.
[Seringe, N.] 1832. Extrait d'un fragment de la physiologie végétale de M. de Candolle, Sur la théorie des assolemens. *Bulletin Botanique* (Lyons) **2**(1): 1-8.
Shirreff, P. 1831. On a combination of grasses for alternative husbandry. *Quarterly Journal of Agriculture* **2**: 242-247.
Smith, J.E. 1809. *An Introduction to Physiological and Systematic Botany.* Second Edition. Longman, Hurst, Long, and Orme, London.
Sprengel, C.S. 1831-1832. *Chemie für Landwirthe, Forstmänner und Cameralisten.* Vandenhoeck und Ruprecht, Göttingen.
Sprengel, K.P.J. 1812. *Bau und der Natur der Gewächse.* K. Kummel, Halle.
Stendahl. 1830. *Le Rouge et Le Noir. Chronique de 1830.* Garnier Frères, Paris.
Sussex, Duke of. 1833. [Address to the Royal Society]. *Proceedings of the Royal Society* **1832-1833** (14), in: *Abstracts of the Papers presented in the Philosophical Transactions of the Royal Society* **3**: 215-235 (1837).
Towers, G. 1830. *The Domestic Gardener's Manual; being an Introduction to Gardening, on Philosophical Principles; to which is added, A Concise Naturalist's Kalendar, and English Botanist's Companion, or Catalogue of British Plants, in their Monthly Order of their Flowering.* Whittaker Treacher and Co., London.
Towers, G. 1833. On the excretory powers of plants. *Quarterly Journal of Agriculture* **4**: 656-667.
T., G.J. [Towers, G.J.] 1834. Investigations on the theory of the rotation of crops. *Gardeners' Magazine* **10**: 12-18.
Towers, G. 1835. On the excretion from the roots of plants. *Quarterly Journal of Agriculture* **6**: 360-368.
Towers, G. 1836. Theory of fallowing. In: *Baxter's Agricultural and Horticultural Gleaner*, pp. 64-71. Simpson, Marshall, and Co., London
[Towers, G.] 1839. *The Domestic Gardener's Manual; being an Introduction to Gardening, on Philosophical Principles; to which is added, A Concise Naturalist's Kalendar, and English Botanist's Companion, or Catalogue of British Plants, in their Monthly Order of their Flowering.* Revised Edition. London.
Valcourt, L.P. 1841. *Mémoires sur l'Agriculture, les Instrumens Aratoires et d'Économie Rurale.* Bouchard-Huzard, Paris.
Vancouver, C. 1810. *General View of the Agriculture of Hampshire, including the Isle of Wight.* R. Phillips, London.
W., J.B. 1835. On the vegetable excretions of plants. *Gardener's Magazine* **11**: 278-280.
W.,S. 1834. Remarks on the excretory powers of plants. *Quarterly Journal of Agriculture* **4**: 880-890.
W.,S. 1835. Excretion of plants. *Quarterly Journal of Agriculture* **6**: 133-138.
Willis, R.J. 1985. The historical bases of the concept of allelopathy. *Journal of the History of Biology* **18**: 71-102.
Willis, R.J. 1996. The history of allelopathy. 1. The first phase 1785-1845: the era of A.P. de Candolle. *Allelopathy Journal* **3**: 165-184.
Willis, R.J. 2002. Pioneers of allelopathy. XII. Augustin Pyramus de Candolle (1778-1841). *Allelopathy Journal* **9**: 151-158.
Wolff, E.T. 1847. *Die chemischen Forschungen auf dem Gebiete der Agricultur und Pflanzenphysiologie.* Joh. Ambrosius Barth, Leipzig.
Yvart, V. 1821. Assolement. In: *Nouveau Cours Complet d'Agriculture Théorique et Pratique, contenant la grande et la petite culture, l'économie rurale et domestique, la médicine vétérinaire, etc., ou Dictionnaire Raisonnée Universel d'Agriculture, Nouvelle Édition*, Volume 2, pp. 38-191. Deterville, Paris.
Yvart, V. 1843. *Assolements, Jachère et Succession des Cultures.* La Librairie Encyclopédique de Roret, Paris.

CHAPTER 8

THE DECLINE OF ALLELOPATHY IN THE LATTER NINETEENTH CENTURY

> Plants, in a state of nature, are always warring with one another, contending for the monopoly of the soil, - the stronger ejecting the weaker, - the more vigorous overgrowing and killing the more delicate. Every modification of climate, every disturbance of the soil, every interference with the existing vegetation of an area, favours some species at the expense of others.
>
> *Flora Indica: A Systematic Account of the Plants of British India.*
> J.D. Hooker (1855)

More or less commensurate with the death of Augustin Pyramus de Candolle in 1841, there was a groundswell of overt opposition to the root excretion theory. The reasons for this were manifold. Renewed interest in plant nutrition, led by Justus von Liebig at Giessen in Germany caused a re-examination of many of the precepts of the functioning of the root, at both an anatomical and physiological level. De Candolle had supposed that roots passively absorbed all solutes, and that the root spongioles were the active organs in this function. Since the early parts of the eighteenth century, there had been controversy about the function of the root, particularly in consideration of its structure. Moldenhawer (1820) had uniquely suggested that root exudations were not excretory in function, but occurred to assist in the absorption of food substances. Murray (1822a, 1822b) claimed that the structure of the root was not well suited to the absorption process, but was better suited for excretion. Experiments by workers such as R.J.H. Dutrochet showed that the spongioles were not active in absorption, but that this function occurred in the distal portions of root tips furnished with root hairs. There was genuine doubt as to whether root excretions actually existed, as careful experimentation by men such as Walser, Braconnot, Daubeny, Wiegmann and Polstorff, and others failed to confirm the ebullient claims of Macaire-Prinsep. There was debate about whether the roots were selective in their uptake of materials and whether organic substances played any role at all in plant nutrition. In the latter half of the nineteenth century there arose an awareness of the presence of microorganisms in the soil, and their importance in decomposition processes, making available certain nutrients for plants, and plant disease. With these points in mind, it is then instructive to examine the breadth of different views on root excretions, which was the principal mode of supposed allelopathic effect supposed in the nineteenth century.

ROOT EXCRETION SUPPORTED

While most French botanical writers had distanced themselves from root excretion theory by 1850, it still had substantial currency for French farmers and agricultural writers (e.g. van dem Broeck 1855). A French landholder, Léon de Rosny (1852) at a meeting of the Société d'Agriculture, des Sciences et des Arts de Boulogne-sur-Mer in northern France wrote:

> Thus we know in a certain manner that roots indiscriminantly absorb all the substances and all the salts dissolved in water; here the water is the great vehicle that transports and conveys them to the roots and stems of the plants; these then appropriate and assimilate the juices that they require, and reject to the outside those which are not useful, and hence excretions, a a necessary result of the nutrition of life; and these excremental exudations constitute a sort of poison for the plants that have produced them, and for those of the same family. Thus the diverse functions of life are provided in plants just as in animals, and neither one nor the other can feed on their own excrements. Hence is the necessity of crop rotation for all types of plants. Fallowing after oats does not destroy the excretions of this cereal, which remain in the soil, and after a year are still harmful to wheat. Clover, on the contrary, absorbs these excretions, which constitute for it a food, and in its turn, it leaves in the soil excrements of a different nature which with its fleshy roots form a potent fertiliser for wheat. This is why a good clover crop after oats is a better preparation for wheat than a fallow in the same situation. (p. 25)

In Germany, the concept of root excretions persisted largely through Carl Sprengel (1787-1859), an agricultural chemist at Göttingen and Brunswick, in Hannover. Sprengel had, to some extent, pioneered the notion that root excretions may be important in crop rotation (Sprengel 1831-1832; see Chapter 7). Although Sprengel's views on mineral nutrients were quite modern – indeed it was he who first formulated the idea of the Law of the Minimum, commonly attributed to Justus von Liebig - he ran contrary to his younger contemporary, Liebig, and continued to maintain that plants excreted their metabolic wastes via both the leaves and roots, and he initially agreed with de Candolle and Macaire-Prinsep that root excretions could affect the growth of other plants.

Justus Ludewig von Uslar

One of the more curious figures to embrace root excretion theory was Justus Ludewig von Uslar (1780-1862; Figure 8.1), a Hannoverian minor noble, with no formal botanical training. He was the son of the noted forester and botanical writer, Julius Heinrich von Uslar, but he studied mining at university. He seemed destined to run the course of his life as a fairly pedestrian, but comfortable, mining administrator for the Hannover government, until he was approached by a British investment company, *The Mexican Company*, to oversee their silver mining operations in southern Mexico. After some tricky negotiations, von Uslar managed to overcome his reluctance to leave the Hannover civil service, by securing a diplomatic mission as well. Thus, from 1827 to about 1837, von Uslar was living in the state of Oaxaca, Mexico, engaged in the supervision of mines, and serving as the Hannoverian Consul General to Mexico. During this period, von Uslar was oblivious to any debate about root excretions, and was apparently completely ignorant of the writings of de Candolle, Macaire-Prinsep and others. However, something, perhaps discussions with a colleague, or observations of Mexican agriculture, set his mind thinking about the subject of the role of root excretions in plant interactions.

Upon his return to Europe in about 1837, von Uslar retired to a farm property in Holstein, near Hamburg, where he quietly pursued his interests and became a corresponding member of the Hamburg Natural Science Society. Over the following six years, he wrote a book, which was ultimately published in 1844 with the title, *Die Bodenvergiftung durch die Wurzel-Ausscheidungen der Pflanzen als vorzüglichster Grund für die Pflanzen-Wechsel-Wirthschaft* (The poisoning of the soil through root excretions of plants as a most excellent reason for plant rotation). This is arguably the first book devoted to the topic of allelopathy, and an annotated edition in English has recently been published (Willis 2004). A so-called second edition (von Uslar 1852) was also published, but this appears to be actually a re-issue with an altered titled by a different publisher, who likely purchased the unsold stock of the 1844 edition.

While the ideas in the book were very similar to those presented by de Candolle, von Uslar claimed that, until he had first drafted his manuscript, he had no knowledge of de Candolle's writings. Indeed, von Uslar was unable to read French, and the full account of de Candolle's *théorie des assolemens*, appeared in volume 3 of *Physiologie Végétale*, which was never translated into German. Of course, there were translations into German of Macaire-Prinsep's work, reviews by German plant physiologists such as L.C. Treviranus, and the works of Carl Sprengel, but again, von Uslar claimed that he only became aware of these later, when he was preparing his work for publication.

Von Uslar's ingenuous, but remarkable, book offered numerous observations on a host of issues relating to allelopathy. He described several examples of soil sickness, which he explained, as had de Candolle, were owing to the accumulation of toxic root excreta. Crops which demonstrated soil sickness included wheat, corn, secale, rice, asparagus, strawberries and spruce. Several plants were indicated as being toxic to other plants, including *Cirsium*, *Euphorbia*, *Brassica*, *Erica*, *Quercus*, *Sambucus*, and *Cichorium*. He suggested that chemical substances may play a role in diverse phenomena such as plant parasitism, e.g. in *Orobanche*, maintenance of diversity in rainforests, and succession. He also proposed that an understanding of root exudates could prove beneficial in intercropping and weed control.

While von Uslar's ideas seemed superficially rather avant garde, they were couched within a framework which was amateurish, somewhat verbose, and sadly, anachronistic. Von Uslar had published his work ten years too late. The book made little impact, sold few copies, and to my knowledge achieved only one review. The reviewer was the acerbic Jacob Matthias Schleiden (1845), who dismissed the work as a relic, and chastised the unfortunate von Uslar for his lack of familiarity with the contemporary authors on agricultural chemistry.

Figure 8.1. A unsigned portrait of Justus Ludewig von Uslar (c. 1805), in Hannover civil service unform, (Courtesy of Gesine von Uslar).

It was the agricultural and gardening communities that most vigorously held on to the idea of root excretion. For example, in England, Jane Loudon (1841) in her *Ladies' Companion to the Flower Garden* stated that oleanders (*Nerium* spp.) required repotting a least once a year because of the accumulation of root excretions, and similarly tulips required a fresh bed each year, and benefited from a rotation of unlike plants. Joshua Trimmer (1842) in his *Practical Chemistry for Farmers and Landowners* provided an amalgam of excretion theories and reiterated the belief that root excretion was most active during flowering:

> Matters rejected by one organ contain the elements which enter into the composition of others, till, being incapable of further transformations, they are thrown off from the system – gaseous matters by the leaves and blossom, solid excrements are deposited in the bark, soluble substances are passed off by the roots. These secretions are most abundant just before the formation and during the continuance of the blossom, and diminish after the development of the fruit. The excrementitious matter which the soil absorbs from the roots is still capable of decay, and being converted into carbonic acid. It becomes in fact humus. (p. 95)

Cuthbert Johnson (1848) mentioned root excretion amongst the theories of crop rotation, but noted somewhat deprecatingly that it concerned some "curious experiments" and "writings of some foreign naturalists". Falkner (1847) accepted that crop rotation could be explained by root excretions as advanced by de Candolle, and added that amelioration of a toxic effect became accelerated on light open soils, in contrast to heavy soils, and was assisted by fallowing and the action of lime, as often seen in continuous wheat cultivation. He added an interesting bit of personal theory, and suggested that the advantage of crop rotation was also related to disrupting the build-up of populations of insects damaging to a particular crop. However in a companion work published in the same volume, Smith (1843) would have none of this, and adopted the emerging view that any negative effect of one plant on another was explicable through understanding their nutrient requirements. Nonetheless, it was widely stated in lay agricultural works that the deleterious effects of the "excrementitious matter" of roots was primarily a problem where there was inadequate drainage (e.g. Stephens 1853, Waring 1854, Phin 1862). The notion that plant roots excreted matter, and that the associated effects were either beneficial or harmful, remained surprisingly entrenched in agriculture, perhaps especially so in regions little concerned with the academic debates of Great Britain's and Europe's learned societies, such as the rural United States, Canada (Dawson 1864) or various other English colonies. For example, Leonard Wray (1848), who had spent sixteen years working in colonial sugar plantations in Jamaica, Bengal (India), and the Straits Settlements (Malaysia) wrote:

> In Europe and all cold climates, this excrementitious matter, voided by plants, is much longer passing into putrefaction than in tropical countries; the necessity therefore of adopting a rotation of crops is much greater in the former than in the latter. All plants void excrement, which when acted on by air and moisture, putrefies and becomes converted into "humus," or vegetable matter in a state of decay. This deposit of organic matter is common to all plants, and exercises a very beneficial effect on land, by furnishing it with a substance capable of being converted into humus, and which is so desirable in soil: but plants cannot long be planted in the same soil without being seriously affected by their own excrement; so much so that at length they altogether fail. Artificial aid, however, induces a more speedy conversion of this matter into humus, than would otherwise take place: this is effected by frequently turning up the soil with the plough or hoe, so as to expose the excrement to the influence of the atmosphere; and by irrigating the land with river water; as the water of rivers and streams contains oxygen in solution, which effects the most rapid and complete putrefaction of the excrementitious matter contained in the soil. (p. 172)

In Switzerland, as for example at the Société des Sciences Naturelles de Neuchâtel, de Candolle's theory remained under consideration, although it was acknowledged that sound evidence was lacking, and other hypotheses were gaining credibility (Ladame 1845, Sacc 1846-1847).

The concept of root excretions outside of agriculture was fraught with difficulty, as in the face of the stable growth of forests decade after decade, or century after century, it seemed discredited. Évon (1846), in discussing the alternation of forest trees, accepted the possible participation of root excretions, and he argued that the diversity of species, as seen for example in natural prairie, allowed the dilution of root excretions in comparison with grasses and herbs under monoculture. Ryland (1843), in discussing the relationship between tree roots and the saprophyte *Monotropa hypopitys*, suggested that the excremental matter of the trees was involved in the relationship.

The Italian botanist Guglielmo Gasparrini (1804-1866) claimed that he had witnessed root excretion and seen the excretory structures in the root hairs of *Poa annua* (Figure 8.2); however, his claims have been regarded as entirely fanciful.

OPPOSITION GROWS

Many of the principal commentators on plant physiology had initially been swayed by the mechanistic logic of the root excretion theory, but then as evidence failed to accrue, they rapidly joined in opposition.

In Germany, Carl Sprengel, who had been a staunch supporter of root excretion theory, became distinctly more cautious in his views in his *Die Lehre vom Dünger* (Sprengel 1839, 1845). Sprengel recalled the views of de Candolle and Macaire-Prinsep, particularly with regard to the toxic effects of weeds on crops, but he was impelled to consider other factors:

> Experiments, in this regard where I used with rye the yellow rattle (*Rhinanthus crista galli*) that becomes so damaging, leave me from this, in no doubt, because I saw that rye plants, which still had not set seeds, were brought into contact (wetted) repeatedly with root excretions of the yellow rattle, this not at all affected the grain formation unfavourably, on the other hand I found by means of a routine chemical analysis that earlier as the rye was becoming ripe, yellow rattle contained exactly those mineral substances in large amounts, that the rye grains also require absolutely for their own formation, from which one may probably conclude that it is just the same with other weeds in most cases that they are thus suppressive. Regarding the poisoning of the soilby the root excretion, I beg to consider against it, that the released substances, as organic bodies, go through a very fast decomposition, so that e.g. the material, which becomes excreted by the potato roots, can affect probably not more injurious to the following rye. Recalling however that the nature effects rarely or never simpler, but usually compound kind are, I want to grant that the root excretions can have a small unfavorable and evenly so a somewhat favorable influence on the growth of the following plants, and which therefore the rules of the crop rotation with on that must be justified. With the crop rotation one has to consider indess also excellently that growth of the plants depends also on the length of their roots; if the earth is in the upper layer of certain, the plants exhausted to the food serving materials, then contain another, more deeply lying layer probably still another genugsame quantity of these materials, which then, reached by which more deeply which are rooted plants needs are sufficient for that. (Sprengel 1845, pp. 28-29)

Perhaps the most important and influential of these commentators was the renowned German agricultural chemist Justus von Liebig (1803-1873), who was based at Giessen (Figure 8.3). Liebig's fundamental work, *Die organische Chemie in ihrer Anwendung auf Agricultur und Physiologie*, appeared firstly in German, but almost simultaneously in English in 1840 as *Organic Chemistry in its Applications to Agriculture and Physiology*, and then in French, Danish, Dutch, Polish and Russian, and went through numerous editions. It is generally considered that Liebig's personal

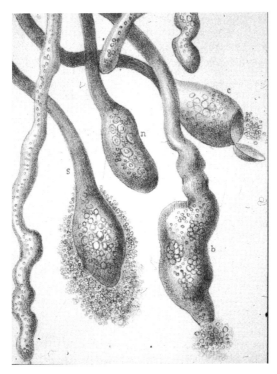

Figure 8.2. Detail of the roots of Poa annua, *showing the excretion of material and a root tip operculum (Gasparrini 1856).*

Figure 8.3. A view of the interior of Liebig's laboratory at Giessen, as drawn by Wilhelm von Trautschold, c. 1840 (Courtesy of the Liebig Museum, Giessen).

knowledge of agriculture was actually quite poor, and his early injudicious support of the ideas of others often led to severe criticism (Browne 1944). The first edition was fundamentally a overview of existing knowledge, and it is commonly noted that Liebig was originally an ardent supporter of de Candolle's theories:

> Of all the views which have been adopted regarding the cause of the favourable effects of the alteration of crops, that proposed by M. Decandolle alone deserves to be mentioned as resting on a firm basis. (Liebig 1840, pp. 161-162).

Liebig's reputation was high, especially in England, and it was Liebig's initial endorsement, that prolonged acceptance of de Candolle's ideas and overrode what was possibly scepticism of Gallic theories (e.g. Squarey 1842). However, one English reviewer (Anonymous 1843a) was scathing in his criticism of Liebig's ill-considered views on root excretions, and furthermore he regarded Macaire-Prinsep as having been a liability to de Candolle's work.

As de Candolle's theories slipped into obscurity in the 1840's, we find that the latter editions of Liebig's *Organic Chemistry* and related works omitted any reference to de Candolle. Liebig's ideas on the importance of mineral nutrients became deeply and firmly embedded in agricultural chemistry. Liebig, however, did admit later the existence of root excretions, notably carbonic acid, and in later editions of *Organic Chemistry*[1] he wrote:

> We have every reason to believe that this secretion takes place over the whole surface, we observe them not only at the trunk, but also at smallest branches, and we must conclude from it that this excretion process also occurs in the roots. An elimination of excrements cannot therefore be denied in plants, although it is possible that they do not take place to the same degree in all plants. (Liebig 1865)

Similarly in England, John Lindley (1799-1865), the prolific author of numerous botanical and horticultural texts, in his early works, e.g. *Theory of Horticulture* (Lindley 1840)[2] endorsed the phenomenon of root excretion, but reserved opinion as to its importance in crop rotation. In the second English edition, published as *The Theory and Practice of Horticulture* (Lindley 1855), Lindley dismissed root excretions as being an artifact or of little significance. In *An Introduction to Botany*, Lindley (1848) wrote:

> Root excretions are now regarded as unimportant, if not apocryphal, except in cases where the roots are wounded. (Volume 2, p.183)

---

[1] The seventh and eighth editions of *Die Chemie in ihrer Anwendung auf Agricultur und Physiologie* (1862, 1865) were greatly enlarged, but were not immediately translated into English, as was customary with previous editions. Liebig's criticism of English agricultural practice had offended the English agricultural establishment and soured his relationship with his English publisher. An abridged version of the seventh edition was subsequently published in English in 1863 with the title *The Natural Laws of Husbandry*.

[2] Comments emanating from Lindley's *Theory of Horticulture* require some explanation. This work was first published in 1840. This book was widely translated, and appeared also in annotated editions in the United States. The second American edition (1852) is based on the English edition of 1840, and thus still maintains the validity of root excretion. The authentic second edition of Lindley's *Theory of Horticulture* was published in 1855 with a modified title, *The Theory and Practice of Horticulture*.

James F.W. Johnston (1796-1855) was a Scotsman whose work on agricultural chemistry was highly regarded, notably also in Europe and in the United States, where his concerns about the practical applications of theory sometimes found him a kinder reception than that of Liebig. In his substantial text, *Lectures on Agricultural Chemistry and Geology* (Johnston 1847), Johnston summarily dismissed the theories of de Candolle:

> Being unsupported by decisive facts and observations, therefore, the hypothesis of Decandolle must, for the present, be in a great measure laid aside, and we must look to some other quarter for a more satisfactory theory of rotation. (p. 855)

Another of the English natural history authors, whose works became adopted as standard texts, was William Benjamin Carpenter (1813-1885). Carpenter was a physician who authored texts in a wide range of fields. His first botanical text appeared as the first volume in the *Popular Cyclopaedia of the Natural Sciences* (Carpenter 1841), but it was his *Vegetable Physiology and Botany* which became a standard textbook in many institutions. As in the case of Roget, Carpenter was not a botanist, but he presented a sound, if slightly embroidered, synthesis of comparatively orthodox views in the science. Thus, in this latter work, Carpenter (1844) readily accepted the views of de Candolle and Macaire-Prinsep.

> Plants, as already stated (§. 119), not only draw various substances from the soil, but impart to it a portion of the juices, which they have formed within themselves. A well-marked instance of this is the oak; which so completely impregnates the soil around its roots with *tannin* (the substance which gives the oak-bark its peculiar power of converting animal skin into leather), that few trees will grow in the same spot from which it has been rooted up; since the agent, even when a very minute quantity of it is dissolved in water, produces an effect like tanning upon the delicate tissue of the spongioles, and destroys their peculiar properties. It is probable that every species of forest-tree produces a similar effect; since it is well known that, when a wood composed of one kind has been cleared by the hatchet or by fire, the new growth which soon springs up, is not of the same, but of a different species. Again, some of the plants which are known as the rankest weeds, secrete from their roots substances equally injurious to plants around them; the Poppy tribe impregnates the soil around it with a substance analagous to Opium, which is easily shown by experiments, to have as injurious an effect upon Plants, as an overdose of this powerful medicine has upon Animals; and the Spurge tribe exudes an acrid resinous matter.
> The excretions of all Plants seem injurious to themselves, as well as to others of the same species grown on the same spot; and in many instances, as in those just quoted, they are injurious to plants of other tribes also. (pp. 141-142)

However, in later editions of Carpenter's textbook (e.g. Carpenter 1883), the theories of de Candolle were mostly abandoned and his approach was deferential to the theories of Liebig:

> It has been supposed that excretions from the roots of one plant are injurious to another, and an attempt has been made to account for the necessity of a rotation of crops on this ground alone. There is no doubt that the products secreted by one plant are often injurious to another. Thus few plants will grow in the soil formed by the leaves of the beech; and the oak, it is said, so completely impregnates the soil around its roots with tannin, that few trees will grow in the spot from which it is rooted up. This does not, however, appear to be the case with any of the plants ordinarily cultivated, and the necessity for the rotation of crops is much better explained by the exhaustion of the soil of the mineral ingredients necessary to the growth of all plants. Thus Liebig proposes to

divide plants into those which require silex, lime, potash, and soda; of course such plants require other constituents, but these are the substances which they need most, which failing in sufficient quantity, they die. (p. 113)

In France, C.F. Brisseau-Mirbel (1815) had originally endorsed the root excretion theory as originated by Brugmans, but like many of his contemporaries he subsequently revised his opinion:

....that such excretions he [De Candolle] supposes to be emanations from the roots – the remains of those juices which the earth and air conjointly supply, and upon which in reality the plant exists. But against even the very fact mentioned by De Candolle, in confirmation of his opinion, that opium, strewn upon the ground, kills plants, and renders the soil henceforth unproductive, we may quote the much more apposite fact, that trees (why not, *a fortiori* corn and grasses) grow and flourish for entire centuries in the midst of excretions from their roots. (quoted by Gyde 1846, p. 276)

The botany textbooks of Achille Richard (1794-1852), a French physician and botanist, became largely the benchmark for those used in France, as they were endorsed by the Faculty of Medecine, University of Paris. Richard's basic text, *Nouveaux Éléments de Botanique et Physiologie Végétale*, was sold over a period of seven decades and was available in most European languages. His texts, with their numerous editions, offer a useful barometer of popular trends in botany of the time. For example, in 1819 he wrote in regard to root excretions (Richard 1828):

It is to this material, which as we have spoken, is different in each species, that are attributed the sympathies and antipathies that certain plants have one for another. One knows, in effect, that certain plants look for each in some way, and live constantly one beside the other; these are the social plants; however, on the contrary, other plants seem not to be able to grow in the same place. (p. 41)

However, by 1852, the subject of root excretions was clearly out of vogue, and there is no mention of them at all in later editions of Richard's text.

The German botanist M.J. Schleiden[3] (1804-1881), who had flayed the work of Justus von Uslar, was the author of an influential book, *Grundzüge der Wissenschaftlichen Botanik nebst einer methodologischer Einleitung als Anleitung zum Studium der Pflanze*. An interesting dichotomy emerges in that. while Scheiden's text achieved critical acclaim, it was ruthless in its treatment of matters with an opposing view, and the only language into which it was subsequently translated was English, whereas Richard's more convivial text became available in most common European languages, except English. Schleiden did not mince his words in summarily dismissing root excretion, and the quote below is taken from the first English edition, *Principles of Scientific Botany* (Schleiden 1849), which was a translation of second edition of the German work:

The worthlessness of the experiment of Prinsep has been pointed out by Meyen (Physiologie, vol. ii. P. 528), Treviranus (Physiologie, vol. ii. p. 117), and Hugo von Mohl. On the other hand, the experiments of Unger and Welser (*sic*), which were performed with all proper care and accuracy, gave a perfectly negative result; so that

---

[3] There is sometimes confusion regarding Schleiden's initials, as in *Principles of Scientific Botany*, his name was wrongly given as J.M. Schleiden.

there can be no doubt that an excretion from the root, such as that believed in by DeCandolle, Prinsep, and Liebig, has no existence at all. (p. 497)

The agricultural chemist J.B. Boussingault (1841, 1843-1844) expressed his reservations concerning de Candolle's theory of crop rotation, as in his experience, many crops grew satisfactorily on the same soil year after, and that this was especially true in regions such as Central and South America, where for example, wheat has been grown for two centuries on the same soil without diminution in yield. Boussingault's major objection was that the root excretion theory seemed to take no account of the fact that root excretions, being organic substances, would readily decompose.

Johann Friedrich Schmalz (1781-1847) was a German agriculturalist who, amongst his works, wrote the *Theorie des Pflanzenbaues* (Schmalz 1840). He indicated that the theory of root excretion in crop rotation had become current in Germany, and that it was believed that the root excretions of a potato crop led to a lower yielding crop of rye sown afterwards. However, Schmalz himself had found little, if any, evidence of any inhibition of a succeeding rye crop, although his experience was with cropping on sandy soils.

The disfavour accorded to de Candollean ideas is amply illustrated by their increasing neglect in works from the mid-nineteenth century dealing with the topic of crop rotation. The writings of de Candolle and Macaire-Prinsep were considered in the review by Gasparin (1851), but were utterly ignored by Heuzé (1862) in a monograph on *assolements*, that even included a discussion of the concept of plant antipathies and sympathies. Gasparin concluded:

> The effect produced by excretions is reduced then to that which we have described in the preceding chapter: the greatest absorption of certain principles compared to others; the impoverishment of the soil relative to the first and the necessity of rendering them and reestablishing their proportions if one wishes to continue the culture.

While the theory of root excretion in crop rotation slid into obscurity, there remained substantial debate concerning whether root excretion played a special role in ridding the plant of harmful substances. Botanists were largely divided in opinion as to whether the root or the leaf was the principal organ responsible for disposing of toxic substances.

In Italy, there was initially limited reaction to de Candolle's crop rotation theory (e.g. Bellani 1834, Anonymous 1837), but then there was an escalation in opposition as Italian work on the functioning of roots demonstrated, similarly to other researchers such as Braconnot, Walser and Gyde, that roots indeed were selective in their uptake of chemical substances. Bellani (1843a, 1843b) presented a lengthy review of work concerning the absorptive functions, and in the same year Trinchinetti (1843) described results of a series of experiments in sand that demonstrated that roots selectively absorb substances. Apparently Macaire-Prinsep replied and addressed some of the issues raised, and furthermore he criticised the methodology of Trinchinetti, although the publication details of Macaire-Prinsep's article are unknown. In particular, Macaire-Prinsep had criticised using sand as a growth medium in that it required the plants to be placed initially into a dry medium, and secondly it obscured any view of excretions. The root excretion theory of de Candolle and Macaire-Prinsep was based on the premise that plants absorb all soil solutes indiscriminately, and then excrete the undesirable materials. Trincinetti argued that plants thus would continuously absorb more and more toxic material, which Macaire-Prinsep agreed was the core of the argument. However, how could one explain the continuous growth of individual trees in the same soil for hundreds of years? Macaire-Prinsep conceded that some toxic substances may decompose or that the seasonality of tree growth may play a mitigating role (Moretti 1845).

It remained for de Candolle's son, Alphonse de Candolle[4], to seemingly bring closure to the matter. Alphone de Candolle (1806-1893), who became an eminent taxonomist in his own right, took over many of his father's botanical duties, and in 1835 was appointed professor of botany at the Université de Genève. He is remembered particularly for his works concerning the biogeography of plants, and the origins of cultivated plants. In the former work, *Géographie Botanique Raisonée*, de Candolle (1855) wrote:

> Thus, the results that are attributed to it, in the experiments of M. Macaire, to the lesions of roots or to some accidental expulsions, holds well perhaps to these causes; but importantly, if in nature these same accidents are not rare? Moreover, the old roots are shed at the surface; they slough off fragments around themselves, a bit like the trunks of trees. There are also the portions of roots which die and which decompose. The proof of it is that near the trunk of a tree and around its oldest roots, there are few small roots, whereas several years before, this region was occupied thickly with root hairs. The detritus of roots contains tannin and other substances, according to the species. Thus it is evident that the prolonged presence of a species alters the soil, by the effect of the irregular shedding of the roots and their ramifications. Along side this incontestable fact, an excretion in the strict sense near the extremity of the roots would have less importance. It would be in all cases so feeble that ordinarily one would not notice it. It remains to be proved that the accidental sheddings and the detritus of roots corrupts the soil for plants of the same species. It would be worthwhile to be conduct an experiment. In that which we are occupied, I restrict myself to state that the prolonged growth of a species in a location is a source of harm, and even a source of exclusion sometimes for the establishment of plants of the same species or very similar species. This local effect is only transitory and probably only of a very little importance in nature. I was obliged to indicate this so as to leave out nothing. (p. 449)

EXPERIMENTAL WORK

Views on the physiology of plants had begun to change quite radically in the early nineteenth century, as the gaseous origin of organic matter in plants was demonstrated by chemists such as Ingenhousz and Senebier. The development of analytical techniques notably in Germany led to a more considered view of plant nutrition. The experiments of Macaire-Prinsep came to be considered as altogether too unnatural to warrant status as evidence in favour of de Candolle's theory of crop rotation, and consequently, there was a clear demand for credible experimental data. For the period of a decade commencing in about 1836, there was dedicated research in France, Germany, Italy, England and Scotland, that attempted to repeat earlier experiments, to demonstrate the existence of root excretions under more natural conditions, or at

---

[4] A.P. de Candolle had a daughter Amella (born 1804), two sons, Alphonse (born 1805) and Benjamin (born 1812), but only Alphonse survived past childhood.

least to show that there was declining yield with repeated culture that was likely associated with a factor such as soil toxicity.

Eduard Walser

Eduard Walser (1805-1872) was a medical student at the University of Tübingen, whose dissertation (Walser 1838, 1840) was directed by Hugo Mohl, and who was awarded a prize by the Faculty of Medicine, for the work in 1836. Mohl set Walser the task of repeating and elaborating the experiments of both Brugmans and Macaire-Prinsep. For example, Walser attempted to grow the combinations of plants that Brugmans had suggested as inimical due to root excretions: *Lolium* and wheat, *Spergula* and buckwheat, *Euphorbia* and flax, but the weeds failed to germinate, or else grew poorly. Walser could not corroborate the visual observation of root excretion as reported by Brugmans, and with regard to Macaire-Prinsep's experiments, he was of the opinion that any dissolved matter that appear in aqueous growth medium was due to leakage from broken roots and/or decomposition of dead cells, and that there tended to be a natural concentrating effect due to water loss from solution due to transpiration. The former was an opinion that was shared by a number of earlier botanists including Link (1807) and Hedwig (1782), and subsequently Treviranus (1838). Treviranus also recalled the dissertation by Backer (1829) who had failed to show that buckwheat was inhibited by *Spergula*.

Henri Braconnot

The first French experimental work to be published that re-examined the crucial experiments of Macaire-Prinsep was that of Henri Braconnot (1780-1855), who early in his career, had endorsed the idea of root excretion (Braconnot 1807). Braconnot (1839) leached the soil of a mature oleander (*Nerium oleander*) grown in a closed pot for three years, and found that the liquid yielded over 3 g of residue, which on analysis consisted almost entirely of mineral salts, not bitter substances that one associates with this plant. Braconnot attempted to repeat Macaire-Prinsep's experiment for obtaining root excretory material from *Chondrilla muralis*, but as this species was unavailable to him, he used lettuce. He concluded that any release of organic material was most likely a consequence of damage done to the root hairs, in preparing the plant for water culture. In experiments with various other bitter plants, *Euphorbia peplus*, *Asclepias incarnata*, and *Papaver somniferum* he failed to retrieve bitter substances from their soil. The only inhibitory substance isolated was acetate. Furthermore, Braconnot showed that Macaire-Prinsep's claim that roots would absorb poisonous substances such as lead acetate, and then later excrete them, seemed easily explained by the passive capillarity of the root tissue. Braconnot was forced to conclude:

> The experiments that I have just presented, are not favourable, as can be seen, to the theory of crop rotation based on root excretions. These excretions, if they really occur normally, are otherwise so obscure or so poorly known, that there is reason to presume that one must have recourse to other causes in order to explain the general system of rotations.

Anselme Payen

Another important Frenchman to look at the interaction of chemical substances in soil and plant roots was the industrial and agricultural chemist Anselme Payen (1835, 1842). Payen is a greatly underrated figure in the history of botany, but he is occasionally remembered as the first person to isolate an enzyme, diastase, and for discovering that plant cell walls are made of cellulose. Payen recognised that tannin, insofar as it was capable of complexing a number of substances, was likely able to affect the seed germination and plant growth. He cites the 1833 observation of Silvestre[5] that trees transplanted into soil containing debris from felled oaks showed poor growth. He germinated seeds of wheat, rye, oats, and corn in a 0.1% solution of tannin, and observed severe inhibition of root and shoot growth. Similarly, the roots of young wheat plants that were exposed to a tannin solution, were strongly inhibited. Payen performed what is likely the earliest piece of histological work relating to allelopathy when he sectioned damaged roots, and observed with the microscope that the damage progresses from the "spongioles" to the vascular tissue, and that damaged cells became brown and lost their integrity. Further studies showed that levels of nitrogenous substances seemed higher in the root extremities, and that the damaging effects of tannin were in some way related to interfering with these substances.

Jean Baptiste Boussingault

The important agricultural chemist Jean Baptiste Boussingault (1802-1887) largely endorsed the concerns of Braconnot, and reported that he also had found little evidence in support of root excretions having any importance in crop rotation (Boussingault 1841), a view that was supported by the influential chemist Berzelius (1843). Boussingault was a pioneer in having performed detailed analyses of the elemental composition of crops in various crop rotation systems from his farm at Bechelbronn, and in the case of artichokes which were grown year after year on the same ground with adequate fertilisation, he could find no evidence of a decline in yield. Although Boussingault still maintained that soil organic matter must have some importance in plant nutrition, de Candolle's theory of crop rotation was unsupportable from his experience, as many crops grew satisfactorily on the same soil year after, and that this was especially true in regions such as Central and South America, where for example, both maize and potatoes had been grown for two centuries on the same soil without diminution in yield. The same seemed true for other crops including indigo and sugar cane. However, Boussingault's major objection to the root excretion theory was that it seemed to take no account of the fact that root excretions, being organic substances, would readily decompose under normal conditions of heat and moisture (Boussingault 1841, 1843-1844).

---

[5] This is the agronomist Augustin François Silvestre (1791-1853?), the son of Augustin François Silvestre (1762-1851).

Alfred Gyde

In 1842, an English surgeon, Alfred Gyde (1781-1858) began a set of experiments, which were similar in many respects to those of Braconnot (Gyde 1845, 1846). The work was published in the *Transactions of the Highland and Agricultural Society of Scotland*, where it earned awards totalling 35 sovereigns as part of its prize essay program. In view of the writings of de Candolle and Macaire-Prinsep, Gyde attempted to assess the composition of root excretions from different families of plants, the relationship of root excretions to the sap, and the role of excretion in ridding the plant of harmful substances. Perhaps the most innovative experiments were those in which Gyde attempted to assess the role of soil, allegedly charged with the excretions of one plant, on another plant. An example of his ingenuity is given in an experiment that attempted to address whether plant roots could excrete toxic substances previously absorbed. A potato plant was placed with its roots split between two containers: one contained potassium ferrocyanide, and the other a ferric solution. Contamination of either solution by the other would have been indicated by the formation of a blue colour (Prussian blue reaction), and in this experiment the only blue colour noted was within the plant itself.

Gyde (1845) attempted to ascertain whether plants actually excreted material from their roots. He was aware that the process of freeing plants from ordinary soil would damage the fine roots and root hairs, and thus compromise his results. Consequently he grew his plants in a sandy preparation which allowed removal of the plants with minimal damage. He found:

1. Plants in flower yielded the greatest amount of root excretions; young plants yielded a medium amount; and plants in fruits yielded least.
2. Plants with fleshy roots, notably legumes, yielded the greatest amounts of material.
3. Generally the odour of the root excreted material was similar to that of the plant shoot (sap), notably with the Brassicaceae and legumes.
4. Plants were exposed to solutions of magnesium sulphate, potassium ferrocyanate, sodium sulphate, muriate of soda and potash, and then placed in distilled water. The plants readily absorbed these substances, but excreted little.
5. A potato plant had part of its root system exposed to an iron solution, and part to potassium ferrocyanate. Mixture of the two salts yields the Prussian blue colour. Blue was observed in the plant tissue, but not in the original solutions. The salts were not excreted from the plant.
6. Flowering kidney bean plants were grown in distilled water. The water containing their root excretions was then applied to selected bean plants growing in garden soil. Plants thus watered were deemed healthier in appearance. In similar experiments with wheat, cabbage, beans, peas, and mustard, there were no adverse effects on subsequent growth.

Gyde (1846) concluded that:

1. Plants are capable of excreting substances, although the strong resemblance of the material to the sap raises suspicions about the origin of the material.
2. Excreted material contained both organic and inorganic material.
3. Excreted material varied from species to species.
4. Excreted material is similar in composition to sap.
5. The mechanism of excretion accords with Dutrochet's model of endosmose and exosmose, and the root spongioles are the likely regions of excretion.
6. Metallic salts administered to healthy generally were not excreted, but caused either death or damage; however, doses were high (~1000 to 10000 ppm).
7. Seeds impregnated with strong solutions of metallic salts generally grew poorly, and seeds planted in soils steeped in similar solutions grew poorly. Bean plants grown in soil watered with collected root excretions or macerated bean plants were unaffected. Gyde presented data from a wheat plot in Gloucester in which the yield was monitored for 12 successive years. While Gyde claims the data show no change in yield, the data are inconclusive.
8. Gyde concluded that the success of certain plants on particular soils is likely related to the specific nutrients found in such soils, and the nutrient demands of the plant. Crop rotation is similarly explained on the basis of available nutrients.

Gyde's results largely swung the balance against acceptance of root excretions for authors such as Johnson (1868), the influential American agricultural writer, who was professor of agricultural chemistry at Yale University.

A.F. Wiegmann and L. Polstorff

One of the more influential works damaging the credibility of root excretion theory was a prize-winning essay by Arend Friedrich Wiegmann (1771-1853), a professor of natural science at Brunswick, assisted by an apothecary, L. Porstorff (Wiegmann and Polstorff 1842, 1843), although the damage came largely through association with evidence so convincing in resolving another controversy. In 1838 an anonymous sponsor had offered a prize was advertised in Germany for the best work which addressed the question: "The so-called inorganic[6] elements, which are found in the ash of plants, and thus in plants, are to be found then in plants if external sources of these are not furnished; and whether these elements are such essential constituents of the plant organism, that they are required for its complete development."

In the eighteenth and nineteenth centuries, it was common practice for learned societies to stimulate research by posing questions publicly, and the subsequent paper which best addressed the issue was awarded a prize, and was usually published. Such had been the case in 1797 when the Berlin Academy of Science, in view of

---

[6] At this time, the term "organic compounds" referred to substances made by living organisms, usually carbon-based, in contrast to today where it refers strictly to carbon-based compounds, be they natural or synthetic.

controversy surrounding vitalism and the issue of the origin of mineral substances in plants, posed a set of questions:

> Of which type are the earthly materials which are encountered by means of chemical analysis of native grain species? Do they come into the grains as they are found, or do they come into being by means of the life force and brought into growth by the workings of the plant?

Schrader and Neumann (1800) were awarded the prize for a suite of two papers which allegedly found that seeds of various grains germinated in relatively inert medium of sulphur powder, contained new nutritive material when the seedlings were ashed. Braconnot (1807) reported similar results and concluded that plants had the capacity to transform elements, and produce an endogenous supply of essential nutrients. These erroneous findings handicapped the advance in understanding of plant nutrition, and eventually were proved as erroneous. Fortunately, the mainstream of botanists, such as Senebier, insisted that the supply of nutrients came entirely from the soil, and to a lesser extent rainfall and the atmosphere. Experiments by Wiegmann and Polstorff comparing the growth of various crop plants in sand and in an artificial soil containing most of the essential nutrients confirmed solidly the need for externally supplied nutrients for satisfactory growth and development. Furthermore, comparison of the ash content of plants grown in an inert platinum pot with the ash content of seeds showed no difference in composition, and thus refuted the notion of elemental transformation or generation. Almost as an addendum, Wiegmann and Polstorff then addressed the question of root excretion theory in relation to soil fertility and crop rotation, and it was reported that work by Wiegmann in 1834 and 1838 which had repeated the key experiments of Macaire-Prinsep, using the plants *Mercurialis* and *Senecio* found no evidence was found to support root excretion, and any movement of substances was satisfactorily explained by capillarity. Wiegmann and Polstorff also refuted the notion of humus as a plant food, as there was little loss of humic material when plants were grown in humus extract over a one month period, in comparison to a plant-free conrol. Wiegmann and Polstorff (1842) concluded:

> The well established observation, that cultivated plants seldom thrive perfectly if they are grown again on the same soil on which similar crops were grown and ripened in the year before, yes, according to the words of the worthy agriculturalist von Schwerz that field peas, if they had formerly occurred and ripened on the same field, after 6 years must not be grown (W.), has likewise been attributed to the effect of root excretions. Similarly it has been stated that that just as an animal cannot thrive upon its excrement, so a plant is unable to thrive upon the exudations of its own kind, but plants of another family can utilise them as food and manure.
>
> Here it has not been considered, that organic substances are destroyed by fermentation, and that inorganic ones, by being plowed under or mixed with other substances of the soil, are rendered innocuous, and finally that trees flourish luxuriantly on their ejecta for several hundred, yes even a thousand years.
>
> The above mentioned observation of farmers and gardeners is much more simply explained by supposing that the soil has been so robbed by the previously harvested crop of the inorganic materials which are necessary for plant development that another crop of the same kind (even when the ground is plowed and newly fertilised with an animal manure deficient in the necessary mineral element) is unable to find the requisite amount of plant food that is necessary for its complete development. (p. 50)

These results, coupled with evidence of the absolute necessity of supplying nutrients to the growing plant, and the more or less simultaneous rise of Liebig's views on plant nutrition solidified the demise of root excretion for years to come.

Charles Daubeny

Of all the British botanists who engaged in the debate concerning de Candolle's theories, the most qualified was Charles Giles Bridle Daubeny (1795-1867; Figure 8.4). Daubeny was Professor of Chemistry at Oxford University, but had many other interests, including botany. In the summer of 1830, he visited Geneva to study with Seringe and de Candolle, although he spoke little French. He attended de Candolle's lectures on plant physiology, as he wished to improve his botanical knowledge. In 1833 the British Association for the Advancement of Science received a report from Lindley (1833) on the current status of botany, which included a favourable hearing of de Candolle's root excretion theory. However, the more conservative members (such as Thomas Knight[7]) were in favour of obtaining better evidence, and thus the meeting resolved (Anonymous 1833):

> That Professor Daubeny be requested to institute an extended inquiry into the exact nature of the secretions of roots of the principal cultivated plants and weeds of agriculture: and that the attention of Botanists and Chemists be invited to the degree in which such secretions are poisonous to the plants that yield them, or to others; and to the most ready method of decomposing these secretions by manure or through other means.

Figure 8.4. Photograph of Charles Daubeny c.1845 (courtesy of the Royal Society).

---

[7] According to his biographer, Knight who corresponded with de Candolle, was opposed to the root excretion theory until better evidence was furnished (Knight 1841).

Daubeny had aspirations of obtaining the Chair of Botany at Oxford, and one suspects that this commission greatly suited him. In 1834 he was appointed Professor of Botany, and he later became the Professor of Rural Economy at Oxford. In any case, Daubeny immediately set to work on experiments as requested by the British Association, and he provided a verbal report in 1834. Daubeny set up a number of plots in which he grew 15 different common crops, either in continuous culture, or in shifting cultivation where a given crop was preceded by a different crop. He was to find that the experiments were to occupy him for the next decade, partly because, as his data accrued, he recognised that there was little evidence to support de Candolle's ideas. In the light of other work critical of de Candolle (e.g. Braconnot), and emergent and more favoured theories concerning mineral nutrition, Daubeny felt it wise to vary his experiments in these directions.

Daubeny's findings were eventually published in 1845 in a lengthy memoir (Daubeny 1845). Firstly, Daubeny reluctantly admitted that after growing crops known to be rich in narcotic substances, such as poppies and tobacco, he was unable to find any trace of morphine or nicotine, or any other extraneous material in the soil. Amongst all the crops tested, the only species that showed a dramatic failure was *Euphorbia lathyrus*; a first crop in 1835 yielded a respectable 18 pounds from a plot of 100 square feet, whereas in 1836 the crop was negligible. Replanting in 1837 also produced a dramatic failure, which was continued in 1838; however, when the plot was planted with other crops, such as flax, barley and beans in 1839, growth of these was normal. Secondly, contrary to his expectation, he was unable to show, to his satisfaction, any difference in crop yields between species that had been cropped continuously on the same soil, and those which had been alternated, as being due to root excretions. Many of the crops did show substantially increased yields when they were grown following a different species. However, as there was little variation in yield among continuously cropped plant over successive years, he ascribed the former to differential use of mineral resources.

It must have been with disappointment that Daubeny published his results, as he had evidently had a deep respect for de Candolle, whose theories he described as ingenious and eloquently expressed. It was during the mid-course of these experiments that de Candolle died in 1841, and Daubeny provided a lengthy biographical notice for English readers (Daubeny 1843), of which de la Rive (1845) remarked:

> Mr. Daubeny was the one whom to me seemed to have best appreciated the character of de Candolle from the viewpoint of science and work. One discovers in his notice the recollections of an intelligent disciple, who has seen close at hand the master and has learned to understand him.

DIFFERENT APPROACHES

Henry David Thoreau

The concept of ecological succession is commonly attributed to Henry David Thoreau (1817-1862), the eloquent American naturalist. He is remembered chiefly for his remarkably insightful account of Concord, Massachusetts, entitled *Walden*. However, much of Thoreau's writings on natural history remained unpublished, including his remarkable essay, *The Dispersion of Seeds*, which was eventually published 130 years after Thoreau's death (Thoreau 1993). An early précis of this work was presented by Thoreau in 1860 in an address, "The Succession of Forest Trees", given to the Middlesex Agricultural Society (Massachusetts), and subsequently published in the *New York Weekly Tribune*, and in several works thereafter (e.g. Board of Agriculture 1861). It is often claimed that with this work, use of the term "succession" in ecology originated, although it was borrowed from agriculture where a sequence of crops in crop rotation was termed a succession (e.g. Low 1834) and the already term had considerable currency in the United States relation to changes in forest vegetation (Caldwell 1808, Peters 1808ba). The eminent English geologist Charles Lyell (1849), during his travels in the United States, drew attention to similar matters:

> Near the house of Hopeton there was a clearing in the forest, exhibiting a fine illustration of that natural rotation of crops, which excites, not without reason, the surprise of every one who sees it for the first time, and the true cause of which is still imperfectly understood. The trees which had been cut down were full-grown pines (*Pinus australis*[8]), of which the surrounding wood consists, and which might have gone on for centuries, one generation after another, if their growth had not been interfered with. But now they are succeeded by a crop of young oaks, and we naturally ask whence came the acorns and how were they sown here in such numbers? It seems that the jay (*Garrulus cristatus*) has a propensity to bury acorns and various grains in the ground, forgetting to return and devour them. The rook, also (*Corvus americanus*), does the same, and so do some squirrels and other Rodentia; and they plant them so deep, that they will not shoot unless the air and the sun's rays can penetrate freely into the soil, as when the shade of the pine trees has been entirely removed. It must occasionally happen, that birds or quadrupeds, which might otherwise have returned to feed on the hidden treasures, are killed by some one of their numerous enemies. But as the seeds of pines must be infinitely more abundant than the acorns, we have still to explain what principle in vegetable life favors the rotation. Liebig adopts De Candolle's theory, as most probable. (Volume I, pp. 246-247)

In his native Concord, Thoreau (1860) attempted to understand why both pines and oak seedlings fared poorly among mature trees of their own kind:

> The shade of a dense pine wood is more unfavorable to the springing up of pines of the same species than of oaks within it, though the former may come up abundantly when the pines are cut, if there chance to be sound seed in the ground.

It is recorded that Thoreau received a copy of Darwin's *Origin of Species* in 1860, which bolstered his own nascent evolutionary opinions. Darwin (1859) himself had nothing to say concerning what might be construed as the chemical interactions of plants, but he was aware of the differential and perhaps deleterious effects of certain plants. For example, regarding a heathland[9] in central England, Darwin (1859) wrote:

> I will give only a single instance, which, though a simple one, has interested me. In Staffordshire, on the estate of a relation where I had ample means of investigation, there was a large and extremely barren heath, which had never been touched by the hand of

---

[8] = *Pinus palustris*.
[9] The site was Maer heath, on one of the properties owned by the Wedgwood family.

man; but several hundred acres of exactly the same nature had been enclosed twenty-five years previously and planted with Scotch fir [=pine]. The change in the native vegetation of the planted part of the heath was most remarkable, more than is generally seen in passing from one quite different soil to another: not only the proportional numbers of the heath-plants were wholly changed, but twelve species of plants (not counting grasses and carices) flourished in the plantations, which could not be found on the heath. (Chapter 3, p. 67)

The new rhetoric of "struggle for existence" and "war of nature"[10] undoubtedly fired the imagination of others. Thoreau greatly fleshed out his arguments in a manuscript work entitled "The Dispersion of Seeds", and this is perhaps the earliest known ecological work to incorporate Darwinian theory. He presaged many of the ideas current today in the ecology of plant-animal interaction. In his lengthy essay, Thoreau discussed the interactions of pines and oaks, and surmised that pines do not succeed well after pines, nor do oaks after oaks, due in part to excretions released by the roots (p. 121), although it should be acknowledged that predecessors such as Lambert and de Candolle had alluded to ideas of excretions playing a role in forest succession thirty years earlier (see Chapter 7). Thoreau was influenced in his considerations of forest ecology by theories surrounding crop rotation, and he was familiar with the phenomenon of "soil sickness". As early as 1808, American country folk had referred to certain problematic pine soils as "pine-sick" (Peters 1808b). Thoreau was also substantially influenced by the British physiologist W.B. Carpenter (1813-1885), whose botanical works initially favoured de Candolle's root excretion theory, including the injurious effects of weeds, and the rationale of crop rotation.

The Black Walnut and Other Trees

While the allelopathic properties of the American black walnut, *Juglans nigra*, came to be reported, rather anomalously in Switzerland in the eighteenth century by Plappart (see Chapter 6), this report remained virtually unnoticed. It was still widely held that the "shade" and/or the drip from trees, notably the beech, ash and walnut, was injurious to plants growing underneath (Johns 1847, Plues 1863, Heath 1881), and that with walnut, "bitter properties of its leaves" were thought to be involved (Anonymous 1843b). J.C. Loudon[11] (1838) had noted that the drip of the locust tree (*Robinia pseud-acacia*) was less injurious than that of other trees, as the leaflets of the compound leaves tended to fold together during wet weather.

Interest in the allelopathic effects of black walnut surfaced in the central United States in the 1870's largely through the meetings and reports of various state horticultural societies (Willis 2000). The first of these reports seems to be that of O.B. Galusha (1870), secretary to the Illinois State Horticultural Society. Galusha, in discussing that "certain kinds of trees are poison to an orchard", related that his neighbour, in planting a row of black walnut trees on one side of his apple orchard, found after 12 years that almost all the adjacent apple trees had died.

---

[10] This phrase actually originated with A.P. de Candolle (1820), and was known to Darwin (1975).
[11] John Claudius Loudon (1783-1843) was married to Jane Loudon (1807-1858).

This was followed in 1874 by publication of discussion, also at the Illinois State Horticultural Society, where Mr. McWhorter and Mr. Douglas agreed that black walnut affected the palatability of grass for stock. Mr Bryant added that he had witnessed black walnut killing apple trees within a radius of about 25 meters, and he added that walnut roots were in some way poisonous. However, others at the meeting were in disagreement, and when Dr Schroeder commented that walnut leaves "contain a great proportion of bitter stuff and they embitter the ground and make it sick", he was confronted with laughter.

Remarkably similar comments appeared elsewhere in the American agricultural press not long thereafter. A Pennsylvania farmer, Mr O. Snowberger, wrote to a Farmers' Club in New York (Anonymous 1871) :

> I feel satisfied that I have seen three apple trees destroyed by black walnuts and I believe they destroy grape vines. I judge it is the water dropping from the walnut leaves that does the work.

In 1881, J.S. Stickney presented a paper on "Timber Culture" before the Wisconsin State Horticultural Society, and listed black walnut as a tree of prime value in terms of rapid growth and ease of culture, but declined to recommend it as a specimen or street tree. When queried, he replied that black walnut suppresses plants growing underneath, but he reserved judgment as to whether this was attributable to leaf leachates or competition. However, Dr Hoy added clearly that:

> The main reason why vegetation does not thrive under these trees is the poisonous character of the drip.

In 1883, an anonymous American[12] report similar to that by Galusha was published in an English forestry journal:

> Some thirty years ago I planted an orchard of about 200 apple trees on one of my farms – open prairie. Having a lot of three to four-year old Walnut trees growing from seed, I planted a few rows of them on the north side as a windbreak. Both did well for some time, and now some of the walnut trees have reached a height of 40 feet. The first row of Apple trees has long since been killed out. The rest of the orchard is doing well; having a large crop of fruit during the past season and is generally fruitful. With my experience, I should as soon think of feeding poison to my stock as planting such trees near enough to Apple trees to subject the latter to their influence. This would seem to deprive arboricultural schemes of the romance with which they have been surrounded in theory.

However, despite the groundswell of interest in the toxic properties of the black walnut, Crozier (1891) dismissed all such claims as unsupported, and wrote:

> The supposition that the injury is caused by poisonous excretions or exhalations from their leaves is wholly unfounded. We must believe the cause of this injury in all cases to be the same as that by which buckwheat, hemp and other strong and rapidly growing plants are able to free the land from weeds and other vegetation, namely the production of shade and the extraction of moisture from the soil. (p. 126)

---

[12] The report is described as Armenian in origin; I take this to be a typographical error.

Gustav Jaeger

Another forgotten individual in the history of allelopathy is the remarkable German naturalist and physician Gustav Jaeger (1832-1917; Figure 8.5). Jaeger had extraordinarily diverse interests ranging from aquatic biology to developmental biology to alternative medicine. He trained in zoology and medicine and became Professor of Zoology and Anthropology at the Hohenheimer Akademie, followed by joint appointments at the Königlichen Polytechnikum, Stuttgart and the Stuttgarter Tierarzneischule (Weinrich 1993). It was Jaeger who founded the Vienna Zoo. Curiously, he became best known in the English-speaking world for his eccentric views on clothing and bedding. He taught that fabric based on vegetable fibres such as cotton and linen were potentially harmful as they were capable of harbouring noxious substances that could be absorbed or could promote disease. He advocated that all clothing, including underwear, and bedding be based on wool, which did not attract such unsanitary problems. His views were taken seriously especially in England, and inspired a line of clothing that was named in his honour; his name still survives today in the London-based fashion house of Jaeger.

While Jaeger's interests were foremost in the sphere of zoology, medicine and anthropology, he did gain an interest in allelopathy. This arose through his eccentric theories of "Duftstoffen", which were explained in his *Entdeckung der Seele*, first published in 1878. Jaeger developed the general theory that all organisms interact through volatile substances that they produce and receive, and that such interactions

Figure 8.5. Photograph of Dr. Gustav Jaeger c. 1880 (courtesy of Jaeger Company, London).

can explain everything from human emotions to disease to homosexuality. In some ways this is reminiscent of the ancient theory of Empedocles (see Chapter 2); on the other hand, from our point it foreshadows the importance of essential oils and other volatile substances in allelopathy, and even the idea of induced-response compounds such as jasmonates. One of Jaeger's former students, H. von Ziegesar, had noted that Jaeger's theories, which had been built chiefly around animal systems, might also explain certain phenomena in agriculture such as soil sickness. Jaeger's theories concerning allelopathy apparently first appeared in 1880 in a series of articles in the now rare magazine, *Neuen deutschen Familienblatt*; however they were more accessible and appeared in their most developed form in the third edition of *Die Entedeckung der Seele* (Jaeger 1885), notably in a chapter titled *Die Seele der Landwirtschaft*, which appeared separately (Jaeger 1884)[13] prior to publication of the book.

Jaeger's writings on agriculture are slight on fact but rich in polemic and speculation. As with many unorthodox theories, it is perhaps too easy to focus on the statements that, with hindsight, seem to be close to the mark, and to ignore less credible ideas, for example, that plant roots seek out plant nutrients in the soil on the basis of smell. In any case, Jaeger argued that, as it seems obvious volatile substances play an important role in the life of the animal, for example in food selection, why does not the same hold true for plants? He believed that the essential nature of the plant is fixed in its proteins, and that as these proteins decompose, particular volatile substances are given off that may characterise the plant's state. Favourable growth leads to the emission of pleasant substances such as plant perfumes. Stressful events can lead to the emission of noxious substances. The volatile substances of a plant can have various effects. For example, it may be responsible for the antipathy and sympathy of plant species. Jaeger (1885) became quite well versed in the history of plant interactions, and he cited several works ranging from Theophrastus to de Candolle and Macaire-Prinsep to Liebig. He went so far as to suggest that congeneric species pairs such as *Achillea atrata* and *A. moschata*, *Primula elatior* and *P. officinalis*, and *Rhododendron alpinum* and *R. hirsutum* while exploiting different but neighbouring niches, may actually repel each other with volatile substances. Thus he adopted the de Candollean view that similar species may repel one another whereas dissimilar species may have no effect or be mutually attractive. He attempted to understand the problem of soil sickness in agriculture, which was well known to German growers of carrots, peas, flax, clover and other crops. Jaeger accepted the possibility that substances released by continuous cultivation of a crop lead to an increased growth of harmful organisms such as nematodes. However, the long-term growth of certain plants on the same soil, as in the case of trees, complicated understanding the issue of soil sickness. Also within his realm of interactions, Jaeger envisaged such substances being active in attracting pollinators and in repelling pathogens and herbivores.

In 1895, Jaeger again touched on the matter of substances in soil sickness. However, this was little more than a brief restatement of his ideas. Reinitzer (1893)

---

[13] This separate publication is not entirely identical to the chapter in *Entdeckung der Seele*, as it has an additional prefatory summary. However, the *Entdeckung der Seele* has an important 24 page addendum to the chapter, that contains most of the discussion regarding root excretions and soil sickness.

had published a note about fatigue substances as an occurrence in the life of cells, and Jaeger took the opportunity to chide Reinitzer for having too narrow a view, and for having ignored the importance of fatigue substances in plant ecology.

It is little known that the Swedish literary giant, August Strindberg (1849-1912), also had an interest in the natural sciences, notably later in life, as he became increasingly reclusive. Amongst the eclectic content of his "Blue Book", he considered the antipathy and sympathy of plants (Strindberg 1907). However, his words amply illustrate how, even in modern times, misinformation begins, for Strindberg has confused the classical sympathy of the rue and fig for that of the rue and the unrelated figwort (*Scrophularia*):

> Why certain plants do not grow together with others cannot be properly explained. Since Roman times, the common grape vine has always been "wed" with the elm tree, with which it sought support; now it is more with the true chestnut tree, as in Savoie, with the poplar and the mulberry in Lombardy. On the other hand, the vine abhorred the cabbage. In rose-green Provence onions are planted in order to increase the smell of the roses. Lily bulbs are regarded like onions, which probably explains the sympathy between the rose and the lily. *Ranunculus* (cocksfoot) and *Nenuphar* (white waterlily) like each other, perhaps because the dampness and muddiness attracts them both. The rue, which gives a lovely smell if one carefully pinches the leaf, but stinks if one crushes it, has a certain preference for figwort. Cyclamens and cabbage hate each other, so much so that both die if they are planted together. Clover and vetch, grain and vetch prosper on the same ground. *Euphorbia peplus* (petty spurge) is found in cabbage gardens, but *E. helioscopia* (sun-spurge) in herb gardens; one looks for *E. dulcis* (purple spurge) on chalk. And so forth.

The nineteenth century headed toward closure with minimal interest in allelopathy. However, there remained considerable interest in root physiology, notably in France, and the topic was reviewed in depth by Duchartre (1868). The principal unresolved question of relevance here was the same question that had troubled de Candolle, whether roots absorbed substances passively or actively from the soil. Some experiments by Chatin (1845) and Reveil (1865), involved exposing plant roots to sublethal doses of known poisons, such as arsenic salts (Figure 8.6).

While some physiologists, such as Chatin (1845) and Bouchardat (1846, 1846a) continued to support root excretion as a means of ridding the plant of unwanted substances, there was more compelling evidence to suggest that the uptake of substances from the soil by roots was indeed selective (Cauvet 1861, 1864), and that the abscission of leaves may serve an excretory function (Roché 1862). Cauvet (1861) was emphatic in writing:

> Roots that are physiologically sound do not excrete poisonous substances absorbed by plants; excretions, such as those claimed by Macaire and Candolle, do not actually exist; all theory based on the existence of these excretions will then necessarily be false.

Despite the dearth of data supporting either the reality of root excretions, or, should they exist, any ecological role for them, there remained scattered, but persistent reports, commonly anecdotal, that related to root excretion or similar theory. Many of these appeared in the *Gardeners' Chronicle*, a British periodical which commenced publication in 1841 and which became the dominant publication of its type during

Figure 8.6. An apparatus as used by Reveil (1865) to investigate the effects of various toxic substances on plants.

the nineteenth century. In the 1841 volume, the editor noted that *Rhododendron arboreum* survived well under the drip of trees when growing in a light, gravelly soil, but did not fare well when so growing in clay soil (Anonymous 1841)[14]. Consequently a correspondent observed that grass and evergreen shrubs such as laurels, box, rhododendrons, yews and hollies grew poorly beneath beech (*Fagus silvatica*) trees perhaps because of the "noxious quality communicated to the rain-water which drips from the beech foliage", although competition was also considered (T.T. 1842). The distinctive annular pattern of mushroom growth and concomitant regions of dead grass, known as "fairy rings" led a Scottish correspondent to suggest that the fungus exerted some sort of toxic effect in the soil (An Inquirer 1845). J.T. Way (1847), a colleague of Daubeny, acknowledged that de Candolle's theory offered a credible explanation of events, but that such theories had lost favour, and fairy rings could be satisfactorily be explained by nutrient depletion. Westerhoff (1859) maintained that the centrifugal growth of mushroom colonies was due to refuge from the noxious effects of excremental matter from mushroom "roots". An American, John Kearsley Mitchell, a professor at the Jefferson Medical College of Philadelphia, also agreed with the toxin theory of fairy-rings, and in 1849 he presented a summary of the chemical interactions of plants, that seemed far ahead of its time, particularly in regard to concepts concerning antibiosis and causes of the peach replant problem (Mitchell 1849):

---

[14] The reference provided by Rice (1983) is in error.

> A curious exemplification of the poisoning of the soil against their own growth is afforded by the fungi which have so lately preyed on the potato crop. In Ireland the potatoes grow much better in the subsequent year, when the diseased potatoes have been left to rot in the soil, than when they are carefully removed.
> We have other analogies for this idea. Macaire, who has given much scientific attention to the effect of plants upon soils, observes, that certain vegetables enrich the earth by their *exuviae*[15], as for example, the leguminous vegetables excrete much mucilage, and thus fertilize it for *gramineae*, but that the *papaveraceae* injure the soil by the deposit of opiatelike substances, and these prevent or render growth imperfect. So is it with the peach and bitter-almond trees, which, as well as other plants that produce prussic acid and the poisonous hydrocyanates, render the soil in which they grow incapable of successive crops of the same kind of trees. A nursery in which young peach trees have been planted, and from which they have been soon removed, will not sustain the same kind of stock for eight or ten years afterwards (note – Manuring the soil from which a peach tree has been removed does not mend the matter; removal of the soil, or long repose, will alone suffice.) Nature thus secures a variety, by a succession of dissimilar vegetations. (pp. 127-128)

A Scottish correspondent to the *Gardeners' Chronicle* (Beobachter 1845) suggested that pine trees may be indirectly inhibited by the root excretions of heath (*Erica* spp.) that seemed to promote the formation of a carbonaceous stratum below the topsoil, that stunted tree root growth. Another item of ecological interest was an early consideration of the specific relationship between plant epiphytes and their hosts[16]. Paul Lévy (1869) noted that in the rainforests of Nicaragua, lianes did not utilise certain host trees even when the lianes were brought in close proximity to the trees, and that similarly there appeared to be specific associations and antagonisms between epiphytes, especially the bromeliads *Tillandsia* spp., with host lianes.[17]

Another correspondent to the *Gardeners' Chronicle* revived discussion of the de Candollean theory of root excretions in crop rotation, particularly with reference to well-known problems caused by clover (T.A. 1845), and in reply the editor remarked that "clover-sickness" was a widespread problem in England, although the causes were unknown (Anonymous 1847). In the United States, Owen (1861) stated the paradox of the "soil sickness" problem, with reference to tobacco cultivation in Kentucky:

> In this "exhausted" tobacco soil, the same thing has occurred as frequently takes place with other crops; with clover, for instance; which, after having been grown with great vigor for a few years ceases to do well on the same land, which is hence said to be "clover-sick;" although the land is far from being exhausted, as is shown by the fact that other crops are produced on it in great abundance. (p. 89)

---

[15] The term "exuviae" usually refers to animal sheddings, and it is possible that Mitchell meant something akin to "effluvia".

[16] The specific "sympathy" of a liane and *Liquidambar styraciflua* in America was noted by the obscure French author Gleïzes (1840). He also stated there was a strong antipathy between betel leaf and durian; however, this report can be dated back to the eigthteenth century botanist Rumphius (see Chapter 4). It is also worth noting that Lévy anticipated allometric relationships in plants in stating that there seemed to be a constant proportion between the diamter and the ultimate length of a liane.

[17] The same theme was of great interest to the twentieth century botanist F.W. Went (see Chapter 11), and is still a subject of allelopathic research (e.g. Talley *et al.* 1996).

Copeland (1859) claimed that the straw of flax was poisonous to vegetation. J.C. Draper (1872), in investigating the relative effects of light and dark on pea seedling development, apparently found evidence of "soil sickness", although data were not given:

> Another interesting fact which lends support to the opinion that the process of growth in seedlings in the dark is very similar to that occurring in those growing in the light, is the character of the excrements thrown out by the roots. It is well known that many plants so poison the soil that the same plant cannot be made to grow therein until the poisonous excretions from the roots of the first crop have been destroyed by oxidation. In the case of peas, this poisoning of the soil takes place in a very marked manner, and I have found that in the pots in which peas have been grown in the dark, the soil is so poisoned by the excrements from the roots that a second crop fails to sprout. Does it not follow that since in the two series with which I experimented, the excrements from the roots possessed the same poisoning action, the process in the plants from which these excrements arose must have been similar? (pp. 127-128)

In 1893 a letter to the editor of the *Gardeners' Chronicle*, J.J. Willis stated that cucumbers could not be grown profitably in the same greenhouse soil for more than three years, and Willis (1894a)[18] elaborated further:

> It seems to be the usual practice after the growth of a Cucumber crop to remove all the surface-soil taking it out to a depth of 12 to 15 inches, and to convey into the Cucumber-house entirely fresh soil. Hence, whatever may be the cause of failure, it cannot be attributed to the surface soil.
>
> Plants set in newly-imported soil have been known to make good healthy growth up to a certain point, and then to show signs of decay, and finally yield a meagre crop of fruit. It appears to me, therefore, that we must look for the primary cause of failure to the subsoil, and it is to that, that I have directed my attention. From observations made of the subsoil in which Cucumbers had been grown, I have found that the roots penetrate to a considerable depth, and fill the subsoil with a mass of fibrous root-matter. A soil, therefore, in which several successive crops of Cucumbers have been grown naturally becomes charged with much decaying vegetable matter, and the supposition is, that the primary cause of failure may be due to excreted substances given off by the roots during growth, which accumulate in the subsoil. Another source of failure must be looked for in exhaustion of some kind of plant-food within the range of the roots, and this exhaustion seems likely to be of potash and nitric acid.

Willis (1894b) subsequently noted that it was the experience of some English farmers growing peppermint (*Mentha piperita*) in Surrey that the land could only be cropped for two or three years from a planting, after which "the soil usually becomes so foul that the quality of oil produced would not be good enough to pay for harvesting." Discussion at the Illinois Horticultural Society raised the point that certain crops, such as oats, adjacent to blackberries or raspberries, had the capacity to cause failure of the berry crop (Webster 1893). In a subsequent discussion concerning the failure of certain fruit trees in northern Illinois, Austin (1895), in a remark that anticipated the work of Pickering (see Chapter 9), noted simply that: "You must not expect to raise trees and grass off the same ground. The grass will destroy the trees."

---

[18] Rice (1983) has unfortunately provided much of this quote out of context, as he has omitted Willis' subsequent discussion of nutrient deficiencies, and more importantly, fungal pathogens.

Towards the close of the nineteenth century, reports also began to emanate from Russia, which during the reign of Nicholas II had begun to emerge from agricultural feudalism. Despite its huge population, Russia lagged behind Europe, and there was a groundswell of interest in agricultural research, led initially by learned societies and enlightened amateurs. By about 1890, largely spurred by unrest about famine, the government began to accept responsibility for agricultural research, and by 1917 there was a network of about 300 agricultural research stations in Russia (Elina 2002). Gortinskii (1966) has given a valuable summary of some of the early Russian reports relating to allelopathy, those of Barabanov, Anzimorov, Bogdanov, Doyarenko, Engelhardt, Gedroits, Kossovich, Levitskii, and Malkovskii.

It is also important here to consider the work of the Rothamsted Experiment Station, which had been founded in 1843 by John Bennett Lawes, in part to stake a claim in England for agricultural research, which was dominated by European workers such as Boussingault and Liebig. Lawes, at a young age, inherited the 400 hectare family estate near Harpenden in Hertfordshire, and his interests in agriculture and chemistry led him to successfully develop the commercial production of the artificial fertiliser superphosphate. Lawes began experiments concerning various systems of cropping and fertilisation in 1837, and in 1843 he employed Joseph Henry Gilbert, a chemist who had trained with Liebig, to oversee such experiments and to supervise an agricultural research farm on the Lawes estate, the Rothamsted Experiment Station, the first such station in the world devoted to agricultural research. Many of these experiments have run more or less continuously to the present time, and results have often served subsequently as benchmarks for more radical results, such as found by the U.S. Department of Agriculture, Bureau of Soils during the period 1901-1920 (see Chapter 10). A major summary of the results of the first four decades of research at Rothamsted was published by Lawes *et al*. (1882), and the work, despite acknowledging the necessity of the alternation of crops, and failure of repeat cropping for certain plants, for example clover[19] (Gilbert 1871), could offer no real support for the effect of root excretions. Lawes and Gilbert (1860), at least in the early years of their Rothamsted work, maintained a surprisingly open mind regarding the role of organic substances in soil, and concluded:

> If the failure of the Clover-plant, when repeated too soon upon the same land, de due at all to the excrementitious matters left by the former crop, it is much more probable that the injury is in some way connected with the organic matters which have been rejected. Unfortunately, we are not yet able, by the aid of chemistry, to distinguish those organic compounds of the soil, which are convertible into the substance of the growing plant and those which are not so. Nor do we know how far the excreted organic matters may be necessary complementary products in the formation of some of the essential constituents of the plant. Experience teaches us that when a crop of Clover is eaten by some sheep folded upon the land, animals dislike the growth which immediately succeeds. It might be inferred, therefore, that, in such a case, the plant had taken up from the soil, certain matters which it had not finally elaborated. Whether these organic substances would in the process of time, be converted into living plant-matter, or whether they would wholly, or in part, be rejected as excrementitious organic compounds,

---

[19] Failure of clover on ground previously planted with clover was a long-standing issue in English agriculture (e.g. Young 1804, Legard 1841, Thorp 1842, Anonymous 1847).

> to undergo in the soil certain chemical changes before being adopted for plant-food, we are not able to determine.

However, while many people questioned the existence of root excretions, a Rothamsted associate, Maxwell Masters (1874), admitted that many problems, such as crop failure, seemed to point to as yet undiscovered causes:

> Mere exhaustion of the soil will not account for the phenomenon in all cases, because a crop will fail on a particular soil after a while, and yet chemical analysis of that soil will reveal the fact that the particular elements required by a given plant are still contained in sufficient abundance in it. Land, for instance, that is "clover sick" – on which, that is, good crops cannot be grown – is by no means necessarily deficient in the constituent required for the growth of the plant, and indeed, in the Rothamsted experiments the constituents in question have been supplied as manure, but without any good result. (p. 41)

In Germany there was also sustained interest in the physiology of root excretion; for example Goebel (1893) found organic acids in water in which *Lepidium* and *Hordeum* roots had been growing. It is seldom realised that Hans Molisch as a young academic had a research interest in root excretions, and was well familiar with the nineteenth century controversy concerning the role of root excretions in plant interactions (Molisch 1887). His experimental results indicated that root excretions could either reduce or oxidise substances and were important in affecting the organic constituents of the soil. Furthermore, it is not appreciated that Friedrich Czapek (1896) considered the question of root excretion at length, particularly with regard to the relationship between injury and the release of substances. He identified numerous inorganic substances that had been excreted from plant roots, and amongst organic substances he identified simple acids such as carbonic, formic and oxalic acids, as did several other later investigators (e.g. Stoklasa and Ernest 1909). Relative indifference to the implications of root excretion was tempered by a burgeoning interest in the nascent discipline of microbiology, which was to greatly modify ecological thinking. Important discoveries came in relation to rhizobia, rhizospheric organisms (L. Hiltner), mycorrhizae (A.B. Frank) and soil bacteria.

A little known work by the American Greene Vardiman Black (1836-1915), remembered primarily as an innovator in dentistry practise, perhaps highlights the continuous thread that seemed to persist in these matters. Black, a remarkable individual, who had little formal education but eventually acquired honorary doctorates in four different disciplines, in his *Formation of Poisons by Micro-organisms* (Black 1884), presented the argument that all forms of life produced injurious waste substances, and, that especially in the case of micro-organisms, these were a significant factor in causing disease (including dental caries). In support of the universality of his thesis, he drew on the ideas and work of de Candolle and Macaire-Prinsep, and he ranks among those pioneering American authors to write of the allelopathic effects of black walnut:

> It is rare to see a very large black walnut tree that has not a clearing around it, wherever it may stand in the forest. This is not on account of its shade, but something eliminated by the tree that is hurtful to other trees. I remember well an effort to raise corn on the south side of a row of walnut trees. The experiment was continued for many years. The corn was injured seriously for many feet distant, where it was never shaded by the trees.
> (p. 118)

REFERENCES

Anonymous. 1833. Recommendations of the Committee: Botany. *Report of the British Association for the Advancement of Science* **2**: 481-482.
Anonymous. 1837. Teoria delle rotazioni (alternazioni della coltura). *Giornale Agrario Lombardo-Veneto e continuazione degli Annali Universali di Agricoltura di Industria et d'Arti Economiche, serie 2* **8**(11/12): 193-195.
Anonymous. 1841. Editor's note. *Gardener's Chronicle* **1**:
Anonymous 1843a. Dr. Justus Liebig, in his relation to vegetable physiology. *The Mirror of Literature, Amusement, and Instruction* **2**(16): 247-248.
Anonymous. 1843b. Walnut. In: *The Penny Cyclopaedia of the Society for the Diffusion of Useful Knowledge. Volume XXVII. Wales-Zygophyllaceae*, pp. 44. Charles Knight, London.
Anonymous. 1847. Clover sickness. *Gardener's Chronicle* **7**: 41.
Anonymous. 1871. Does black walnut destroy fruit trees? *Pacific Rural Press*, 1 July : 403.
Anonymous. 1883. Poison in walnuts. *Forestry; a Magazine for the Country* **7**: 234.
An Inquirer. 1845. Fairy rings. *Gardener's Chronicle* **5** (October 25, 1845): 722.
Austin. 1895. Discussion on questions from the question box. *Transactions of the Illinois State Horticultural Society, new series* **29**: 232-233.
Backer, G. 1829. *De Radicum Plantarum Physiologia, eurumque Virtutibus Medico , Plantarum Physiologia Illustrandis*. H. van Munster, Groningen.
Bellani, A. 1834. Delle rotazioni agrarie. *Giornale Agrario Lombardo-Veneto e continuazione degli Annali Universali di Agricoltura di Industria et d'Arti Economiche, Série 2* **1**(1-3): 195-203.
Bellani, A. 1843a. Sulle funzioni delle radici nei vegetabili. Parte prima. *Giornale dell'I. R. Istituto Lombardo di Scienza, Lettere e Arti* **7**(21): 281-303. [also in *Giornale Agrario Lombardo-Veneto e Continuazione degli Annali di Agricoltura* **20**: 177-203; 245-260]
Bellani, A. 1843b. Sulle funzioni delle radici nei vegetabili. Parte seconda. *Giornale dell'I. R. Istituto Lombardo di Scienza, Lettere e Arti* **8**(22): 13-82. [also in *Giornale Agrario Lombardo-Veneto e Continuazione degli Annali di Agricoltura* **20**: 261-276; *Giornale Agrario Lombardo-Veneto e Continuazione degli Annali di Tecnologia, di Agricoltura, di Economica e Domestica, ecc., serie seconda* **1**: 81-98]]
Beobachter. 1845. The highland pine. *Gardener's Chronicle* **5** (February 1, 1845): 69-70.
Berzelius, J. 1843. *Rapport Annuel sur les Progrès de la Chimie, presenté le 31 mars à l'Academie Royale des Sciences de Stockholm*. Fortin, Masson et Cie., Paris.
Black, G.V. 1884. *The Formation of Poisons by Micro-organisms: a Biological Study of the Germ Theory of Disease*. P. Blakiston, Son & Co., Philadelphia.
Bouchardat, A. 1846. *Recherches sur la Végétation Appliquées à l'Agriculture*. Chamerot, Paris.
Boussingault, J.B. 1841. De la discussion de la valeur relative des assolements, par les résultats de l'analyse élémentaire. *Annales de Chimie et de Physique,* Série 3 **1**: 208-246.
Boussingault, J.B. 1843-1844. *Économie Rurale, Considerées dans ses Rapports avec la Chemie, la Physique et la Météorologie*. Béchet Jeune, Paris.
Braconnot, H. 1807. Sur la force assimilatrice dans les végétaux. *Annales de Chemie* **61**: 187-246.
Braconnot, H. 1839. Recherche sur l'influence des plantes sur le sol. *Annales de Chimie et de Physique*. Série 2 **72**: 27-40.
Brisseau-Mirbel, C.F. 1815. *Élémens de Physiologie Végétale et de Botanique*. Magimel, Paris.
Broeck, V. van dem. 1855. *Catéchisme Agricole*. Librairie Agricole de H. Tarlier, Brussels.
Browne, C.A. 1942. Justus von Liebig – man and teacher. In: *Liebig and After Liebig: a Century of Agricultural Chemistry* (ed. F.R. Moulton), pp. 1-9. American Association for the Advancement of Science, Washington, D.C.
Caldwell, C. 1808. Letter re: Changes of timber and plants. Races of animals extinct. By Richard Peters. *Memoirs of the Philadelphia Society for Promoting Agriculture* **1**: 301-307.
Candolle, A. de. 1855. *Géographie Botanique Raisonée, ou Exposition des Faits Principaux et des Lois Concernant la Distribution Géographique des Plantes de l'Époque Actuelle*. Victor Masson, Paris.
Candolle, A.P. de. 1820. Géographie botanique. *Dictionnaire des Sciences Naturelles* **18**: 359-422.
Carpenter, W.B. 1841. *Vegetable Physiology. Popular Cyclopaedia of Natural Sciences*. Volume 1. Tanner Brothers, London
Carpenter, W.B. 1844. *Vegetable Physiology and Botany*. Orr, London

Carpenter, W.B. 1858. *Vegetable Physiology and Systematic Botany*. G. Bohn, London.
Cauvet, D. 1861. *Études sur le rôle des racines dans l'absorption et l'excrétion*. Strasbourg. (Also in *Annales des Sciences Naturelles, 4ème série* **15**: 320-359.
Cauvet, D. 1864. De l'excrétion des matières non assimilables par les végétaux. *Bulletin de la Société botanique de France* **11**: 201-211.
Chatin, A. 1845. Étude de physiologie végétale faites au moyen de l'acide arsénieux. *Comptes Rendus de l'Académie des Sciences* **20**: 21-29.
Coleman, J. 1855. On the causes of fertility or barrenness of soils. *Agricultural Society Journal* **16**: 169-206.
Copeland, R.M. 1859. *Country Life: A Handbook of Agriculture, Horticulture, & Landscape Gardening*. John P. Jewett & Co., Boston.
Crozier, A.A. 1891. *Popular Errors About Plants*. Register Publishing Co., Ann Arbor.
Czapek, F. 1896. Zur Lehre von den Wurzelausscheidungen. *Jahrbücher für wissenschaftliche Botanik* **29**: 321-390.
Darwin, C. 1859. *On the Origin of Species by Means of Natural Selection*. John Murray, London.
Darwin, C. 1975. *Charles Darwin's Natural Seletion, being the Second Part of his Big Species Book Written from 1856 to 1858* . Edited by R.C. Stauffer. Cambridge University Press, Cambridge.
Daubeny, C. 1834. On excretions from the roots of vegetables. *Report of the British Association for the Advancement of Science* **3**: 598.
Daubeny, C. 1843. Sketch of the writings and philosophical character of Augustin Pyramus Decandolle, Professor of Natural Histry at the Academy of Geneva. *Edinburgh New Philosophical Journal* **34**: 197-246.
Daubeny, C. 1845. Memoir on the rotation of crops and on the quantity of inorganic matter abstracted from the soil by various plants under different circumstances. *Philosophical Transactions of the Royal Society* **135B** (2): 179-252.
Daubeny, C. 1845a. On the chemical principles involved in the rotation of crops. *Report of the British Association for the Advancement of Science* **1845** (2): 33-34.
Dawson, J.W. 1864. *First Lessons in Scientific Agriculture for Schools and Private Instruction*. John Lovell, Montreal.
Draper, J.C. 1872. Growth or evolution of structure in seedlings. *American Journal of Science and Arts*, 3rd series **4**: 392-398.
Duchartre, M. 1868. *Rapport sur le Progrès de la Botanique Physiologique. Recueil de Rapports sur les Progrès des Lettres et des Sciences en France*. L'Imprimerie Impériale, Paris.
Dutrochet, H. 1837. *Mémoires pour Servir à l'Histoire Anatomique et Physique des Végétaux et Animaux*. Baillière, Paris.
Elina, O. 2002. Planting the seeds for the revolution: the rise of Russian agricultural science, 1860-1920. *Science in Context* **15**: 209-237.
Evon, M.N. 1846. De l'alternance des végétaux. *Journal d'Agriculture Pratique et de Jardinage*, Deuxième Série 3: 593-603.
Falkner, F. 1847. *The Farmer's Treasure; a practical treatise on the nature and value of manures, founded from experiments on various crops*. D. Appleton & Co., New York.
Galusha, O. B. 1870. O. B. Galusha report *ad interim* – apples. *Transactions of the Illinois State Horticultural Society*, New Series **4**: 78-86.
Garreau, L. and Brauwers, 1858. Recherches sur les formations cellulaires. L'accroissement et l'exfoliation des extrémités radiculaires et fibrillaires des plantes. *Annales des Sciences Naturelles, Botanique,* Séries IV **10**: 181-192.
Gasparin, Comte de 18xx. *Cours d'Agriculture*. Volume 5. Dusacq, Librairie Agricole de la Maison Rustique, Paris.
Gasparrini, G. 1856. *Ricerche sulla natura dei succiatori e la escrezione delle radici ed osservazioni morfologiche sopra taluni organi della Lemna minor*, Dura, Naples.
Gilbert, J.H. 1871. Notes on "clover-sickness". *Journal of the Royal Horticultural Society* **3**: 86-95.
Gleïzes, J.-A. 1840. *Thalysie, ou, La Nouvelle Existence*. L. Delessart, Paris.
Gortinskii, G.B. 1966. 'Allelopathy and experiments of Soviet studies from the early thwentieth century' [in Russian]. *Byulletin Moskovski Obshchestva Ispytarelei Prirodi, Otdeleniye Biologicheskii* **71**(5): 128-133.
Gyde, A. 1845. On the radical excretions of plants. *Transactions of the Highland and Agricultural Society of Scotland* **1**: 75-82.

Gyde, A. 1847. On the radical excretions of plants. *Transactions of the Highland and Agricultural Society of Scotland* **2**: 273-292.
Heath, F.G. 1881. *Autumnal Leaves*. Sampson Low, London.
Hedwig, J. 1782. *Fundamentum Historiae Naturalis Muscorum Frondosorum*. Siegfried Lebrecht Crusius, Leipzig.
Heuzé, G. 1862. *Les Assolements et les Systèmes de Culture*. L. Hachette et Cie., Paris.
Hooker, J.D. and Thomson, T. 1855. *Flora Indica: A Systematic Account of the Plants of British India*. W. Pamplin, London.
Jaeger, G. 1884. *Die Seele der Landwirtschaft, oder die Lehre vom Dünger, der Bodenmüdigkeit und den stofflichen Bedingungen des Pflanzentriebs*. Ernst Günthers Verlag, Leipzig
Jaeger, G. 1885. *Entdeckung der Seele*. Dritte stark vermehrte Auflage. Volumes 1-2. Ernst Günthers Verlag, Leipzig.
Jaeger, G. 1895. Über Ermüdungsstoffe der Pflanzen. *Berichte der Deutschen Botanischen Gesellschaft* **11**: 70-72.
Johnson, C.W. 1848. *British Husbandry; Exhibiting the Farming Practice in Various Parts of the United Kingdom*. Volume 2. Robert Baldwin, London.
Johnson, G.W. 1845. *The Principles of Practical Gardening*. Robert Baldwin, London.
Johnson, S.W. 1868. *How Crops Grow*. Orange Judd & Company, New York
Johnston, J.F.W. 1844. *Lectures on Agricultural Chemistry and Geology*. William Blackwood and Sons, Edinburgh.
Johnston, J.F.W. 1847. *Lectures on Agricultural Chemistry and Geology*. Second edition. William Blackwood and Sons, Edinburgh.
Knight, T.A. 1841. *A Selection from Mr. Knight's Physiological and Horticultural Papers*. Longman, Orme, Brown, Green, and Longmans, London.
Ladame. 1845. Quelques observations sur l'épuisement des sols par le culture. *Bulletin de la Société des Sciences Naturelles de Neuchâtel* **1**: 197-200.
Lawes, J.B. and J.H. Gilbert. 1860. Report of experiments on the growth of red clover by different manures. *Journal of the Royal Agricultural Society* **21**: 178-200.
Legard, 1841. *Report of the Yorkshire Agricultural Society* **1841**: 136.
Lévy, P. 1869. Note sur les lianes. *Bulletin de la Société Botanique de France* **16**: 279-285. An English translation appeared in the *Gardeners' Chronicle* March 19, 1870, pp. 383-384.
Liebig, J. 1840. *Organic Chemistry in its Applications to Agriculture and Physiology*. Taylor and Walton, London.
Liebig, J. 1865. *Die Chemie in ihrer Anwendung auf Agricultur und Physiologie. Band I. Der Chemische Process der Ernährung der Vegetabilien*. F. Vieweg, Braunschweig
Lindley, J. 1833. On the principal questions at present debated in the philosophy of botany. *Report of the British Association for the Advancement of Science* **3**: 27-57.
Lindley, J. 1840. *The Theory of Horticulture: or an Attempt to Explain the Operations of Gardening upon Physiological Principles*. Longman, Orme, Browne, Green, and Longmans, London.
Lindley, J. 1848. *An Introduction to Botany*. Fourth edition. Longman, Brown, Green, and Longmans, London.
Lindley, J. 1855. *The Theory and Practice of Horticulture: or, an attempt to explain the chief operations of gardening upon physiological principles*. Longman, Brown, Green and Longmans, London.
Link, D.H.F. 1807. *Grundlehren der Anatomie und Physiologie der Pflanzen*. J.F. Donckwerts, Göttingen.
Loudon, J.C. 1838. *Arboretum et Fruticetum Britannicum*. Longman, Orme, Brown, Green, and Longmans, London.
Loudon, J.W. 1841. *The Ladies' Companion to the Flower Garden, being an Alphabetical Arrangement of All the Ornamental Plants Usually Grown in Gardens and Shrubberies*. William Smith, London.
Low, D. 1834. *Elements of Practical Agriculture, Comprehending the Cultivation of Plants, the Husbandry of the Domestic Animals, and the Economy of the Farm*. Bell & Bradfute, Edinburgh.
Lyell, C. 1849. *A Second Visit to the United States of North America*. John Murray, London.
McWhorter, Douglas, Bryant, and Schroeder. 1874. Discussion. *Transactions of the Illinois State Horticultural Society*, New Series **8**: 66-67.
Masters, M.T. 1874. The battle of life among plants. *Popular Science Review* **12**: 35-45.
Meyen, F.J.F. 1838. *Neues System der Pflanzen-Physiologie*. Volume 2. Haude und Spanersche Buchhandlung, Berlin

Mitchell, J.K. 1849. *On the Cryptogamous Origin of Malarious and Epidemic Fevers*. Lea and Blanchard, Philadelphia. Also published in 1859 in: *Five Essays*. J.B. Lippincott & Co., Philadelphia.
Moldenhawer, J.J.P. 1820. *Beytrage zur Anatomie der Pflanzen*. C.L. Waser, Kiel.
Molisch, H. 1887. Uber Wurzelausscheidungen und der Einwirkung auf organische Substanzen. *Sitzungsberichte der Kaiserlichen Akademie der Wissenschaften in Wien. Mathematisch-Naturwissenschaftliche Classe. Abteilung 1* **96**: 84-109.
Moretti, G. 1845. Disamina delle oppozioni fatte del signor Macaire di Ginevra ad alcuni capi della Memoria del dottore Trinchinetti sulle funzioni delle radici dei vegetabili. *Giornale del'I. R. Istituto Lombardo di Scienze, Lettere e Arti* **10** (28/29): 19-21.
Murray, J. 1822a. On the physiology of roots in plants. *Edinburgh Philosophical Journal* **7**: 328-331.
Murray, J. 1822b. Further remarks connected with the physiology of the fibres of roots. *Edinburgh Philosophical Journal* **8**: 37-41.
Murray, J. 1833. *The Physiology of Plants*. J. Murray, London.
Murray, J. 1838. *The Economy of Vegetation*. Raffe and Fletcher, London.
Nietner, T. 1839. Kurzer Umriss der Rotation oder des Wechsels der Pflanzen. *Verhandlungen des Vereins zur Beförderung des Gartenbaues in den Preuss. Staaten* **14**: 158-162.
Owen, D.D. 1861. *Fourth Report of the Geological Survey of Kentucky, Made During the Years 1858 and 1859*. J.B. Major, Frankfort, Kentucky
Payen, A. 1835. Mémoire sur la composition chimique des raciness des plantes et l'action du tannin sur ces organes. *Annales des Sciences Naturelles. Partie Botanique* **3**: 5-20
Payen, A. 1842. *Mémoires sur les Developpements des Végétaux*. Imprimerie Royale, Paris.
Peters, R. 1808a. Departure of the southern pine timber, a proof of the tendency in Nature to a change of products on the same soil. *Memoirs of the Philadelphia Society for Promoting Agriculture* **1**: 27-40.
Peters, R. 1808b. Note re: Letter from Rembrandt Pearl. *Memoirs of the Philadelphia Society for Promoting Agriculture* **1**: 300.
Phin, J. 1862. *Open Air Grape Culture: A Practical Treatise on the Garden and Vineyard Culture of the Vine, and the Manufacture of Domestic Wine*. C.M. Saxton, New York.
Plues, M. 1863. *Rambles in Search of Wild Flowers, and How to Distinguish Them*. Journal of Horticulture and Cottage Gardener Office, London.
Reinizter, F. 1893. Über Ermüdungsstoffe der Pflanzen. *Berichte der Deutschen Botanischen Gesellschaft* **9**: 531-537.
Reveil, P.O. 1865. *Recherches de Physiologie Végétale de l'Action des Poisons sur les Plantes*. Adrien Delahaye, Paris.
Rice, E.L. 1983. *Pest Control with Nature's Chemicals: Allelochemics and Pheromones in Gardening and Agriculture*. University of Oklahoma Press, Norman.
Richard, A. 1828. *Nouveaux Éléments de Botanique et Physiologie Végétale*. Quatrième édition. Béchet Jeune, Paris.
Richard, A. 1838. *Nouveaux Éléments de Botanique et Physiologie Végétale*. Sixième édition. Béchet Jeune, Paris.
Richard, A. 1852. *Précis de Botanique et Physiologie Végétale*. Béchet Jeune, Paris.
Roché, M.-E.-E-H. 1862. *De l'Action de Quelques Composés du Regne Minéral sur les Végétaux*. Paris.
Rosny, L. de. 1852. Rapport sur l'assolement triennal et l'assolement quadriennal. *Compte Rendu de la Séance Publique de la Société d'Agriculture, des Sciences et des Arts de Boulogne-sur-Mer* **1852** (20 mars): 16-29.
Ruffin, E. 1839. Application of the principles of the rotation of crops. *Farmer's Register (Shellbanks, Virginia)* **7**: 609-613.
Ruffin, E. 1855. New views on the theory and laws of rotation of crops, and their practical application. *The Southern Planter, Devoted to Agriculture, Horticulture, and the Household Arts* **12** (10): 289-305.
Ryland, T.G. 1843. On the mode of growth of *Monotropa Hypopitys*. *The Phytologist: a Popular Botanical Miscellany* **2**::329-330.
Sacc, F. 1846-1847. L'épuisement des sols. *Bulletin de la Société des Sciences Naturelles de Neuchâtel* **2**: 68-86
Schrader, J.C.C. and J.S.B. Neumann. 1800. *Zwei Preisschriften über die eigentliche Beschaffenheit und Erzeugung der erdigen Bestandtheile in den verschiedenen innländischen Getreidearten*. Friedrich Manrer, Berlin.

Schleiden, M.J.. 1845. Pflanzenchemie. *Neue Jenaische Allgemeine Literatur-Zeitung* **4**(162-164): 645-653.
Schleiden, M.J. 1849. *Principles of Scientific Botany; or, Botany as an Inductive Science*. Longman, Brown, Green, and Longmans, London.
Schmalz, F. 1840. *Theorie des Pflanzenbaues mit Beispielen and der Erfahrung im Grossen erläutert und bestätigt*. Gebrüder Bornträger, Königsberg.
Silvestre 1833 (see Payen 1835, 1842)
Smith, J.A. 1843. *Productive Farming; or a familiar digest of the recent discoveries of Liebig, Johnston, Davy, and other celebrated writers on vegetable chemistry; showing how the results of tillage might be greatly augmented*. D. Appleton & Co., New York. (also published in conjunction with Falkner 1847).
Sprengel, C.S. 1831-1832. *Chemie für Landwirthe, Forstmänner und Cameralisten*. Vandenhoeck und Ruprecht, Göttingen.
Sprengel, C. 1839. *Die Lehre vom Dünger oder Beschreibung aller bei der Landwirthscaft gebräuchlicher vegetabilischer, animalischer und mineralischer Düngermaterialen, nebst Erklärung ihrer Wirkungsart*. Immanuel Müller, Leipzig.
Sprengel, C. 1845. *Die Lehre vom Dünger oder Beschreibung aller bei der Landwirthscaft gebräuchlicher vegetabilischer, animalischer und mineralischer Düngermaterialen, nebst Erklärung ihrer Wirkungsart. 2. vermehrte und verbesserte Ausgabe*. Immanuel Müller, Leipzig.
Squarey, C. 1842. *Popular Treatise on Agricultural Chemistry: Intended for the Use of the Practical Farmer*. Ridgway, London.
Stephens, H. 1853. *The Farmer's Guide to Scientific and Practical Agriculture*. Leonard Scott, New York.
Stickney, J.S. 1881. Timber culture. *Transactions of the Wisconsin State Horticultural Society* **11**: 156-168.
Stoklasa, J. and A. Ernest. 1909. Beiträge zur Losung der Frage der chemischen Natur der Wurzelsekrets. *Jahrbücher für wissenschaftlichen Botanik* **46**: 55-102.
Strindberg, A. 1907. *En Blå Bok. I*. Björck & Börjesson, Stockholm.
T.A. 1845. Rotation of crops. *Gardener's Chronicle* **5**: 159.
T.T. 1842. Beech-trees. *Gardener's Chronicle* **2**: 253.
Thoreau, H.D. 1860. *The Succession of Forest Trees. New York Weekly Tribune* **1860** (6 October).
Thoreau, H.D. 1993. *Faith in a Seed: The Dispersion of Seeds and other Late Natural History Writings* (ed. B.P Dean). Island Press, Washington D.C.
Thorp, W. 1842. On the failure of red clover. *Journal of the Royal Agricultural Society* **3**: 326-36.
Treviranus, L.C. 1835-1838. *Physiologie der Gewächse*. Adophus Marcus, Bonn.
Trimmer, J. 1842. *Practical Chemistry for Farmers and Landowners*. J.W. Parker, London.
Trinchinetti, A. 1843. Sulla facolta' assorbente delle radici de vegetabili. *Giornale dell'I. R. Istituto Lombardo di scienze, lettere e arti.* **7** (19): 21-83.
Unger, F. 1836. *Über den Einfluss des Bodens auf die Vertheilung der Gewachse nachgeweisen in der Vegetation des nordostlichen Tirol's*. Rohrmann und Schweigart, Vienna.
Uslar, J.L. von 1844. *Die Bodenvergiftung durch die Wurzel-Ausscheidungen der Pflanzen als vorzüglichster Grund für die Pflanzen-Wechsel-Wirthschaft*. Georg Blatt, Altona.
Uslar, J.L. von 1852. *Die Wurzeln der Pflanzen oder die Bodenvergiftung durch die Wurzel-Ausscheidungen der Pflanzen als vorzüglicher Grund für die Pflanzen-Wechsel-Wirthschaft*. Robert Kittler, Hamburg.
Walser, E. 1838. *Untersuchungen über die Wurzel-Ausscheidung*. Gustav Bähr, Tübingen.
Walser, E. 1840. Recherches sur les sécretions des racines. *Annales des Sciences Naturelles, Seconde Série, Botanique* **14**: 100-119.
Waring, G.E. 1854. *The Elements of Agriculture: a Book for Young Farmers*. Orange Judd Company, New York.
Way, J.T. 1847. On the fairy-rings of pastures, as illustrating the use of inorganic manures. *Journal of the Royal Agricultural Society* **7**: 549-552.
Webster. 1893. Question Box. Question No. 1.-Can crops of any kind be grown near to small fruits without injury to them? *Transactions of the Illinois State Horticultural Society, new series* **27**: 243-244.
Weinreich, H. 1993. *Duftstoffe-Theorie: Gustav Jaeger (1832-1917) Vom Biologen zum "Seelenriecher"*. Wissenschaftliche Verlagsgesellschaft, Stuttgart

Westerhoff, R. 1859. *Verhandling over de kol- of hekselringen, ool wel tooverkringen genaamd; voorgelzen in de vergadering van het Genootschap ter Bevördering der Natuurundige Wetenschappen te Groningen, op Woensdag den 3 Februarij 1859, door R. Westerhoff*. C.M. van Hoitsema, Groningen.

Wiegmann, A.F. and Polstorff, L. 1842. *Ueber die anorganischen Bestandtheile der Pflanze, oder Beantwortung der Frage: Sind die anorganischen Elemente, welche sich in der Asche der Pflanzen finden, so wesentliche Bestandtheile des vegetabilischen Organismus, dass dieser sie zu seiner völligen Ausbildung bedarf, und werden sie den Gewächsen von Aussen dargeboten?* Friedrich Viewig und Sohn, Braunschweig

Wiegmann, A.F. and Polstorff, L. 1843. Ueber die anorganischen Bestandtheile der Pflanze, oder Beantwortung der Frage: Sind die anorganischen Elemente, welche sich in der Asche der Pflanzen finden, so wesentliche Bestandtheile des vegetabilischen Organismus, dass dieser sie zu seiner völligen Ausbildung bedarf, und werden sie den Gewächsen von Aussen dargeboten? *Flora* **26**: 21-35.

Willis, J.J. 1894a. Failure in cucumber culture. *Gardener's Chronicle, 3rd series* **16**: 98-99.

Willis, J.J. 1894b Peppermint culture. *Gardener's Chronicle, 3rd series* **16** (November 17 1894): 594.

Willis, R.J. 1996. The history of allelopathy. 1. The first phase 1785-1845: the era of A.P. de Candolle. *Allelopathy Journal* **3**: 165-184.

Willis, R.J. 2000. *Juglans* spp., juglone and allelopathy. *Allelopathy Journal* **7**: 1-55.

Willis, R.J. 2004. *Justus Ludewig von Uslar, and the First Book on Allelopathy*. Springer, Dordrecht.

Wolff, E.T. 1847. *Die chemischen Forschungen auf dem Gebiete der Agricultur und Pflanzenphysiologie*. Joh. Ambrosius Barth, Leipzig.

Wray, L. 1848. *The Practical Sugar Planter: A Complete Account of the Cultivation and Manufacture of the Sugar-Cane*. Smith, Elder & Co., London.

Young, A. 1804. *The Farmer's Calendar*. R. Phillips, London.

CHAPTER 9

SPENCER PICKERING, AND THE WOBURN EXPERIMENTAL FRUIT FARM, 1894-1921

> All men who hitherto expressed opinions on this point have been entirely wrong.
>
> S.U. Pickering as quoted by Hall (1920)

BACKGROUND

One of the key figures in the revival of interest in allelopathy in the twentieth century was the Englishman, S. U. Pickering (Willis 1994, 1997). Percival Spencer Umfreville Pickering[1] (1858-1920; Figure 9.1) was born into an upper middle-class family, and as a youth he had the luxury of pursuing his interest in chemistry within a private laboratory at home. He eventually attended Oxford and had a relatively distinguished academic career, which culminated in an academic appointment in chemistry at Bedford College, Oxford, and ultimately in his becoming Professor of Chemistry there in 1886. In 1878 Pickering had lost his right eye, which he had initially damaged as a youth in a chemistry accident.[2] In any case, Pickering suffered continuing poor health, and he also became disillusioned through the indifferent reception of his chemical research, which focused largely on the nature of aqueous solutions. He resigned from Bedford College in 1887, although he maintained a private interest in chemistry until about 1896. He recuperated routinely in the country at Harpenden in Hertfordshire, and to relieve any sense of idleness, he became a part-time labourer at the nearby Rothamsted Agricultural Station. He eventually decided that the lifestyle suited him, and he bought a small property in Harpenden in 1885, where he learned the rudiments of farming and horticulture, and which became his permanent home from 1902 onwards.

---

[1] Although christened as such, Pickering never used the name Percival.
[2] It was widely reported that the loss of his eye occurred due a second chemistry accident (e.g. H., A. or Harden 1926), but his colleague and friend E.J. Russell revealed that knowledge of the real cause, an errant tennis ball (R, E.J. or Russell 1921), was suppressed likely out of vanity. Similarly, portraits of Pickering usually show only his left profile.

Pickering's experiences presented him with the idea of establishing an experimental farm for scientific research on the growth of fruit trees. Pickering did not have the resources to carry this out himself, but he had been a schoolmate with Herbrand Arthur Russell (1858-1940, Figure 9.2), who had succeeded as the 11th Duke of Bedford in 1893. Bedford inherited a vast estate at Woburn surrounded by 11 miles of wall, had a huge private income of some £200,000 per annum, and maintained a strong interest in natural history, albeit mainly zoology (Bedford 1959). The Duke was a disdainful and aloof individual, and perhaps is remembered most favourably because of his flamboyant wife who took up aviation at the age of 62 and was endearingly known to the public as the "Flying Duchess". Nonetheless, the Duke had a strong and historical sense of public duty, and he agreed to grant Pickering a parcel of 20 acres at Woburn to establish the first experimental farm in England, indeed the world, devoted solely to the study of fruit trees, and furthermore agreed to pay all costs. The Woburn Experimental Fruit Farm was thus born in June 1894. Within a year, a weedy paddock had been transformed into an experimental farm containing 500 experimental plots for fruit trees, a manager's house, nursery, strawberry beds and hedges (Anonymous 1895; Figure 9.3). The fruit farm is not to be confused with the nearby Woburn Experimental Farm, which had been established in 1876 by the 9th Duke of Bedford.

The results of Pickering's work over the next 25 years appeared mainly in a series of eighteen reports[3] that were issued by the Woburn Experimental Fruit Farm. Pickering published his most interesting findings also in scientific journals, and reprints of these, as available, were commonly appended to the Woburn reports. As the research was funded by Bedford, his name usually appears as first author on most of the reports, although he considered it beneath his station in life to actually soil his hands. The farm itself had a manager and several labourers; however, Pickering was the solitary researcher at the Fruit Farm (see Figure 9.4). As with his chemistry research, his work was characteristically meticulous, and often went against the mainstream, despite the cost[4]. Hall (1920) noted that Pickering seemingly prided himself in his unorthodoxy and had "a disconcerting habit of making discoveries which contradicted the common form." It seems that Pickering and Bedford had a mutual respect; they both were austere, dedicated and somewhat anachronistic gentlemen. In 1919 Bedford and Pickering co-authored a book, *Science and Fruit Growing*, that summarised the work with fruit species at Woburn. During the heydays of the Woburn Experimental Fruit Farm, Pickering worked seemingly tirelessly, and a document printed in 1904 indicated that he had 960 different experiments in progress at that time (Woburn Experimental Fruit Farm, 1904).

---

[3] The Eighteenth Report, published in 1921, was posthumous and edited by E.J. Russell.

[4] Russell (1921) related that Pickering, while a student at Oxford, unwisely published a chemistry paper that directly contradicted the work of his tutor.

Figure 9.1. A photographic portrait of Percival Spencer Umfreville Pickering (1858-1920) taken by Walter Stoneman in 1917 (NPG Acqn. No. x43876). Courtesy of the National Portrait Gallery, London.

Figure 9.2. Photograph of the Eleventh Duke of Bedford from Science and Fruit Growing (Bedford and Pickering 1919).

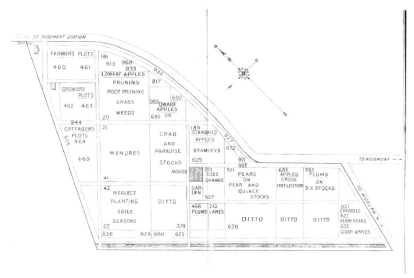

Figure 9.3. Plan of the experimental plots at the Woburn Experimental Fruit Farm as of 1904 (Woburn Experimental Fruit Farm 1904).

The last few years of the fruit farm were difficult ones, as the First World War greatly restricted resources. Furthermore, the Duke of Bedford had invested nearly £2,000,000 in imperial Russian bonds, and the Russian Revolution in 1917 rendered these worthless (Bedford 1959). By 1918 the Duke of Bedford was forced to withdraw his financial support; however, a benefactor in the form of the Lawes Agricultural Trust, which funded the Rothamsted Experiment Station at Harpenden under the direction of Pickering's long-time colleague E.J. Russell, agreed to fund Pickering such that he could finish his experimental work. Pickering's health was declining, and his death in December 1920 brought a conclusion to the Woburn Experimental Fruit Farm. Pickering's pride and his frustration in failing to achieve greater recognition for his work during his lifetime became particularly evident after death. While he had shunned the camera during his lifetime, his final funeral rights were recorded on movie film, and he was buried on the Devonshire coast with his head facing the dawn (Topical Film Company 1920). Furthermore, the terms of his will provided funds for a memorial volume, celebrating his life's work in chemistry and horticulture, which was eventually published in 1927 (Lowry and Russell 1927).

Pickering's work at the Woburn Experimental Fruit Farm covered numerous aspects relating to fruit culture, including planting, pruning, soil conditions, manuring, and diseases and pests. Three facets of Pickering's work relate to allelopathy: 1) the effect of grass cover on the growth fruit trees, 2) the effect of various crops on other crops, and 3) the effect of heat on soils. A summary of the work can be found in Bedford and Pickering (1919) and Lowry and Russell (1927).

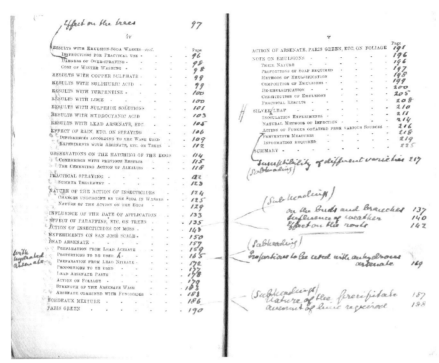

Figure 9.4. Pages from Pickering's copy of the proofs of the Sixth Report of the Woburn Experimental Fruit Farm *(1906), with corrections and additions by Pickering.*

THE EFFECT OF GRASS ON FRUIT TREES

The negative effect of grassing on fruit trees seems to be have been widely experienced by various fruit farmers in temperate climates, but was rarely investigated. For example, a discussion at Northern Illinois Horticultural Society in 1875 revealed that both timothy and blue grass were thought deleterious to apple trees (Anonymous 1875). Pickering's work on the effect of grass on young fruit trees was first reported in 1897, and again in 1900, when it was observed that a grass (a mixture of *Alopecurus pratensis, Cynosurus cristatus, Dactylis glomerata, Festuca duriuscula, Phleum pratense, Poa nemoralis* and *Lolium perenne*) undergrowth, even in comparison with weeds, caused both discoloration and stunting of young apple trees (Bedford and Pickering 1897, 1900). The effect was most pronounced during the first few years, and then the trees recovered gradually (Figure 9.5). Different varieties of apple trees normally showed an increase of growth ranging between 896-1200% after four years, but grass-affected trees showed an increase of only 56-64%. The cause of the effect was unknown at this stage, although it was suggested firstly that the grass was accelerating evapotranspiration and causing moisture stress, and then that it was possibly restricting aeration of the soil. Another result, which had parallels with the work that was to emanate from the USDA Bureau of Soils, was that manuring had little effect on tree growth, at least in the first few years.

The hypothesis that the effect of grass on apple trees was allelopathic was first raised in 1903 (Bedford and Pickering 1903, Pickering 1903). Pickering also suggested, with typical caution, that the effect may not be direct poisoning, but may be some indirect effect, where, for example, toxic substances arise through bacteria or affect bacteria beneficial to the trees (Pickering 1907a). These ideas were again discussed in the Fifth Report (Bedford and Pickering 1905), and Pickering became further convinced that the inhibitory effects were due to some toxin. He reported that grassed apple trees showed different autumn coloration, and also tended to yield more fruit, an outcome common in inhibited fruit trees. Remarkably, at first, Pickering could find no evidence of toxic substances in leachates from the grass or grassy soil (1907b).

Figure 9.5. Advanced seedlings of Bramley variety apple trees grown with a grass cover (left) and without (right). Reproduced from Bedford and Pickering (1900).

Experiments were established that demonstrated that the effect varied considerably with different cultivars of apple, and with other *Rosaceae* fruit trees such as plum, cherry and pear (Bedford and Pickering 1911, Pickering 1913). Pickering reported that in such experiments, most plants during the very early stages of grassing actually do better, but then growth becomes progressively inhibited (Pickering 1912). In most cases, there was a marked inhibition of vegetative growth and crop yield (Bedford and Pickering 1911). Pickering also found that grassy cover inhibited coniferous trees. Other experiments demonstrated that tree growth was inhibited regardless of at what stage of tree growth the grass was added. There seemed to be a correlation between severity of inhibition and proximity of the grass cover. The effects of grass cover were assessed in different soils, for example in a deeper soil at Harpenden, and a sandier soil from Milbrook, and in general the effects were greater in heavier soils such as at Woburn. Pickering assessed the effects of different individual grasses, and the most inhibitory were the robust grasses such as *Festuca pratensis*, *Alopecurus pratensis Lolium* spp. and *Dactylis glomerata*, although all had a negative effect (Bedford and Pickering 1911). An interesting finding was that the addition of clover to the grass cover made little difference to inhibition of growth, but alleviated the typical yellowing of the foliage, evidently by supplying nitrogen. Pickering tested numerous hypotheses regarding the cause of the inhibition, including high concentrations of carbon dioxide, but all his work seem to point toward the cause being a toxin, that under favourable conditions was readily oxidised (Pickering 1920). Pickering devised a number of leachate experiments that demonstrated that the inhibitory effects were associated with substances emanating from the grass roots (Bedford and Pickering 1911); this work led eventually to the somewhat more elegant methodology described below.

THE EFFECT OF VARIOUS CROPS ON OTHER CROPS

The extensive experimentation with the effect of grass on trees led eventually to exploring the general idea that virtually any plant may have an inhibitory effect on another plant (Pickering 1917, 1919). Pickering set up a number of experiments in which one variety of plant was grown in a perforated annular ring pot, which fitted atop a conventional pot with plants grown within the center (Figure 9.6). In this way the effects of leachates of clover were tested on the growth of tobacco, tomatoes, and mustard, and the effects of *Festuca pratensis*, clover, mustard and *Dactylis glomeratus* on themselves. Pickering even reported that apple seeds had a negative effect on grass growth, albeit small (Pickering 1920).

With regard to the effect of one plant on another, Pickering summarised it as follows:

> The nature of the action appears to be substantially that which we have long held it to be – a toxic action – and it now proves to be but a special instance of a general phenomenon, which must have a wide bearing on the question of the growth of all plants, and on that of soil fertility in general. Briefly, the explanation is that a toxic substance is formed in the soil by the growth of any plant in it, originating probably in the decomposition of the debris of roots during growth, this substance being toxic, not only to other plants of a different nature but to the plants themselves which form it: but, by the action of air and moisture, probably with the assistance of bacteria, this toxin becomes oxidised and converted into plant food, thus eventually rendering the soil more fertile than it would be in the absence of vegetation. In time, therefore, the deleterious action of grass, or any other crop, becomes converted into a beneficial action, which takes due effect on any plants growing in the soil unless these have been so injured by the previous toxic action as to be beyond recovery. In the case of hard-wooded plants, such as trees, where recovery from stunting is very difficult, the injury is generally permanent, but with soft-wooded plants recovery usually occurs.
>
> (Bedford and Pickering 1914, pp. 46-47)

Tobacco.
Without grass in trays.   With grass in trays.

Tobacco.
Without clover in trays.   With clover in trays.

Festuca pratensis.
Festuca in trays.   No festuca in trays.

Figure 9.6. The effect of various plants on other plants or the same plant. Top: the effect of grass (Festuca pratensis) on tobacco. Middle: the effect of clover on tobacco. Bottom: the effect of Festuca pratensis *on itself.* (Bedford and Pickering 1914b).

Thus Pickering took his toxin theory to the extreme in claiming that every plant continuously produces toxic substances. He claimed to have demonstrated this in a series of experiments in which plants were grown in a fixed volume of soil, either in isolation or in root contact with one another (Bedford and Pickering 1919, Pickering 1920). When all plants were of equal age there were no differences in the plants' growth; however, when plants were interplanted with a second, delayed group of plants, there was an unaccountable relative decrease in growth of the second group of plants, which Pickering unconvincingly interpreted as due to the large amount of toxin produced by the older plants. He added that the theory also helps explain the commonly observed phenomenon that the outer rows of crops often show greater growth.

THE EFFECT OF HEAT ON SOILS

In the course of attempting to elucidate the nature of inhibitory soils, Pickering conducted a number of experiments in which soils were heated to temperatures ranging from 60° to 200°C, or were sterilised with organic solvents (Pickering 1907a, 1907b, 1907c, 1908, 1910a, 1910b, Bedford and Pickering 1908, 1911, 1919). Russell and his co-workers at Rothamsted, and workers at the USDA Bureau of Soils had pioneered similar work, and Pickering was particularly impressed with the concept of elevated temperatures or chemical treatments having the capacity to kill selectively the soil bacteria and protozoa that preyed on them.

Generally, Pickering found that heating or sterilizing the soil caused the soil to become inhibitory to seed germination and plant growth, which he interpreted as being the result of the accumulation of a soil toxin through breakdown of organic matter, due to the diminished soil microflora. He regarded this as evidence that the production of toxins in soil is a normal phenomenon that is counterbalanced by the actions of the soil microflora. Further evidence was provided, as heated soil starved of oxygen retained its toxic character longer (Pickering 1910a). Soil that had been heated or sterilised, whilst initially inhibitory, lost its inhibitory properties and frequently became strongly growth promoting after several weeks, evidently due to an increase in soluble nutrients and the gradual loss of toxicity through oxidation. For example, the growth of three grass species in soil heated to 125°C eventually increased in growth by a mean of 321% compared to unheated soil, although the soil was initially inhibitory (Pickering 1910b). It is interesting that grasses responded very favourably to soils heated up to 150°C, whereas dicots showed maximal growth with soils heated at 100°C and were inhibited in soils heated at higher temperatures (Pickering 1910b). Pickering was unable to identify the toxin(s) generated by heating soil, but he characterised it as non-acidic and perhaps nitrogenous, and not the compounds (e.g. dihydroxystearic acid, picoline carboxylic acid) isolated by Schreiner and his co-workers at the USDA Bureau of Soils (Pickering 1907c, Bedford and Pickering 1919). Pickering actually acquired some of the infamous Takoma soil from Milton Whitney, and he described it as peculiar, in having an unusually high proportion of insoluble organic matter, and in generating an offensive odour and becoming distinctly poisonous on heating (Pickering 1908).

IMPACTS

Over the years, Pickering's results have been largely forgotten, although the integrity of his work has never been in question. However, there are several facets of Pickering's endeavour that caused it to have less impact than he might have hoped. Firstly, the soil at Woburn is essentially a shallow sandy loam over heavy clay, and one that is rarely used commercially in the district for the production of fruit. Secondly, Pickering tended to use apple trees that were different from those used by most commercial growers; Pickering's apple varieties were commonly grafted onto clonal "Paradise" rootstock of uncertain origin, whereas the commercial growers tended to use a hardier seed-raised crabapple or free stock, which was evidently less affected by a grassy cover (Lowry and Russell 1927). Finally, Pickering never had any great rapport with the local fruit growers, who obviously had immense practical experience. It seems that Pickering and Bedford were tolerated, but the farmers took exception to being told to change their ways, especially by an Oxford don. When closure of the Woburn Experimental Fruit Farm became imminent in 1918, the fruit farmers did nothing to prevent its demise. One correspondent to a local farm journal wrote, assumedly following the death of Pickering: "While regretting the cause, the writer was glad on the whole that the station was closed." (Lowry and Russell 1927). As strange as it may sound, it was the Russian Revolution that precipitated the demise of a remarkable British episode in the history of allelopathy.

Pickering's experiments and theories attracted considerable attention at the time, both in England and abroad. A prolix and laudatory review of *Science and Fruit Growing* by an Oxford don (Keeble 1920) only highlighted the gulf between academic and practical horticulture. As indicated above, Pickering was received by the local farming community with deference. The effect of grass on apple trees was tested by some other fruit-growing institutions, e.g. the National Fruit and Cider Institute at Long Ashton (Bristol), the Fruit and Cider Institute in Gloucestershire, and the Harper-Adams College (see Bedford and Pickering 1919, p. 307-8), and while results were not as dramatic as those seen at Woburn, there was, nonetheless, a reduction in growth. Researchers at the nearby Rothamsted Experimental Station, including A.D. Hall[5] and E.J. Russell, had great respect for Pickering, but were never fully convinced by his findings, as they could not be duplicated elsewhere (Anonymous 1912, Brenchley 1917, Hall 1908, Hall *et al.* 1913, Russell, Russell 1912a, Russell 1912b, Russell 1914a, Russell 1914b, Russell 1918, Russell 1936). This must have been exasperating for Pickering as Russell, in particular, was often invited to observe or comment on Pickering's experiments at Woburn. The Rothamsted Experimental Station, in particular, had a long-term history of investigating de Candollean ideas of crop rotation (see e.g. Lawes and Gilbert 1889), and generally had found little evidence in their favour[6]. Although Pickering's experiments were carefully designed, his results were

---

[5] A.D. Hall (1864-1942) was Director at Rothamsted from 1902-1912, and was succeed by E.J. Russell (1872-1965) who was Director from 1912-1942.

[6] Despite this, de Candolle's theories still were to be found in textbooks of Pickering's time. For example Martin (1919) wrote, "It seems in many cases the deleterious substances are more poisonous to the roots of plants of the same kind, and this may help explain the value of crop rotation."

criticised for being too variable or inconclusive (B.[7] 1921, Johnson 1919, Lowry and Russell 1927).

There were some attempts to repeat Pickering's experiments overseas, especially those concerned with the effect of grass on trees. The most noteworthy of these investigators was Albert Howard (1873-1947), who served as Economic Botanist, Indian Department of Agriculture, Pusa 1905-1924, and was Director of the Institute of Plant Industry at Indore, India (about 500 km NE of Mumbai [formerly Bombay]), 1925-1931. Howard is chiefly remembered as a pioneer in biodynamic farming, in developing and advocating techniques for manufacturing humic materials from agricultural waste, the so-called Indore Process. While at Pusa, Howard investigated the effect of grasses on fruit trees during the years 1914-1924. It had been found that the grassing of peach, guava, litchi, mango, loquat, lime, custard apple, and plum trees caused severe retardation of growth, and even death in some cases, of the trees, of which the custard apple was the most sensitive and the guava the least sensitive (Howard 1925, 1940). Detailed studies of the root systems, and subsequent measurements of the carbon dioxide levels in the soil volume led Howard to conclude that the chief effect of the grass was asphyxiation of the fruit tree roots, which were generally ill adapted to penetrate the mat of the grass surface roots. The low levels of oxygen also caused nitrogen depletion, as nitrification was inhibited, and this phenomenon was also reported by Lyon *et al.* (1923). Howard reported that grass seemed to cause no obvious effect on forest trees in cultivation, as their root systems were still commonly active when the grasses were dormant, and were much better at competing for oxygen in the surface soil layers. In summary, Howard found no reason to ascribe the inhibitory effects of grasses to a toxin.

Another researcher in India who followed Pickering's work with interest was Fletcher (1907), who, using water culture methods, could not verify Pickering's findings regarding the effects of heat treatment on growth media. His results more closely paralleled those of the United States Department of Agriculture, Bureau of Soils, and he found that the suppressive effects of *Sorghum vulgare* on *Sesamum indicum* could be neutralised by agents such as tannic or pyrogallic acids or carbon black.

Pickering's findings also came under scrutiny in the United States. Alderman and Middleton (1925), at the Minnesota Agricultural Experiment Station, attempted the investigate the effect of ten different crops on tomato, with methodology similar to that used by Pickering, but generally the presence of a crop above the tomato roots led to increased growth and yield. Cubbon (1925) also conducted experiments similar to those of Pickering, but used rye as sod and grape plants as the woody species. Cubbon found a distinct inhibition in grape plant growth in his grassed plots and he discounted soil moisture and nitrate levels as factors. Cubbon concluded that there must be a growth inhibitor associated with the rye plants.

Interest in allelopathic interpretation of plant interaction sparked by Pickering's work was not simply confined to agriculture. In England, Farrow (1917) and Jeffreys (1917) described the inhibitory effect of decomposing bracken (*Pteridium aquilinum*) fronds and pine litter on plants such as *Calluna vulgaris*, *Carex arenaria*, *Nardus*

---

[7] This was B.T.P. Barker, Director of the National Fruit and Cider Institute, Long Ashton.

stricta and *Deschampsia flexuosa*, and Jeffreys demonstrated that the effects could be relatively specific; for example, whereas the grass *Deschampsia flexuosa* was severely inhibited by bracken fronds, a similar grass, *Holcus mollis*, was not. Both Eardley-Wilmot (1910) and Woodruffe-Peacock (1917) revived the old de Candollean notion that plants, as they grow, produce toxic waste products which prove more inimical to their own species, thus fostering the natural succession and mixture of species. Revived interest in toxin theory caused Bayliss (1911, 1926) and Shantz and Piemeistel (1917) to revisit the intriguing "fairy-ring" effect in mushrooms, as raised by Way (1847) and others many years earlier. Bayliss concluded that fungi such as *Marasmius* and *Clitocybe* do produce staling substances or autotoxins that also have the effect of destroying grass roots; however, in the following years there is luxuriant growth due to surplus nutrients. Also in the U.S. it was believed that substances leached from trees may disrupt "fairy rings" (Coville 1898, Reed 1910), but Shantz and Piemeistel (1917) in their extensive studies could find no evidence of autotoxins.

Although Pickering's findings regarding toxins in soils were little studied by others following 1920, they left a greater impact on contemporary ecological thought than is generally recognised. Influential ecologists such as E.J. Salisbury (1929) and C.E. Lucas (1947) accepted the reality of toxins in soils as envisaged by Pickering. Salisbury (1886-1978), who was born in Harpenden and possibly knew Pickering, described in his presidential address to the British Ecological Society in 1929, as a matter of course, how the runners of common creepers such as *Lamiastrum galeobdolon*[8] must negotiate toxins in soil in the spread of the plant. Even Russell, who had been enduringly sceptical of Pickering's findings, but having witnessed the course of many of Pickering's experiments, was forced to admit:

> In consequence we must be prepared to consider possible toxic effects of one plant *growing alongside of it*, and the part such effects may play in determining natural plant associations and in explaining some of the bad effects of weeds. (Russell 1921, p. 247)

REFERENCES

Alderman, W.H. and J.A. Middleton. 1925. Toxic relations of other crops to tomatoes. *Proceedings of the American Society for Horticultural Science* **22**: 307-308.

Anonymous. 1875. Renewing old orchards. *Transactions of the Illinois State Horticultural Society. New Series.* **8**: 182-187.

Anonymous. 1895. The Woburn Experimental Fruit Farm. *Nature (London)* **52**: 508-510.

Anonymous. 1912. Fruit trees in grass land. *The Gardeners' Chronicle*, 13 January 1912 (no. 1307): 17-18.

B. [Barker, B.T.P.] 1921. Toxic root-interference in plants. *Nature (London)* **106**: 666-667. Also in: *Agricultural Journal of India* **16**: 325-329.

Bayliss, J.S. 1911. Observations of *Marasmius oreades* and *Clitocybe gigantea* as parasitic fungi causing fairy rings. *Journal of Economic Biology* **6**: 111-132. check

Bayliss Elliott, J.S. 1926. Concerning fairy rings in pastures. *Annals of Applied Biology* **13**: 277-288.

Bedford, Duke of, 11th [Herbrand Arthur Russell] and Pickering, S.U. 1897. *Report on the Working and Results of the Woburn Experimental Fruit Farm since its Establishment. First Report.* Eyre and Spottiswoode, London.

Bedford, Duke of, 11th [Herbrand Arthur Russell] and Pickering, S.U. 1900. *Report on the Working and Results of the Woburn Experimental Fruit Farm. Second Report.* Eyre and Spottiswoode, London.

---

[8] Formerly known as *Galeobdolon luteum*.

Bedford, Duke of, 11th [Herbrand Arthur Russell] and Pickering, S.U. 1903. *Third Report of the Woburn Experimental Fruit Farm.* Eyre and Spottiswoode, London.
Bedford, Duke of, 11th [Herbrand Arthur Russell] and Pickering, S.U. 1905. *Fifth Report of the Woburn Experimental Fruit Farm.* Eyre and Spottiswoode, London.
Bedford, Duke of, 11th [Herbrand Arthur Russell] and Pickering, S.U. 1908. *Ninth Report of the Woburn Experimental Fruit Farm.* The Amalgamated Press, London.
Bedford, Duke of, 11th [Herbrand Arthur Russell] and Pickering, S.U. 1911. *Thirteenth Report of the Woburn Experimental Fruit Farm.* The Amalgamated Press, London.
Bedford, Duke of, 11th [Herbrand Arthur Russell] and Pickering, S.U. 1914a. *Fourteenth Report of the Woburn Experimental Fruit Farm.* The Amalgamated Press, London.
Bedford, Duke of, 11th [Herbrand Arthur Russell] and Pickering, S.U. 1914b. The effect of one crop on another. *Journal of Agricultural Science* **6**: 136-151. Also reproduced in Appendix to Bedford and Pickering (1914a).
Bedford, Duke of, 11th [Herbrand Arthur Russell] and Pickering, S.U. 1919. *Science and Fruit Growing.* MacMillan and Co. Limited, London.
Bedford, Duke of, 13th [John Arthur Russell]. 1959. *A Silver-plated Spoon.* Cassell, London.
Brenchley, W.E. 1917. The effects of weeds upon cereal crops. *New Phytologist* **16**: 53-76.
Coville, F.V. 1898. The fairy-ring mushroom. *Plant World* **2**(3): 39-41.
Cubbon, M.H. 1925. The effect of a rye crop on the growth of grapes. *Journal of the American Society of Agronomy* **17**: 568-577.
Eardley-Wilmot, S. 1910. *Forest Life and Sport in India.* Edward Arnold, London.
Farrow, E.P. 1917. On the ecology of the vegetation of breckland. V. Observations relating to competition between plants. *Journal of Ecology* **5**: 155-172.
Fletcher, F. 1907. Root action and bacteria. *Nature (London)* **76**: 270.
H., A. [Harden, A.] 1926. Percival Spencer Umfreville Pickering – 1858-1920. *Proceedings of the Royal Society, A.* **111**: viii-xii. This also appears in Lowry and Russell (1927), *The Scientific Work of the late Spencer Pickering, F.R.S.* (ed. Lowry, T.M. and Russell, J.), pp. v – ix. The Royal Society, London.
Hall, A.D. 1908. Agricultural chemistry and vegetable physiology. *Annual Report on Progress in Chemistry* **5**: 242-257.
H., A.D. [Hall, A.D]. 1920. Obituary – Spencer Pickering, F.R.S. *Nature (London)* **106**: 509-510.
Hall, A.D., Brenchley, W.E. and Underwood, L.M. 1913. The soil solution and the mineral constituents of the soil. *Philosophical Transactions of the Royal Society of London B* **204**: 179-200. Also in (1914) in *Journal of Agricultural Science* **6**: 278-301.
Howard, A. 1925. The effect of grass on trees. *Proceedings of the Royal Society of London. Series B* **97**: 284-321.
Howard, A. 1940. *An Agricultural Testament.* Oxford University Press, London.
Jeffreys, H. 1917. On the vegetation of the four Durham coal measure fells. IV. On various other ecological factors. *Journal of Ecology* **5**: 140-154.
Johnson, J. 1919. The influence of heated soils on seed germination and plant growth. *Soil Science* **7**: 1-87.
Keeble, F. 1920. Science and Fruit Growing. *Edinburgh Review* **1920** (2): 263-282.
Lawes, J. and Gilbert, J.H. 1895. The Rothamsted experiments. Being an account of some of the results of the agricultural investigations conducted at Rothamsted, in the field, in the feeding-shed, and the laboratory over a period of fifty years. *Transactions of the Highland Agricultural Society, 5th Series* **7**: 11-354.
Lowry, T.M. and J. Russell. 1927. *The Scientific Work of the late Spencer Pickering, F.R.S., with a Biographical Notice by Prof. A. Harden, F.R.S.* The Royal Society, London.
Lucas, C.E. 1947. The ecological effects of external metabolites. *Biological Reviews* **22**: 270-295.
Lyon, T.L., A.J. Heinicke and B.D. Wilson. 1923. The relation of soil moisture and nitrates to the effects of sod on apple trees. *Memoir Cornell Agricultural Experimental Station* **63**:
Martin, J.N. 1919. *Botany for Agricultural Students.* John Wiley and Sons, Inc., New York.
Pickering, S.U. 1903. The effect of grass on apple trees. *Journal of the Royal Agricultural Society* **64**: 365-376.
Pickering, S. 1907a. Root action and bacteria. *Nature (London)* **76**: 126-127.
Pickering, S. 1907b. Root action and bacteria. *Nature (London)* **76**: 315-316.
Pickering, S.U. 1907c. Studies on germination and plant-growth. *Journal of Agricultural Science* **2**: 411-34. Also reproduced in Appendix to Bedford and Pickering (1908), pp. I-XXIV.

Pickering, S.U. 1908. The action of heat and antiseptics on soils. *Journal of Agricultural Science* **3**: 32-54. Also reproduced in Appendix to Bedford and Pickering (1908), pp. XXV-XLVII.

Pickering, S.U. 1910a. Studies of the changes occurring in heated soils. *Journal of Agricultural Science* **3**: 258-276. Also reproduced in Appendix to Bedford and Pickering (1911).

Pickering, S.U. 1910b. Plant-growth in heated soils. *Journal of Agricultural Science* **3**: 277-84. Also reproduced in Appendix to Bedford and Pickering (1911).

Pickering, S.U. 1912. The effect of grass on trees. *Nature (London)* **89**: 399.

Pickering, S.U. 1913. Horticultural research. III. The action of grass on trees. *Science Progress* **7**: 490-503.

Pickering, S.U. 1917. The effect of one plant on another. *Annals of Botany (London)* **31**: 181-7.

Pickering, S.U. 1919. The action of one crop on another. *Journal of the Royal Horticultural Society* **43**: 372-80.

Pickering, S.U. 1920. *Seventeenth Report of the Woburn Experimental Fruit Farm (under the control of the Lawes Agricultural Trust)*. The Amalgamated Press, London.

R., E.J. [Russell, E.J.] 1921. Percival Spencer Umfreville Pickering – Born 1858; died December 5th, 1920. *Journal of the Chemical Society* **119**: 564-569.

Reed, H.S. 1910. An interesting *Marasmius* fairy ring. Plant World **13**(1): 12-14.

Russell, E.J. 1912a. *Soil Conditions and Plant Growth*. Longmans, Green and Co., London.

Russell, E.J. 1912b. The effect of grass on fruit trees. *Nature (London)* **88**: 486-487.

Russell, E.J. 1914a. The effect of one growing crop on another. In, Bedford, Duke of, 11th [Herbrand Arthur Russell] and S.U. Pickering, *Fourteenth Report of the Woburn Experimental Fruit Farm*, pp. 51-68. The Amalgamated Press, London.

Russell, E.J. 1914b. The nature and amount of the fluctuations in nitrate contents of arable soils. *Journal of Agricultural Science* **6**: 18-57.

Russell, E.J. 1918. The effect of one growing plant on another. *The Gardeners' Chronicle* **63**: 23-24.

Russell, E.J. 1921. *Soil Conditions and Plant Growth*. Fourth Edition. Longmans, Green and Co., London.

Russell, E.J. 1936. Interactions between roots and soils. The growing plant: its action on the soil and its neighbours. In: *Proceedings of the 6th International Botanical Congress*, pp. 1-3.

Salisbury, E.J. 1929. The biological equipment of species in relation to competition. *Journal of Ecology* **17**: 197-222.

Shantz, H.L. and R.L. Piemeistel. 1917. Fungus fairy rings in eastern Colorado and their effect on vegetation. *Journal of Agricultural Research* **11**: 191-245.

Topical Film Company. 1920. *Mr Spencer Pickering, Burried (*sic*) Under Rock on Devonshire Coast with Head Towards the Dawn*. 35mm black and white film, Topical Budget 485-2. Available British Film Institute.

Way, J.T. 1847. On the fairy-rings of pastures, as illustrating the use of organic manures. *Journal of the Royal Agricultural Society* **7**: 549-552.

Willis, R.J. 1994. Pioneers of allelopathy: Spencer U. Pickering (1858-1920). *Allelopathy Journal* **1**: 69-77.

Willis, R.J. 1997. The history of allelopathy. 2. The second phase (1900-1920): The era of S.U. Pickering and the U.S.D.A. Bureau of Soils. *Allelopathy Journal* **4**: 7-56.

Woburn Experimental Fruit Farm. 1904. *Catalogue of Experiments, August 1904*. Timaeus Printer, Bedford.

Woodruffe-Peacock, E.A. 1917. Toxic atrophy. *Quarterly Journal of Forestry* **11**: 88-93.

CHAPTER 10

THE USDA BUREAU OF SOILS AND ITS INFLUENCE

> To suppress what one conceives to be the truth, because it does not accord with the views of colleagues, is an enormity hardly conceivable to liberal-minded men.
>
> *Report of the Committee on President's Address, Association of Official Agricultural Chemists* (1908)

Without doubt, the most tumultuous period in the history of allelopathy was that associated with the United States Department of Agriculture (USDA), and its Bureau of Soils, during the first two decades of the twentieth century. During this period, the academic debate on allelopathy became charged with emotion, the commentary eventually became libelous, and academic and political reputations were on the line, as were the budgets of numerous agricultural institutions and departments.

ORIGINS

The Bureau of Soils, based in Washington, D.C., began as the Division of Agricultural Soils of the Weather Bureau in 1894, as studies of climate were seen as instrumental in understanding soils (Weber 1928). Milton Whitney (1861-1927; Figure 10.1), who was to head the Bureau of Soils, had trained as a chemist at Johns Hopkins University, and had served briefly in various scientific roles in Connecticut, North Carolina, South Carolina and Maryland. Whitney's work on the effects of the properties of soils had begun as early as 1886 during his early years at the Research Farm of the North Carolina Experiment Station, West Raleigh, North Carolina, and he was a pioneer in using pots in this type of research. He joined the Weather Bureau in 1892, and in that year he published a paper, "The Physical properties of soils in relation to moisture and crop distribution", which was to become the cornerstone of his theories on soil fertility. He had accumulated a following of agriculturalists in the eastern U.S. and he had the ear of the Secretary of Agriculture, J. Sterling Morton, and more importantly the Assistant Secretary of Agriculture, Charles Dabney, who had been Whitney's mentor in North Carolina (Helms *et al.* 2002). In 1894, Whitney, aged 34, was appointed as first head of the newly created Division of Agricultural Soils and he continued in office until his death in 1927

(Baker *et al.* 1963, Weber 1928). In 1895, the Soils Divisions was shifted to the Department of Agriculture, and in 1901 it was transformed into an autonomous bureau by the U.S. Department of Agriculture (USDA). Following Whitney's death in 1927, the Bureau of Soils was subsumed into the Bureau of Soils and Chemistry of the USDA, which itself became defunct in 1938.

Milton Whitney was very much the Southern gentleman, and was renowned for his ever-present cigar[1]. He was a persuasive individual, who was just as comfortable talking with land-owners as he was dealing with politicians. The Bureau of Soils had been charged with the responsibility of investigating the relation of soils to climate, and the texture and composition of soils. As a result, the Bureau of Soils is best known today for its extensive and popular series of descriptive soil maps produced beginning in 1899. However, Whitney through his early research on the quantities of minerals in the soil solution, had a strong interest in the causes of soil fertility/ infertility, and began to promote his views, which became increasingly unorthodox, particularly with regard to the role of fertilisers. In 1901 he stated that the principal role of fertilisers was to obtain an improvement in the texture of the soil (Whitney 1901). He showed little hesitation in using the power of his office to advance his ideas, and he doggedly stuck to his views. Staff who questioned the research program were rapidly subordinated or dismissed, and while Whitney was shrewd in recruiting talented scientists, many of the best quickly sought other posts once conflict with Whitney was apparent[2].

Whitney had at his disposal staff and laboratories to investigate the properties and constituents of soil. Amongst the first Bureau researchers in this area were several some very able soil scientists: F. D. Gardner, L. .J. Briggs, T. H. Means, F. K. Cameron and F. H. King. By 1903, The Bureau of Soils had a staff of over 120, including about 80 soil scientists or experts. The Bureau of Soils had a number of briefs, including producing a comprehensive soil map of the United States, investigating the increasing alkalinity in soils associated with irrigated regions, improving soil management and soil reclamation, and improving soil fertility. In many parts of the United States, agricultural soils had ceased to be productive due to years of neglect and/or poor agricultural practise. It is also of historical interest that synthetic nitrogen fertilisers were early investigated by the Bureau of Soils, and that the use of such fertilisers was ultimately accelerated by the sudden demand for nitrates in the manufacture of explosives, as the United States became embroiled in the First World War.

As far as the history of allelopathy is concerned, there are two fairly distinct phases of work associated with the Bureau of Soils: 1) differences in crop productivity caused primarily by physical attributes of the soil, especially soil moisture,

---

[1] Some have said that Whitney's brandished his cigar as if a symbol of power, and but it was also a link to the American rural south. Apparently Whitney claimed he could distinguish different tobacco soils based upon the smell of the resultant cigar smoke .

[2] For example Lyman J. Briggs, a physical chemist, was seen as a likely threat, and was dismissed for seemingly unfounded or trivial reasons. Others such as F.H. King, B.E. Livingston, and H.S. Reed flourished elsewhere after unusually short service with the Bureau of Soils.

Figure 10.1. Photograph of Milton Whitney, (courtesy of the National Archives at College Park, Maryland).

chiefly from about 1901-1908, and 2) the study of organic substances isolated from soil, from about 1905-1919. The more immediately pertinent of these is the latter period, during which numerous toxic organic substances from soil were isolated and studied, hence realising an ambition stated by de Candolle. However, in order to understand the context of this research, and its impact upon agriculture and soil science, it is essential to examine the prior phase, as this greatly polarised the scientific community with regards to the theories and work of the USDA Bureau of Soils.

THE EFFECT OF SOIL MOISTURE ON CROP PRODUCTION

The pivotal moment that began the controversy that was to embroil the Bureau of Soils for the next twenty years was publication in 1903 of the *U.S. Department of Agriculture Bureau of Soils Bulletin* **22**, entitled "The chemistry of soil as related to crop production." (Whitney and Cameron 1903), which was available for the princely sum of 5 cents[3]. As Jenny (1961) has remarked, "bedlam broke loose" in 1903 with publication of this controversial publication. The premise of the research was virtually the same as that Whitney had presented a decade earlier, although now it had the full imprimatur of the USDA. Whitney and Cameron (Whitney and Cameron 1903; Anonymous 1903; Cameron 1910, 1911, 1912;) promoted the highly contentious idea that it is essentially the physical properties of the soil, such as moisture, that determines productivity. It was held that many nutrients, such as phosphate, have a relatively low solubility in water and exist in most soils at near saturation; and as salts are absorbed by the root, new mineral material dissolves [4]. Whitney and Cameron (1903) advanced the view that the "soil solution", that is aqueous extracts of soil, despite variation in apparent soil fertility, showed little variation in nutrient content and the authors concluded that:

> The exhaustive investigation of many types of soil by very accurate means of analysis, under many conditions of cultivation and of cropping, in areas yielding large crops and in adjoining areas yielding small crops, has shown that there is no obvious relation between the amount of the several nutritive elements in the soil and the yield in the crops; that is to say, that no essential chemical difference has been found between the solution produced in a soil yielding a large crop of wheat and that in a soil of the same character in adjoining fields giving much smaller yields. The conclusion logically follows that on the average farm the great controlling factor in the yield of crops is not the amount of plant food in the soil, but is a physical factor the exact nature of which is yet to be determined. (p. 63)

The first review of the bulletin appeared in September 1903 in the *Scientific American* (Anonymous 1903a). and whilst it was essentially positive, it also indicated that the status quo had been upset. A brief review in the *Forestry Quarterly* suggested ironically that agriculture had merely lagged behind forestry in adopting the primacy of the physical condition of the soil as a measure of productivity (Anonymous 1903b). The ideas had some appeal to the nascent field of plant ecology, and in 1904 it was recorded by B.E. Livingston, who was subsequently in the employ of the Bureau of Soils, that the theory made sense in relation to the ecology of native vegetation, as the low concentrations and consequently small differences in soil nutrients seemed inadequate to explain the distributions of different species (Livingston 1905a).

E.W. Hilgard (1833-1916; Figure 10.5), who was Professor of Agriculture and Botany at the University of California, and the first director of the California

---

[3] While the Bulletins of the Bureau Soils were likely printed in reasonable numbers and were inexpensive (as of 1908 the price range was between 5 and 15 cents), they are extremely difficult to obtain today apart from access in public libraries where copies were lodged, in contrast to the soil survey work of the Bureau of Soils, which appears regularly in the antiquarian trade. It is possible, in view of the controversy that surrounded the soil fertility work, that following Whitney's death in 1927, any surplus copies were destroyed by the USDA. This repudiation of the work of the Division of Soil Fertility Investigations seems supported by the omission of any reference in Whitney's official obituary by the USDA (Anonymous 1927) and the lack of any relevant records in the National Archives.

[4] It may be of interest that the idea of inherent long-term soil fertility of cropped soils was suggested in 1849 by the Englishman, the Rev. Samuel Smith in his pamphlet *A Word In Season*. He stated that wheat could be grown year after year without manuring; his results were sufficiently startling that Lawes and Gilbert at Rothamsted repeated his experiments, but without success.

Agricultural Experiment Station at Berkeley, was the then senior figure in the American soil science establishment. Hilgard was well familiar with Whitney's ideas, and the two has already skirmished in agricultural journals over both principles and points of detail in response to Whitney's papers published in the 1890's[5]. Hilgard had become alarmed to see the young Whitney, whom he had viewed initially as a fairly harmless upstart, emerge intact as a major and powerful player in the field of soil science, backed by the U.S. Government, something Hilgard had never managed[6]. Hilgard apparently loved a stoush, and he had already fought and won numerous battles in California to develop his small agricultural research empire. Hilgard wasted little time in sharply criticising Whitney and Cameron's work (Hilgard 1903, 1904). His palpable irritation was based not simply on his dissatisfaction with the USDA pushing a theory he saw clearly lacking in substance, but also on the patent disregard of his own related pioneering work, and the dubious methodology used in the preparation of soil solutions and analyses. Hilgard used his influence amongst his colleagues, and persuaded others to enter the fray (e.g. Myrick 1904).

The work was also challenged from across the Atlantic, and the two senior English soil scientists, A.D. Hall (1904) and E.J. Russell (1905), were compelled to express their strong reservations. However, more importantly, at home there was a groundswell of opposition stirring among the American agricultural colleges and allied institutions. Hilgard, in particular, had been agitating that the viability of state experimental research stations was threatened by Whitney and the soil mapping program. In November 1903 there was a scheduled meeting of the Association of American Agricultural Colleges and Experiment Stations to be held in Washington, D.C. Hilgard had prepared a draft resolution calling for an enquiry into the Bureau of Soils work on soil fertility. However, Hilgard, now 70 years old, was unable to attend the meeting due to poor health. Hilgard's resolution was presented eventually by C.L. Penny, and although it passed, it was not supported with the same vigour, had Hilgard been able to attend*. In any case, the USDA did respond to the situation, and the Secretary of Agriculture, James Wilson, appointed a tame investigative panel, which, to Hilgard's chagrin, included Whitney himself. The outcome was a foregone conclusion.

Even though the Bureau of Soils was represented at the conference (Whitney, Cameron and F.H. King attended), Whitney and Cameron's theory of soil fertility was attacked in two major presentations: one by Hilgard, who had his paper read by C.E. Thorne (Director of the Ohio Experiment Station), and one by the Professor of Agronomy from the University of Illinois, Cyril G. Hopkins (1866-1919). Both

---

[5] Hilgard's initial irritation may have been twofold, as not only were there acute differences in theory, but Whitney's paper of 1892 in essence had trumped that of Hilgard's, entitled "On the relations of soils to climate", which appeared immediately before as *Bulletin* **3** of the same series.

[6] In 1889 Hilgard had been first choice for the newly created position of Assistant Secretary of Agriculture, but he declined in favour of a more lucrative offer from the University of California. Consequently Dabney was appointed, and thus Hilgard himself was indirectly responsibility for the course of development of soil science within the USDA.

Figure 10.2. Photograph of Eugene Woldemar Hilgard, c. 1905, published by E.O. Cockayne, Boston. (Courtesy of the University of California, Berkeley)

papers eventually appeared in print later among the proceedings of the conference (Hilgard 1904, Hopkins 1904b), but both also appeared in print as separates almost instantly after the conference. Hopkins, who was also director of the University of Illinois' agricultural experiment station, distrusted the USDA, and felt that Whitney would use his influence to have publication of both his paper and that of Hilgard suppressed. Hopkins' address was distributed widely as *Circular* **72** of the University of Illinois Agricultural Experiment Station in late 1903 (Hopkins 1903); Hilgard (1903) had submitted his address to *Science* (Hilgard 1903). It was Hopkins who caused the greater consternation to the Bureau of Soils. Hilgard, both because of distance and age, was seen as more of a nuisance than a threat to the USDA. Hopkins' attack, now in print, was fairly venomous, and the Bureau of Soils was forced to respond. The loyal F.K. Cameron, who was a physical chemist, acted as Whitney's spokesperson on technical issues. Cameron's reply (Cameron 1904) was equal to the task, although he did concede some ground on the utility of fertilisers. Hopkins

(1904a) was clearly irritated with the Bureau's two-faced approach, and he replied, largely maligning Cameron's integrity.

Back on the West Coast, the overall handling of Hilgard's concerns had left him incensed. He set in motion a national petition among his colleagues, which in essence questioned the competence of Whitney. Once again, Hilgard became frustrated; support became lukewarm, and he was forced to retreat, as an appropriations bill seeking improved funding for agricultural research, the Adams Bill (approved March 1906), was before the Congress, and the full support of the Secretary of Agriculture was required. Hilgard retired in 1906, and seemingly left the battle for others to pursue. In his penultimate publication, his tome on soils (Hilgard 1906), Hilgard continued to snipe at Whitney, although F.K. Cameron, Whitney's co-author in *Bulletin* **22**, was largely spared of criticism.

F.H. King (1848-1911) was another respected soil scientist who became disaffected, and indeed embittered, with the Bureau of Soils. He was a professor at the University of Wisconsin, and joined the Bureau of Soils in 1902 to assume the position of Chief of the Division of Soil Management. King had collaborated with Cameron and Whitney on the material that was ultimately presented as *Bulletin* **23**, although his role was apparently not known at the time amongst his peers (Jenny 1961). King was furious that his dissenting views were not allowed to be aired, and he insisted that his name be removed from the work. The manuscript for King's main work, *Bulletin* **26** (King 1905), was at first considerably longer, but almost half was censored by Whitney, as it conflicted strongly with the ideas expressed in *Bulletin* **22**. Finally, Whitney demanded King's resignation, and in early 1904 King left the Bureau of Soils and returned to the University of Wisconsin. However, King refused to be silenced and published his views privately (albeit with the permission of the Secretary of Agriculture) and pointedly titled as *Investigations in Soil Management, Being Three of Six Papers on the Influence of Soil Management Upon the Water-Soluble Salts in Soils and the Yield of Crops* (King 1904). The three papers in question were those which Whitney had suppressed. Hilgard was delighted to find an ally in adversity, in King, and in 1904 he published another article in *Science*, which defended King's suppressed work, and he unleashed some extraordinary invective:

> But worse than the ill-founded hypotheses of the head of one of the most important bureaus of the Department of Agriculture, which, moreover, receives and spends one of the largest appropriations in the budget of that department, is the return to medievalism indicated in the case before us. It is not only that of a deliberate attempt to suppress the truth, but it indicates on the part of the morally responsible head of that bureau a more than child-like confidence in the permanent success of the obscurantist regime such as is practiced and defended by Pobyedonosteff[7]. Yet it is doubtful that even the latter, or the puissant head of the Russian Empire himself, would undertake to pass the censor's black brush over inductive scientific papers like these of King.

---

[7] Konstantin Pobedonostev (1827-1907) was a greatly influential and reactionary adviser in the service of tsars Alexander III and Nicholas II, remembered for his anti-democratic views and for thwarting any hint of government reform in Russia.

Throughout all of this, King was placed in an awkward position. Proximity to Hilgard, whose attacks on Whitney and even the Secretary of Agriculture, James Wilson (Hilgard 1904a) were becoming uncomfortably personal, was becoming a liability, and furthermore, a promising young colleague and friend from the University of Wisconsin, Oswald Schreiner, had recently joined the Bureau of Soils. King (1908) ultimately accused the Bureau of intellectual dishonesty in suppressing data and he ridiculed many of the experiments and their interpretations. He was joined by Leather (1907) in criticising the methodology of the agar bioassay (Schreiner and Reed 1907a).

Bulletin **22** was followed immediately by another bulletin by Whitney and Cameron (1904); however, the content of this was far less contentious as it essentially described the methodology and results of a novel bioassay using the transpiration of wheat seedlings. Contrary to their earlier findings, the authors found that aqueous extracts of poor yielding soils were strongly inhibitory, and they were subsequently forced to re-evaluate the significance of root excretion. This bulletin was translated into Spanish (1907) and was published in conjunction with a translation of *Circular* **18** (Gardner 1905), which detailed the methodology of the pot studies.

Curiously, it was recalcitrant lawn soil from Whitney's home neighbourhood of Takoma Park, Maryland that propelled Whitney onward. Evidently Whitney was aware of the difficulty of growing lawn on the suburban Takoma Park soil, and the problem was handed to the plant physiologist Burton Livingston, who was the first head of the new Soil Fertility Division. Livingston *et al.* (1905) found that soil was inhibitory to seedling growth, likely due to trace amounts of an acidic organic compound; however, the inhibitory nature of the soil could be countered by various amendments including tannic acid, pyrogallol, manure, mulch and other organic treatments likely rich in tannins, such as oak and sumac leaves, and some mineral salts. Six months after publication of this bulletin, Livingston resigned on 1 January 1906; however, in 1907, *Bulletin* **36** appeared, providing further data derived from other unproductive soils and collected under Livingston's direction (Livingston *et al.* 1907). Although Livingston is stated as the principal author, it is not altogether clear how much of the bulletin he actually wrote. The first part of this bulletin was used to review the early literature of root excretion and resurrect de Candolle's theories, and one gains the impression that Howard S. Reed (1876-1950), who joined the Bureau of Soils in 1906, likely as a replacement for Livingston, may have been an unacknowledged contributor to *Bulletin* **36**. At least, he was a significant figure in stimulating interest at the Bureau of Soils in de Candollean theory, as the historical content of a subsequent review (Reed 1908a)[8] was very similar to that of the introductory pages of *Bulletin* **36**. Early acceptance of the revived de Candollean theories spread elsewhere within the U.S.D.A., and a January 1906 Farmers' Bulletin originating from the Bureau of Plant Industry cited root excretions as contributing to soil infertility, and stated that a virtue of fertilisers was in destroying soil toxins (Spillman 1906).

---

[8] Reed had a passion for history and antiquarian books, and also wrote, later in his career, an account of Ingenhousz's work and *A Short History of the Plant Sciences*.

In spring 1906[9], another highly controversial publication appeared; it was entitled simply *Soil Fertility*, and it offered the content of an address given by Whitney to the Rich Neck Farmers' Club, Queen Anne County, Maryland. It presented Whitney's views in lay terms, and was very convincing. The address was published within the popular *USDA Farmers' Bulletin* series[10], and consequently tens of thousands of copies were distributed. The article was also translated into French by Henri Fabré in 1907. Whitney's ideas were developed into a more global theory that the interaction of the root with the soil largely controls the fertility of soil (Whitney 1906, 1909), a view which essentially revived the theories of de A.P. de Candolle (1832). Whitney touched on many issues which he knew were of great interest to farmers, but made the following points particularly in reference to toxic substances:

1) plants, like animals and bacteria, continuously produce waste products, which if allowed to accumulate, can become harmful.

> We must clean out the soils as we do the stalls in our stables. If we do not, the substances given off by the plants, or the substances that are formed from those substances by the action of bacteria, will produce acid substances, will produce what we call toxic or poisonous matters, that will themselves seriously affect if not kill the crop. (p. 13)

2) The harmful effect of weeds is due to their toxic effects on other plants.
3) The role of humus is to convert toxic organic substances into a harmless form.
4) The role of fertiliser is in assisting the process of converting soil toxins into harmless material.
5) The rotation of crops is effective because each succeeding crop is not injured by the excreta of the previous crop.

> There is another way in which the fertility of soil can be maintained, viz., by arranging a system of rotation and growing each year a crop that is not injured by the excreta of the preceding crop; then, when the time comes round for the first crop to be planted again, the soil has had ample time to dispose of the sewage resulting from the growth of the plant two or three years before. (p. 21)

Had Hilgard been a younger man, there would have been an unrelenting attack on Whitney following publication of this document, but Hilgard's contempt is recorded only in his correspondence (Jenny 1961).

The mantle of chief opponent to the soil fertility theories of the Bureau of Soils passed to Cyril G. Hopkins (Figure 10.3). In 1906 Hopkins was president of the staid Association of Official Agricultural Chemists (AOAC). At its annual meeting in November 1906, Hopkins stunned the membership by delivering a lengthy address

---

[9] This bulletin became widely circulated, and remained in print for several years; many copies bear the date of the reprint, 1909. The 1906 edition usually has an appendix concerning the "wire–basket" pot method.

[10] The *USDA Farmers' Bulletins* were distributed free of charge, often via politicians, and in large numbers. The series began in 1889, and by 1907 a total of over 55 million copies of roughly 300 different bulletins had been distributed. Specialist publications, such as the *Bureau of Soils Bulletins*, were primarily distributed to libraries and research institutions, and were only available for purchase to interested individuals upon request to the Superintendent of Documents.

which lambasted the work of Whitney and Cameron, and which was little short of slander. This was all the more remarkable as Hopkins was known as a very fair and principled man (Anonymous 1922); something had clearly irritated him beyond redemption. The matter became strained, and indeed Hopkins' presidential address was not included among the proceedings of the AOAC (*USDA Bureau of Chemistry Bulletin* **105**, 1907), as was customary. Furthermore, the AOAC had been compelled to convene a select committee to consider the appropriateness of Hopkins' remarks, given his office, and its findings, too, were not issued among the published proceedings of the following year (*USDA Bureau Chemistry Bulletin* **116**, 1908). At best, the committee issued a letter supporting Hopkins' right to speak out, and this was sent to *Science* (Woll 1908). Hopkins ensured that the report itself was published at the University of Illinois (see Davenport 1908?). A dividing line between the USDA and many agricultural scientists had clearly been drawn. The Assistant Secretary of Agriculture, W.M. Hays[11], was called on to vindicate Whitney's views (Hays and Whitney 1907). It is little appreciated that while Hays had established his reputation in plant breeding, he also had a long-standing interest in crop rotation and continuous cropping in relation to soil fertility, and many years earlier he had initiated numerous experiments in both North Dakota and Minnesota, although his report ignored any work by the Bureau of Soils (Hays *et al.* 1908).

Once again, Hopkins exploited the publication system of the University of Illinois Agricultural Experiment Station to ensure that his views were in the public domain. Firstly, he published his suppressed address to the AOAC as *University of Illinois Agricultural Experiment Station Circular* **105** (Hopkins 1906). This was followed by *Circular* **123** (Davenport 1908), which attempted to put on public display various documents in the debate, and *Circular* **124** (Hopkins 1908), the contents of a lecture given to the American Society of Agronomy at Cornell University, Ithaca, New York, of which the aim was to expose the extraordinary testimony of Whitney and Cameron before the Committee on Agriculture of the United States House of Representatives in January 1908.

This Committee on Agriculture had been established to review the work of the Department of Agriculture, in view of the Agricultural Appropriations Bill of 1908 that sought an increase in agricultural research funding (United States House of Representatives 1908). It is perhaps indicative of Whitney's character that he requested more funding for the Bureau of Soils from the Committee than the Secretary of Agriculture had recommended. The work of the Bureau of Soils, especially that concerning soil fertility, was subject to lengthy and pointed questioning. The interviews with Whitney, but more notably Schreiner and Cameron, were quite astounding, and comments were often treated sceptically by the committee. The politicians basically wanted assurance that a farmer could bring a soil sample for analysis, and be given recommendations for successful soil improvement. Schreiner's testimony presented

---

[11] Willet Martin Hays (1859-1928) was a professor of agriculture at the University of Minnesota, best known for his work in plant breeding and as founder of the American Breeders Association (forerunner of the American Genetics Association), who became Assistant Secretary of Agriculture in 1905.

Figure 10.3. Photograph of Cyril Hopkins (Courtesy of Dr Robert Hoeft, Department of Crop Sciences, University of Illinois)

the Bureau's views on soil toxins and their interaction with fertilisers, but some members were clearly irritated when he stated that a farmer could not expect any simple prescriptions for soil amendment from the Bureau. Schreiner later attempted to regain ground, and stated that a farmer could be assisted after consideration of the soil, climate, cultural practice, crops grown, and suitable expertise in the field on "utilization". When queried as to whether he believed "for the most part fertilisers are useful in destroying the toxic condition of the soil", Schreiner was somewhat evasive, and replied that "it is an important effect and has been one that has been practically overlooked". Cameron, when interviewed, affirmed that all soils always have enough nutrients for plant growth, and that basically the only mediating factor was whether the soil was sufficiently moist to allow saturation of solution of the mineral salts. He confirmed that the role of fertilisers was as an antitoxin or to improve the mechanical qualities of the soil, or possibly something to do with microorganisms. The Committee was very much aware that the most controversial idea promulgated by the Bureau of Soils was that fertilisers essentially had little part in improving the nutritive quality of the soil, and even if it were true, it would have been anathema to farming constituencies and chemical industries. Whitney, during his questioning, had also stated that the main benefit of applying fertilisers was to negate the action of soil toxins and changing the mechanical properties of the soil, ideas that were repeated in an overview of work by the Bureau of Soils (Whitney 1909). In view of the furore that surrounded Whitney, he prepared a press release, which included the statement: "The Bureau of Soils find that the decline in yield is due generally to the accumulation of organic products in the soil which are not eliminated through proper cultural methods as fast as they have accumulated and that the failures that are reported are, therefore, due to improper methods of cultivation and crop rotation" (see Hopkins 1910a).

Amongst other issues, Hopkins maintained that Whitney (1906) had misrepresented data from the classic experiments of Lawes and Gilbert at Rothamsted in order to further his own theories. Hopkins (1910b) tirelessly endeavoured to sabotage Whitney's unorthodox views on fertilisation, and he continued to question Whitney's academic honesty (Hopkins 1912). Hopkins (1911) even wrote a children's book concerning soil science, entitled *The Story of the Soil*, in which a callow youth undermines the views of the Bureau of Soils, and ultimately Hopkins wrote:

> Nevertheless the erroneous teaching so widely promulgated by the federal Bureau of Soils is undoubtedly a most potent influence against the adoption of systems of positive soil improvement in the United States, because it is disseminated from the position of highest authority. Other peoples have ruined other lands, but in no other country has the powerful factor of government influence ever been used to encourage the farmers to ruin their own land. (p. 239)

More vitriolic rhetoric flowed from *The Farm That Won't Wear Out* (Hopkins 1913), which was the privately published book form of a series of articles published prior in the long-running American magazine, *Country Gentleman*. Hopkins seems to have been stirred in particular by publication of the *Bureau of Soils Bulletin* **55**, a substantial work with the authoritative title, *Soils of the United States*, and authored by Whitney (1909) himself. Whitney continued to advance the idea that soils and their nutrients were essentially inexhaustible:

> The soil is the one indestructible, immutable asset that the Nation possesses. It is the one resource that can not be exhausted; that can not be used up. (p. 66)

Whitney (1909) had further stated that:

> The important thing is that we now understand the nature of the soil; how it supplies the nutrient constituents for the crops and how it maintains this supply; how crops may affect each other when grown in succession on the soil; how cultivation affects the conditions resulting from the crop, and lastly, we are beginning to understand how fertilizers come into this scheme and themselves act on or change toxic conditions in the soil, rendering the soil again sweet and healthy for the growing crop. (p. 79)

Hopkins' retaliatory and well-nigh libellous remarks were artfully cast, but he plainly regarded Whitney as both irresponsible and contemptible (Hopkins 1913):

> Is it not in order to ask Congress or the president of the United States how long the American farmer is to be burdened with these pernicious, disproved and condemnable doctrines poured forth and spread abroad by the Federal Bureau of Soils?
>
> It is true that these erroneous teachings have been opposed or ridiculed in Europe; they have been denounced by the Association of Official Agricultural Chemists of the United Stated, and rejected by every land-grant college and agricultural experiment station that has been heard from, including those in forty-seven states; and yet this doctrine, emanating from what should be the position of highest authority, is the most potent of all existing influences to prevent the proper care of our soils. (pp. 73-74)

It is also here relevant that in 1908 A.M. Peter, Chief Chemist and Head of the Chemistry Division of the Kentucky Agricultural Experiment Station was sufficiently concerned that he sought to survey the impact of the Bureau of Soils ideas on soil fertility within numerous North American agricultural teaching and research institutions (Peter and Cameron 1909). F.H. Cameron received one of these circulars, and took great exception to the views that were allegedly ascribed to him, and his lengthy reply was published along side Peter's letters. In 1909 Peter collated the results of the 104 respondents, of which only 2 accepted the views of the Bureau of Soils without reservation; of the rest, about half accepted some truth in the work, notably that on toxic substances in soils, whereas the other half reject or oppose the findings.

In retrospect, Livingston (1923) apologetically recalled the bitterness of those early years, and his words ring as true today as eighty years ago:

> People took sides violently, almost as though this were a question of religious faiths, and the desire for truth was in some instances made secondary to the desire to conserve old beliefs.

There appears to be little public record of any embarrassment or reaction to the controversy by the U.S. administration beyond the congressional hearings, and Whitney retained office until his death in 1927, despite many appeals for his removal. It is apparent that Whitney was nonetheless still accountable to higher authority, and commencing in 1907 the Secretary of Agriculture, James Wilson, evidently directed Whitney to have any potentially damaging publications, such as those emanating from the Laboratory of Fertility Investigations, vetted firstly by the Assistant Secretary of Agriculture, W.M. Hays.

It must be said that the majority of agricultural scientists of the era chose to selectively ignore or distance themselves carefully from the views promulgated by the Bureau of Soils. For example, L.H. Bailey, the influential American agricultural writer in his *Cyclopedia of Farm Crops* (Bailey 1922) cited the root excretion theory, but was careful in presenting it as being more or less endemic to the Bureau of Soils. Similarly, Lyon *et al.* (1916) praised Schreiner's identification of more than thirty organic compounds in soils to date, but were judicious in discussing the implications of these substances in plant growth. Many prominent soil scientists had associations with the Bureau of Soils, but as in the case of King, many became disaffected by Whitney's overbearing and one-eyed view of soil fertility. For example, Frank D. Gardner who had been employed during the early years of the Laboratory of Fertility Investigations, in a subsequent comprehensive text on farming (Gardner 1916), omitted any reference to the work of the Bureau of Soils. On the other hand, the work of the Bureau of Soils did enjoy some favourable treatment in the popular press, as it addressed the long-standing issue of land degradation and the causes of unproductive soils, which had attracted considerable public attention and concern (Cameron 1902, Beal 1911, Bruère 1915,).

SOIL FERTILITY INVESTIGATIONS

Until at least 1901, root excretions formed no part of Whitney's views on the causes of soil infertility (e.g. Whitney 1901), but it was subsequently found that aqueous extracts of poor yielding soils were commonly inhibitory to the transpiration of wheat seedlings (Whitney and Cameron, 1904). The function and growth of wheat seedlings, of the Russian variety *Chul*, in response to various soils, soil extracts and isolated compounds, was to become the basis of much work over the next several years.

The Laboratory of Fertility Investigations was soon established at the USDA Bureau of Soils and a young chemist, Oswald Schreiner (1875-1965; Figure 10.2), who had just completed a Ph.D. studying sesquiterpenes at the University of Wisconsin, was recruited to it in 1903. In 1904 this became a separate Division, and B.E. Livingston (1875-1948) became its first head. Livingston was a young and promising plant physiologist from the University of Chicago, and had collaborated in research with the Bureau of Soils in addressing criticism of the validity of transpiration in wheat seedlings as a measure of growth performance (Livingston 1905c). Following Livingston's sudden departure in early 1906, after only a few months service, Schreiner became Head. In 1907, a well-qualified Canadian who had worked as an agricultural chemist in Hawaii, Edmund C. Shorey (1865-1939) joined the Division of Fertility Investigations. The Division benefited from a series of government appropriations in agriculture targeting soil fertility research, and in 1912, there were 11 research scientists based there: E.C. Shorey, M.X. Sullivan, B.E. Brown, J.J. Skinner, F.R. Reid, E.C. Lathrop, J.H Beattie, A.M. Jackson, H. Winckelmann, D.J. McAdam, and Schreiner himself, and extensive facilities and support staff (Figures 3 and 4).

Although the work of the Bureau of Soils concerning soil fertility became clouded in controversy, Schreiner was an important figure in the history of allelopathy, as he was essentially the first person to isolate and identify individual organic compounds, and, in particular, phytotoxic compounds from unproductive soils (Willis 1996). Indeed, until the work of Schreiner and his colleagues, there was meagre information on organic compounds in soils, and the earlier literature simply described amorphous substances such as humic and fulvic acids, or related artifices of alkaline extraction procedures.

Ultimately there were three guiding principles governing the research of Schreiner and his colleagues (Schreiner and Shorey 1909a): 1) Aqueous extracts of soils reflect the fertility or infertility of the soils from which they originate, 2) living roots excrete organic substances that can affect the fertility or infertility of soil, and 3) the death and decay of plant parts contribute to the factors causing the infertility of soil. The results of the Division of Soil Fertility Investigations may be broadly divided into six topics: plant physiological experiments, soil sickness, soil organic compounds, effects of organic substances in soils, soil heating, and ecological work although these are somewhat arbitrary and largely interrelated.

Plant Physiology

Interest in organic substances had begun with the finding that aqueous extracts of certain infertile soils were inhibitory to seedling growth. Although the principle involved was early characterised as acidic, it took some time for the substances to be identified. In the interim, there was considerable experimental work conducted to demonstrate the effect of effect of putative inhibitors, and the performance of roots under various conditions, initially largely due to the influence of the plant physiologists Livingston and Reed. Physiological work showed that roots were believed to

Figure 10.4. Photograph of Oswald Schreiner (1875-1965) as a young man (c. 1902), (courtesy of the American Institute of the History of Pharmacy).

influence the "oxidising power" of the soil (Schreiner and Reed 1907b, 1909; Schreiner and Sullivan 1910, 1911a, 1911b), which was viewed as essential in the breakdown of soil organic matter, and secondly, roots were thought to necessarily excrete into the soil waste substances, which were toxic or subsequently became toxic and also altered the oxidation process. Roots were also found to have reducing power, and this activity, like oxidisation was most active near the root tip (Schreiner and Sullivan 1909b). Schreiner and Sullivan (1907) found that even the juice from germinating wheat seeds was toxic and the idea of toxic waste metabolites became a general theme (Schreiner and Reed 1908b, Schreiner and Sullivan 1908a, 1908b). Before his departure from the Bureau of Soils, Reed became interested in the phenomenon of hormesis, and his summary of the beneficial effects of toxic substances at low concentrations was published within L.H. Bailey's *Cyclopedia of American Agriculture* (Reed 1907b). A related contribution was his review of an unusual French book, actually a published thesis (Reed 1908b), which explored antagonistic and antitoxic effects, and which had implications for the soil work at the Bureau of Soils.

Soil Sickness

"Soil sickness" or "soil fatigue", that is the failure of crops grown successively on the same soil, had been little studied in the United States. Whitney was well familiar with the degradation of soils in the southern United States, especially those associated with tobacco and cotton culture. It was research on the soil from a Tennessee cotton-sick soil that led ultimately to the isolation and identification of dihydroxystearic acid (see below). The "soil fatigue" of wheat and cowpeas (*Vigna unguiculata*) was examined experimentally by growing them repeatedly on the same soil in glasshouses (Schreiner and Shorey 1908a, Schreiner and Sullivan 1909a). Cowpea-sick soil would not grow a reasonable crop of cowpeas, but could grow wheat or potatoes; furthermore, the soil, when extracted with a large volume of water, allowed the improved growth of cowpea, and the extract yielded a crystalline phytotoxin (Schreiner and Sullivan 1909a). Like his predecessors, Schreiner justified the idea of toxic waste substances through analogies with animal and microbial metabolism (Schreiner and Shorey 1909c, Schreiner and Sullivan 1909a). Later, it was conceded that the chief source of the soil toxins was likely the decomposition of plant organic matter and/or the metabolic activity of microorganisms (Sullivan 1912). Skinner (1913) suggested that oil crops such as *Sesamum* may leave in the soil oily material that affects the growth of succeeding crops[12].

Soil Organic Compounds

Schreiner and his associates started the laborious task of identifying organic constituents in soil, particularly unproductive soils, at a time when chromatography was unknown. Early work indicated that soil extracts from unproductive soils would stunt plant growth and blacken root tips, and that the inhibitory effects could be alleviated through dilution, adsorption onto agents such as carbon black and boiling (Breazeale 1906, Livingston *et al.* 1905, 1907) and through the action of mineral salts (Breazeale 1905). Some inhibitory compounds were distillable. Schreiner and Reed (1907a, 1907c) developed an innovative, although flawed bioassay that used the chemotropic response of wheat seedling roots to root-contaminated media such as agar.

There was considerable pressure to demonstrate that soil toxins actually existed, and the first toxic substance isolated from soil and identified was picoline carboxylic acid (Figure 10.5a), and this was followed closely by dihydroxystearic acid (Figure 10.5b). Today these compounds are known respectively as 3-pyridinecarboxylic acid and dihydroxyoctadecanoic acid[13]. A survey of many soil types found that dihydroxystearic acid occurred widely, that a third of all topsoil samples (60) tested contained the compound, and that these were mainly from infertile soils (Schreiner and Lathrop 1911a, 1911b, 1911c), although the authors did concede that the presence of dihydroxystearic acid may not actually cause soil infertility, but may be coincident with conditions which cause poor soils (Schreiner and Lathrop 1911a).

---

[12] *Sesamum* was recorded in the ancient Chinese literature has being damaging to other plants (see Chapter 4).

[13] There are several isomeric forms, and assumedly the form found was 9,10-dihydroxyoctadecanoic acid, a compound produced by various soil fungi; this is the compound illustrated here.

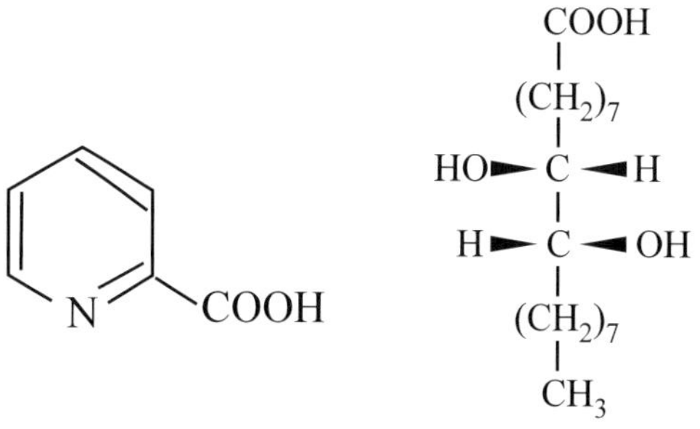

Figure 10.5. Structure of picoline carboxylic acid (a) and dihydroxystearic acid (b).

Schreiner and his associates subsequently identified many more organic compounds in soils (Schreiner 1912, 1913a, Schreiner et al. 1907, Shorey 1913), including terpenes, sterols (Schreiner and Shorey 1909a, 1909b, 1909c, 1911a), hydrocarbons (Schreiner and Shorey 1909a, 1911c), fatty acids (Schreiner and Shorey 1910d, 1911c), phenolic acids and aldehydes[14] (Schreiner and Skinner 1914a, 1914b, 1914c; Shorey 1914; Skinner 1914, 1918a, 1918b, 1918c, 1918d, 1918e; Skinner and Noll 1916; Walters 1917) and nitrogenous compounds (Schreiner and Lathrop 1911c; Schreiner and Reed 1908b; Schreiner and Shorey 1907, 1910b, 1910c; Schreiner and Skinner 1912a) (see Table 1). A useful summary of Schreiner's methods for extracting different groups of organic constituents from the soil was given by Thomas (1914). An unheralded aspect of the Bureau of Soils work was very early studies on soil enzymology (Sullivan 1912). The chemical aspects of the work in identifying organic constituents in soil were highly regarded at the time, and Schreiner, Lathrop and Skinner were all recipients of prestigious Longstreth Medals, awarded by the Franklin Institute. Views of some of the facilities at the Bureau of Soils are in Figures 10.6. and 10.7.

Effects of Organic Substances in Soils

In anticipation of their discovery in soil, many component organic compounds, including amino acids, phenolics, terpenes, and aldehydes were evaluated for their effect on plant growth (Schreiner et al. 1907, Skinner 1918a) and it was generally found that the inhibitory effects of most substances could be ameliorated through the action of boiling, adsorbents such as carbon black, or the addition of substances such as ferric hydrate, nitrate, calcium carbonate or pyrogallol (Schreiner 1913b; Schreiner and Reed 1907a, 1908a). Indeed, many patents were taken on a soil "antitoxin" preparation which allegedly counteracted the effects of excreta of microorganisms and plants (Coates 1910). Of the hundreds of substances investigated, of principal interest were two compounds, picoline carboxylic acid (Schreiner and Shorey 1907,

---

[14] Schreiner (1913) took some pride in that soil salicylic aldehyde was first discovered in rose garden soil from Mount Vernon, Virginia, the home of George Washington.

1908c; Shorey 1906) and dihydroxystearic acid, which were found to be inhibitory in bioassay (Schreiner and Shorey 1908b, 1909a). Picoline carboxylic acid, whilst inhibitory at 100 ppm, was stimulatory at 1-50 ppm, the range containing the known soil concentrations. It was thus discounted as a significant factor in unproductive soils, although a precursor, uvitonic acid, was shown to be far more toxic, but had not been isolated from soils. Thus, interest initially centred on dihydroxystearic acid, a compound which had been isolated from a "soil sick" Tennessee cotton soil in relatively large yields, at 50 ppm, and bioassay indicated inhibition of wheat seedling transpiration at 20 ppm (Schreiner and Shorey 1908b).

Later work focussed on the effects of aldehydes found in soils, such as vanillin and salicyclic aldeyde, and related compounds (Skinner 1914, 1915a, 1915b, 1915c, 1918a, 1918b, 1918c, 1918d, 1918e; Skinner and Beattie 1916; Skinner and Noll 1916); for example, Skinner (1914) reported that salicyclic acid applied at concentrations of 10 ppm in water culture or 25 ppm or more in soil experiments could harm plant growth. Reed and Williams (1915), realising that little was known about the effects of these compounds on soil microorganisms, assayed a range of naturally occurring organic compounds for effects on nitrogen fixation, but despite the unusually high concentrations used (250-2000 ppm), most compounds were not toxic with the notable exceptions of hydroquinone and salicylic aldehyde, and some were stimulatory. Amongst nitrogenous compounds tested, compounds including nicotine, picoline and urea acted as inhibitors, but may have acted as alternative sources of nitrogen for the microorganisms. Commonly, compounds that were known to promote seedling growth were inhibitors of nitrogen fixation, and those known to be potent inhibitors to seedlings had little effect on nitrogen fixers.

An interesting development was work on the interrelationship between organic substances and soil fertilisation (Schreiner 1911, Schreiner and Reed 1908a; Schreiner and Skinner 1911, 1912b, 1914a, Skinner and Noll 1916). It was found generally that the toxic effects of compounds such as coumarin, vanillin, salicylic aldeyde and quinone could be overcome by the application of fertiliser salts of phosphorus, potassium or nitrogen, especially the latter. Moreover, there was a differential effect such that that different nutrients could overcome different toxins. This work was continued by Skinner and Reid (1919), who found that the effects of α-crotonic acid, another toxic substance, initially isolated from soil by Walters and Wise (1916), could be lessened through addition of phosphate. Thus, at this early time it was appreciated that toxic substances in the soil may affect the uptake of various soil nutrients. Some work by Cameron (1911) suggested that oxidative organic compounds such as pyrogallol were just as effective in improving soils as were conventional fertilisers, although the cost was prohibitive (see also Wheeler 1911). A view of the Bureau of Soils was that a key to soil fertility was in allowing oxidative processes, including those stimulated by fertilisers, to alter toxic compounds formed by plants in the soil. A useful summary of the work of Schreiner and his colleagues was presented to the American Philosophical Society in 1913, and Schreiner, evidently mindful of the controversy that had preceded, adopted a tempered approach in reconciling the relationship between soil toxins and fertilisers (Schreiner 1913).

Ecological Work

The work at the Division of Soil Fertility Investigations was primarily concerned with agricultural problems, but other soils were also considered on occasion. The authors speculated that root excretions were involved in natural ecosystems and hypothesised that phenomena such as the maintenance of "oak openings" in oak savannah may be due to the excretions of toxins by grasses (Schreiner and Reed 1907a, 1907c). Work at the USDA Bureau of Soil demonstrated that various leachates of oak, pine, chestnut, tuliptree, dogwood, maple and cherry were inhibitory to wheat seedling transpiration or growth (Jensen 1907, Livingston *et al.* 1907, Schreiner and Skinner 1911), but this work was never followed up. Similarly Reed (1907a) speculated that the unusually poor understorey growth beneath Kentucky coffee-trees (*Gymnocladus canadensis*) was due to toxic substances leached from the bark of the tree. Fungal fairy rings were also seen as examples of toxic effects and Schreiner and Reed (1907c), like Westerhoff (1859), who fifty years earlier, had viewed the striking fungal growth as due to the avoidance of and patterning were all related to root excretion by plants. It is noteworthy that Schreiner and Shorey (1910) described one of the earliest instances of water repellency, in a California soil, which they attributed to the accumulation of a varnish-like material coating the soil particles.

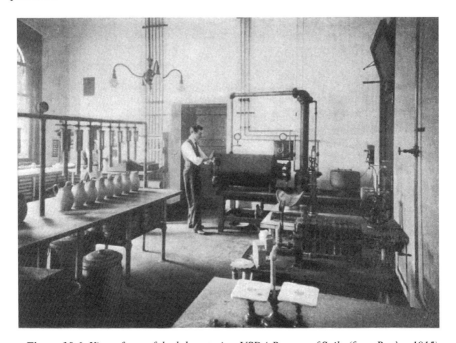

Figure 10.6. View of one of the laboratories, USDA Bureau of Soils (from Bruère 1915).

228　　　　　　　　　　　History Of Allelopathy

Figure 10.7. Glasshouse experiment at the USDA Bureau of Soils in which organic compounds were tested on wheat seedlings (from Bruère 1915).

Effects of Heat on Soils

Similarly to Pickering in England, Schreiner began investigating the effects of heat treatment on soil, and Schreiner and Lathrop (1911a, 1912a, 1912b) found that autoclaving an unproductive Maryland soil at 135°C increased both the number and the amounts of detectable organic constituents, including dihydroxystearic acid.

METHODOLOGY

While the soil fertility work of the USDA Bureau was controversial because of its theoretical basis, and consequently fell into disrepute, one must not lose sight of the fact that significant advances were made, as in the soil chemistry discussed above. Also, an often forgotten aspect of the soil fertility work was the development of standard and often innovative methods, many of which still form the basis of research in allelopathy today. Whitney himself had largely pioneered the use of standard pots in his soil fertility experiments during the 1890's.

Early in the soil fertility work at the USDA Bureau of Soils, there were various attempts to develop a quick, inexpensive and easily repeatable method of extracting water with its dissolved substances (the soil solution) from soil. Even dry soils could be induced to yield small quantities of moisture if necessary via a special centrifugation technique developed by Briggs. However, Whitney and Cameron (1903) for most soils used a standard technique of simply taking 100 g soil, and stirring it for three minutes with 500 ml distilled water. The soil-water mixture was then allowed to stand for 20 minutes, was decanted, and the supernatant was filtered to yield the aqueous extract.

As the soil solution work pointed increasingly toward inhibitory substances in soil, a standard bioassay using wheat seedlings was employed to assess soil fertility/infertility (Livingston 1906, Livingston *et al.* 1907, Whitney 1906). In the same vein, the researchers realised that toxins in the soil may be at such low concentrations as to defy detection with the analytical techniques then available, but that seedlings themselves could be used as bioassay agents to detect the presence of toxic substances (Schreiner and Reed 1907). It was known that substances at low concentration would promote root growth; however, roots tend to curve towards nutrients and away from harmful substances. To eliminate gravitropism as a factor in root curvature, Schreiner and Reed (1907) employed the use of a rotary klinostat.

Another innovation was the use of agar as a plant growth medium to trap substances released from roots (Schreiner and Reed 1907). The agar could be harvested, gently melted, and the reused as a growth medium in bioassay. Schreiner and Reed performed some novel experiments which demonstrated that plants tend to grow away from neighbouring roots evidently in response to some chemical factor which could be negated by the presence of adsorptive material such as carbon black (Livingston *et al.* 1905, Schreiner and Reed 1908).

AGRICULTURAL EXPERIMENT STATIONS AND RELATED INSTITUTIONS

Government financed Agricultural Experiment Stations first appeared in England and in Europe in the mid-part of the nineteenth century. In response to trends in Europe, agricultural expansion, and concerns about soil degradation, the United States government instigated a number of initiatives beginning in 1862. Firstly, the Morrill Land-Grant College Act of 1862 provided funds for the establishment of agricultural education institutions, and as a consequence, a small number of ancillary agricultural experiment stations were soon established. In 1887 the Hatch Experiment Station Act provided federal funds, to be matched by the states, for the establishment of state experiment stations, commonly tied to local state colleges, and consequently there has been an agricultural experiment station in every state and major territory. In 1888 the Office of Experiment Stations was created as a branch of the U.S. Department of Agriculture (USDA), and the experiment stations were required to report through the USDA.

The controversy surrounding the work of the Bureau of Soils, created enormous unease at the state experiment stations, for a variety of reasons. The vast majority of workers at the experiment stations were opposed to the soil fertility theories being promulgated by the Bureau of Soils. The Bureau of Soils soil mapping program was seen by many as of dubious value and undermined, at the local level, the work of the experiment stations. Administrators of the experiment stations were very mindful of Whitney's influence at the federal level, and as station funding was heavily dependent upon Washington, they were reluctant to be overly critical of Whitney and the

Table 10.1. Principal substances isolated from various soils by the Bureau of Soils.

Substance	References
Acrylic acid	Schreiner 1913
Adenine	Schreiner and Skinner 1912, Schreiner 1913
Agroceric acid	Schreiner and Shorey 1909a
Agrosterol	Schreiner and Shorey 1909a, 1909b
Arginine	Schreiner and Shorey 1910, 1910b, 1910c
Choline	Schreiner and Skinner 1912, Schreiner 1913
Creatinine	Schreiner and Skinner 1912, Shorey 1912
α-crotonic acid	Walters and Wise 1916
Cytosine	Schreiner and Shorey 1910, 1910c
Dihydroxystearic acid	Schreiner and Shorey 1908c, 1909a
Guanine	Schreiner and Skinner 1912, Schreiner 1913
Hentriacontane	Schreiner and Shorey 1910
Histidine	Schreiner and Shorey 1910
Histidine	Schreiner and Skinner 1912, Schreiner 1913
Hypoxanthine	Schreiner and Shorey 1910, 1910c
Lignoceric acid	Schreiner and Shorey 1910
Lysine	Schreiner 1913
Mannite	Schreiner 1913
Monohydroxystearic acid	Schreiner and Shorey 1910
Nucleic acid	Schreiner 1913
Oxalic acid	Schreiner 1913
Paraffinic acid	Schreiner and Shorey 1910
Pentosan	Schreiner and Shorey 1910
Pentosan	Schreiner 1913
Pentose	Schreiner 1913
Phytosterol	Schreiner and Shorey 1910, 1911a
Picoline carboxylic acid	Schreiner and Shorey 1908b, 1909a
Resins	Schreiner 1913
Rhamnose	Schreiner 1913
Saccharic acid	Schreiner 1913
Salicylic aldehyde	Shorey 1913, Schreiner and Skinner 1914b
Succinic acid	Schreiner 1913
Trimethylamine	Schreiner 1913
Trithiobenzaldehyde	Schreiner 1913
Vanillin	Shorey 1914
Xanthine	Schreiner and Shorey 1910, 1910c

Bureau of Soils, for fear of jeopardising funding. At the same time, the agricultural experiments rarely followed the fertility work of the Bureau of Soils, despite the vast array of relevant resources.

Examination of work done at the USDA Bureau of Soils took place at university agricultural departments and state agricultural experiment stations. As early as 1895 workers at the Rhode Island Agricultural Experimental Station of the Rhode Island State College had suggested that there were toxic substances in soil (Wheeler *et al.* 1895, Wheeler and Hartwell 1900), and in the early years of the twentieth century, the Rhode Island Agricultural Experiment Station, under the direction of H.J. Wheeler conducted experiments in collaboration with the USDA Bureau of Soils. These were able to confirm the toxicity of soil extracts from infertile Rhode Island soils and that the addition of salts and other agents lessened the toxicity of these extracts (Wheeler and Breazeale 1905), as reported by the Bureau of Soils for Takoma (Maryland) soil.

While the Agricultural Experiment Stations progressively distanced themselves from the views of the Bureau of Soils, it was inevitable that common problems for investigation would present themselves. A prominent example of this is the work of the Agricultural Experiment Station on "the influence of crop plants on those which follow". Although the experiments began in 1907, and obviously related to the work of the Bureau of Soils, there is absolutely no reference to the work of Whitney or Schreiner. The work in Rhode Island was begun by H.J. Wheeler, but was continued by Burt Hartwell. Hartwell evidently preferred to distance himself from the USDA, and in 1928 was suspended, and ultimately resigned, as director of the station, for refusing to send results to the USDA, and hence for jeopardising funding.

The research at Rhode Island was based on sixteen similar linear plots (9.14 by 59.01 m) that were grown in a three-year cycle (Hartwell and Damon 1918). For the first two years plots were planted with one of the following crops: onion, potato, mangel (beet), ruta-baga (turnips), cabbage, buckwheat, corn, millet, oats, rye, carrot, redtop, timothy, squash, alsike clover, and red clover. In the third year, all plots were grown with the same crop, which in 1910 was onion. In 1913 the experimental crop was buckwheat, and in 1916 it was alsike clover. With the onion crop, there were large differences in yield: preceding roots crops, such as mangel, cabbage, ruta-baga, and potatoes, resulted in low total onion yield (72, 88, 99, 112 bushels respectively), whereas preceding forage plants, such alsike clover and timothy resulted in much higher total onion yield (415 and 524 bushels) respectively. Results with buckwheat showed a parallel range in yield, but different trends: the lowest yield (0.27 bushels) was after timothy, and the highest yield (1.59) after squash. Results with alsike clover in 1916 were much less variable, and ranged from 2.60 bushels after carrots to 4.33 bushels after redtop. Various analyses of plant and soil nutrients indicated that it was not always the crops that removed large amounts of nutrients that caused the lower yield of succeeding crops, but the authors baulked at providing any hypotheses.

Subsequent work (Hartwell *et al.* 1919) sought to elucidate the nutrient relationships and pH factors in these experiments, and there were some evidence to suggest that buckwheat roots may cause a suppressive effect on succeeding crops. Hartwell *et al.* (1927) further examined nutrient effects, and factors such as pH and aluminium toxicity. The final paper in the series was written well after the departure of Hartwell, and during a period which had largely forgotten the tensions raised concerning the topic of root excretions in the first decade of the century, and the authors (Odland *et al.* 1934) were prepared to admit root excretions as one of a number of factors possibly involved in the effects of one crop on another.

Work on soil aldehydes was taken up by several investigators (Davidson 1915; Fraps 1915; Funchess 1916; Upson and Powell 1915) and this work led to an understanding of the degradation of these compounds in soil (Gardner 1924, 1926; Robbins 1916, 1917; Robbins and Lathrop 1919). Fred (1912) found that "toxic" soil could not be amended through sterilising agents such as ether. Upson and Powell (1915) at the University of Nebraska investigated the effects of the putative toxins vanillin, salicyclic aldehyde, coumarin, quinone and dihydroxystearic acid in soil experiments but found little inhibition of seedling growth. Fraps (1915) found that dihydroxystearic acid administered at 500 ppm to corn and sorghum plants in pots had little effect. Vanillin and coumarin administered at 100 ppm affected only one of eight test species and was found to disappear from soil within two weeks. Fraps could find no evidence that fertilisers overcame the effects of these supposed toxins and all evidence pointed toward poor growth in unproductive soils being due to a lack of nutrients. Fraps concluded that results from pot experiments differed markedly from those in aqueous culture. Funchess (1916) also reported on the effects of various supposed soil toxins in pot experiments with oats and corn and his results concurred, generally, with those of other workers (Dandeno 1910; Fraps 1915; Upson and Powell 1915).

Lyon and Wilson (1921) attempted to confirm whether plant roots do indeed release organic matter and whether any of these substances do become oxidising or reducing. The authors confirmed that plants lose significant amounts of organic matter (1.5 to 2.7 per cent of the plant dry weight) through their roots but they could not support the contention that oxidising substances were excreted.

Thomas (1914a) at Columbia University, New York, attempted to follow Schreiner's methodology and compared the organic constituents of burnt and unburnt soil; preliminary results suggested that heat decreased the amount of dihydroxystearic acid. Lodge and Smith (1912) in Massachusetts experimented with the effects of heat sterilisation on bacterial and plant growth and concluded that different soils respond differently to treatment and that any observed effect is chemical or physical in origin, not biological (vide the work of Russell and his co-workers). Seaver and Clark (1910) suggested that the markedly increased growth of the pyrophilous fungus *Pyronema* on heated soils was due to the removal of soil toxins, such as those found by the USDA Bureau of Soils. Seaver and Clark (1912) and Wilson (1914) found that soils heated to relatively low temperature (120°C) enhanced growth and retarded fungal growth, whilst higher temperatures, had the opposite effects. Johnson (1916, 1919) investigated the effects of heating soil on plant germination and growth but used a wider range of temperatures than Schreiner and Lathrop (1911a, 1912); he found that the greatest retardation occurred with temperatures of 250°C and there was no evidence of dihydroxystearic acid.

There was other independent work on root excretion that was contemporary with the USDA Bureau of Soils. It was suggested that the sometimes distorted growth of roots in closed glass tubes might be due to the accumulation of root excretions (Newcombe 1902). L. B. Dandeno completed his Ph.D. at Harvard University on the subject of the effect of various solutions on plant growth; his early work demonstrated that secretion at the root tip may help to counter soil toxicity, and toxicity in many instances can be countered by insoluble agents which are finely divided, e.g. sand (Dandeno 1904). He subsequently took a position at Michigan Agricultural

College and became interested in the causes of soil fertility and fertility. He alluded that plant may affect one another through chemical means (Dandeno 1905). Dandeno seemed unaware of, or at least ignored, Schreiner and Shorey's more recent work, as he stated that there was no information available on plant substances affecting soil fertility (Dandeno 1908, 1910). Dandeno grew various combinations of seedlings, e.g. squash and corn, together in water culture and witnessed enhanced growth in early stages (18-36 h) and then depressed growth. He believed that plant roots can excrete waste material and that, after about 48 h, bacteria and fungi, feeding on these substances and in particular root cap cells, begin to produce compounds inhibitory to plant growth (Dandeno 1908, 1909a, 1909b). Dandeno further suggested that the oxidising power of roots of a species is most effective in neutralising the toxicity of its own excreta.

The lack of seedling growth beneath trees in northern forests led to several investigations based on hypotheses stimulated by the work of the USDA; for example, Robinson (1909) speculated that soil from hemlock (*Tsuga canadensis*) groves contained substances inhibitory to the growth of seedlings but her results contradicted this idea.

TOXICITY IN BOGS

B.E. Livingston, although only briefly associated with the Bureau of Soils, was a stimulus for work concerning the toxic effects found sometimes in bogs. He had reported in 1905 that water collected from a bog was inhibitory to the growth of the green alga *Stigeoclonium*, and that the degree of inhibition could be lessened through dilution (Livingston 1905b). Similarly, Coville (1910) later suggested that the swamp blueberry (*Vaccinium corymbosum*) was devoid of root hairs to minimise damage by bog water. Toxicity in bog water was further explored by Alfred Dachnowski (1905, 1908a, 1908b, 1909, 1912a, 1912b) who bioassayed water and soil from an Ohio bog with the growth of gemmae from the thallose liverwort *Marchantia* and with the transpiration of numerous crop seedlings. Generally, water and soil collected from the centre of the bog were found to be the most inhibitory, and the inhibitory properties could be countered with adsorbent agents. Dachnowski speculated that the accumulation of toxins in bog environments was responsible for the low productivity and the curious xerophytic character of many of the plants found there, and believed that such substances may affect plant nutrition (Dachnowski 1912a). Henry Cowles, who was a colleague of Livingston at the University of Chicago, had a long-standing interest in the dynamics of wetland vegetation, and speculated that the stagnant nature of bogs would allow the accumulation of toxins, and that these may be involved in the successional sequences of wetland plants and the xerophytic nature of many wetland plants (Cowles 1911). His student George Rigg subsequently investigated the phytotoxic decay products of water-lilies (*Nymphaea* spp.) both in Illinois and Washington. Preliminary work suggested that several substances were involved; however, inhibitory effects, albeit less so, were also found in decaying potatoes and turnips (Rigg 1916b). Rigg (1913) had also found that bog water collected from the Puget Sound region in Washington was inhibitory to the root hair development of *Tradescantia*, which he subsequently used in bioassay. Rigg (1916a) also published a review concerning the causes of the poor plant growth commonly associated with bogs, and Rigg *et al*. (1916) concluded that the inhibitory properties of bog water were not due to osmotic effects, but were likely due to toxic substances. Another University of Chicago alumnus, Edgar Transeau (1914) reported that bog water was toxic to the growth of *Rumex acetosella*.

INFLUENCES OVERSEAS

The theories and results of the USDA Bureau of Soils were controversial but the identification of many compounds hitherto unrecorded in soils was probably the greatest single contribution of the Laboratory of Fertility Investigations and stimulated interest from as far away as Australia (Guthrie 1910, 1913) and South Africa (Stead 1918). There was certainly increased awareness of the chemical and biological complexity of soil (Anonymous 1911, Bailey 1907, Chamberlin 1911, Harshberger 1911, Jodidi 1913, Lipman 1918, Stubbs 1913), which led to many research projects elsewhere, especially those involving putative toxic compounds in soil and the effect of heat on soil constituents.

In England and in Europe the work of the Bureau of Soils received considerable exposure (Dietrich *et al.* 1911; H. 1903; Hall 1907, 1908a, 1908b, 1909, 1911; Immendorff 1910; Lemmerman 1909, Pratolongo 1915; Ulpiani 1910, Voelcker 1904) but the work was often regarded as uneven or eccentric. This, coupled with difficulty of access, allowed a limited impact. As with Pickering's work, the conclusions reached ran counter to orthodox views and were not supported by other experimental data (Anonymous 1914a; Hall 1910; Russell 1911a, 1911b, 1912c), although Voelcker (1903) had reported that extracts from infertile soils at the Woburn Experimental Station could be harmful to grain crops such as wheat, barley and oats. Both Russell (1908) and Hall (1910) were critical of the wheat seedling transpiration bioassay as a measure of crop performance. On the other hand, Schreiner's subsequent work with dihydroxystearic acid was regarded at least briefly by Russell (1911a, 1912b) as relatively convincing. Hall (1908b), who visited the U.S. in 1908, regarded the Bureau of Soils theories as overly simplistic and lacking in field evidence but he did concede that crop residues may affect other crops. One reviewer even suggested that the principal function of Whitney and Cameron's work was not meant to be scientific, but was to deter farmers from buying certain fertiliser mixtures that were intrinsically of little value (Miller 1913). Hall (1909) argued that Whitney's theory ignored a host of factors including pH effects, the complexity of phosphorus salts, and the role nitrates in soil, and concluded:

> As it stands at present Whitney's theory must be regarded as lacking the necessary experimental foundation; no convincing evidence has been produced of the fundamental fact of the excretion of toxic substances from plants past the autotrophic seedling stage, nor is there direct proof of the initial supposition that all soils give rise to soil solutions sufficiently rich in the elements of plant food to nourish a full crop, did not some other factor come into play. If, however, we give the theory a wider form, and instead of excretions from the plant understand débris of any kind left behind by the plant and the results of bacterial action upon it, we may thereby obtain a clue to certain phenomena at

> present imperfectly understood. The value of a rotation of crops is undoubted and in the main is explicable by the opportunity it affords of cleaning the ground, the freedom from any accumulation of weeds, insect, or fungoid pests associated with a particular crop, and to the successive tillage of different layers of the soil, but for many crops there remains a certain beneficial effect from a rotation beyond the factors enumerated. (p. 296)

It should be said that the phenomenon of soil infertility following the repeated culture of certain crops was not in dispute, and many researchers believed the problem was tied to soil bacteria, whose importance in soil had begun to be appreciated in the late nineteenth and early twentieth centuries. As with the work of Pickering, and workers at the Bureau of Soils, there were extensive data that demonstrated that either heat or various organic solvents such as chloroform, carbon disulphide or toluene, could act to reduce or alter soil bacterial populations, and consequently increase soil fertility (Anonymous 1912b, Fred 1912, Jamieson 1913, Russell 1910). There were several variations on this theme and researchers at Rothamsted championed the "phagocytic theory" whereby increases in soil protozoa decreased the population of beneficial soil bacteria (Russell and Hutchinson 1909, Russell and Golding 1911, Russell and Petheridge).

The remedial effects of organic solvents on some infertile soils were seen by some as evidence in support of the phagocytic theory, although others such as Robert Greig-Smith (1866-1927) disagreed and interpreted the effect as the removal from soil particles of waxy substances, which he had named agricere. Greig-Smith was a Scottish born chemist who eventually migrated to Australia in 1900, where he assumed the position of Chemist and Bacteriologist to the Linnean Society of New South Wales. The work of Greig-Smith, in contrast to that of Pickering and Whitney and his co-workers is rarely cited, likely because it was mostly was published in Australia. Greig-Smith held that bacteria can cause their own demise through the accumulation of toxic excreted substances, and that this applied equally in soil and with soil bacteria. He collected orchard soil from the Hawkesbury region in New South Wales, and found that aqueous extracts were toxic to soil bacteria. The toxin could be destroyed by heat, sunlight, and storage. Fertile soils yielded extracts which were less toxic.

Greig-Smith, in extracting soils with organic solvents, found that 500 g of soil could yield 0.39 g of material containing waxes, fatty materials and resins, which he named "agricere". It was considered that agricere was derived from the degradation and transformation of plant materials, likely from the cuticle, and was similar to the substances discovered independently was Schreiner and Shorey. Greig-Smith believed that agricere played a significant role in soil fertility, as it waterproofed soil particles, and made soil nutrients less available to bacteria and plant roots. Removal of agricere with organic solvents could thus increase soil fertility; however, he believed that some agricere was required in soils to keep in check bacteria which otherwise would lead to soil toxicity (Greig-Smith 1910). The infertility of soils treated with sewage was seen as due to the unusually high levels of agriceric materials which prevented soil aeration (Greig-Smith 1912b). Russell's theory concerning the role of protozoa was countered as the addition of protozoa to soil had little effect on bacterial numbers (Greig-Smith 1912c).

Greig-Smith performed numerous experiments with soils during the period 1910-1918 in order to elucidate the nature and significance of the so-called soil "bacteriotoxins", and their interactions with agricere; a brief summary of his work was published in 1918 (Greig-Smith 1918). He characterised soil bacteriotoxins as non-volatile and water soluble; they were carried by rain to the subsoil (Greig-Smith 1911a, 1911b, 1913b). Bacteriotoxins were subsequently recorded in run-off water (Greig-Smith 1914). Heat was found to destroy the bacteriotoxins, but also generated different soil toxins (Greig-Smith 1911a, 1911b), or "heat toxins" as reported by Pickering and others. Hear and solvent treatments were found to have different effects on soil toxicity, particularly if the soils were rich in agricere (Greig-Smith (1913a). Greig-Smith (1913b) identified two types of soil bacteriotoxins: a thermolabile type from topsoil, and a thermostable type from subsoil. He found that the optimal toxicity was expressed with a 1:1 soil:water mix, and that drying of the soil allowed degradation of the toxins, but incubation of moistened soil re-established the soil toxicity (Greig-Smith 1913c). Soil toxicity became greatest at 28° C and when the soil was at 25% soil moisture capacity. The addition of dextrose accelerated toxin production (Greig-Smith 1915).

Greig-Smith's work received only modest support. W.B. Bottomley (1911a, 1911b), in England, reported that rotted manure extracts were inhibitory to seed germination and seedling growth, and could inhibit nitrogen-fixing bacteria; the toxicity could be alleviated by heating. In, C.M. Hutchinson (1912) performed similar work to that of Greig-Smith, and claimed that soil bacteriotoxins were destroyed by sunlight, air, heat and organic solvents. Hutchinson (1926) subsequently retained the view that bacteria were important in controlling soil fertility, especially in anaerobic soils, either through the addition of organic compounds which clog the soil, or through the production of toxins that act on plants or susceptible microorganisms such as nitrifying bacteria.

Greig-Smith's work with bacteriotoxins ended with disappointment. In 1918 after lengthy experimentation, he was forced to admit that he could not find any organisms responsible for producing bacteriotoxins, and that his observations on fluctuations in bacterial numbers were most easily explained by changes in soil physical conditions such as pH (Greig-Smith 1918a). Furthermore, H.B. Hutchinson and Thaysen (1918) produced relatively conclusive data that refuted most of Greig-Smith's and C.M. Hutchinson's contentions. They concluded that heating of soil extracts acted through diminishing their nutritive value for bacteria, and bacteriotoxins were shown to exist under artificial conditions, but were of little significance in natural systems.

As mentioned previously, the First World War led to grave concerns about the supply of imports, especially agricultural commodities, as countries such as England, in particular, had become overly reliant upon overseas staples including wheat and fertilisers. A result of this, there was re-evaluation of agricultural land usage and means of improving soil fertility without chemical fertilisers. Advances in the understanding of soil bacteria in nitrogen fixation and nitrification had led to a program in the U.S. Department of Agriculture, where cultured nitrogen fixing bacteria dried onto cotton wool were distributed by mail to farmers for use in soil inoculation to improve fertility. A similar program, on a smaller scale was initiated by W.B. Bottomley, a professor at Kings College, University of London, but was relatively unsuccessful, due to the limited longevity of the bacteria. Bottomley, in investigating soil amendments, discovered that peat allowed to decompose aerobically for 14 days (bacterised peat) yielded humic substances that promoted plant growth, which he subsequently named "auximones". Further investigations showed that these substances could affect numerous types of plants, as well as the organisms involved in nitrogen fixation, nitrification and denitrification. Bottomley's work was never really fully understood, but it generated considerable public interest. Bottomley eventually abandoned this work in frustration, as in the face of having received commercial offers from Germany, he was eventually denied funding by English Board of Agriculture and Fisheries, because of variable results.

The work of Whitney and his colleagues received probably the most sympathetic hearing in France (André 1922, Demolon 1922, Foussat 1911, Parisot 1911, Pouget and Chouchak 1907, Rousset 1908, Zolla 1907, 1908), as translations of key works were published in France (Cameron and Bell 1907, Whitney 1907), and there was parochial interest in the revival of de Candollean theories. Massart (1912) envisaged the new findings regarding plant secretions as paving the way toward understanding the causes of localised plant distributions. However, despite this rampant enthusiasm, further experimentation and documentation were recommended by the mainstream scientists (Zolla 1908). Amongst those who took up the challenge were Pouget and Chouchak (1907) and Marin Molliard (1913, 1915). Pouget and Chouchak (1907) investigated inhibitors associated with lucerne (*Medicago sativa*) "soil sickness". Soil collected from an old lucerne field was extracted with water; the extract was dried and added back into fresh soil. In pot trials, the amended soil proved inhibitory to Lucerne seedling growth, whereas control soil and soil prepared with charred extract or an alcoholic extract showed good growth. In laboratory studies, Molliard (1913, 1915) found that the aqueous medium used for growing pea seedlings was inhibitory to the growth of both successive pea and corn seedlings. Later, Lumière (1920) suggested that the well known flush of herbaceous growth which accompanies spring may be due not so much to the change in physical soil conditions but to the oxidation, decomposition and dilution of soil toxins resulting from the previous year's growth. Forest soil which had been repeatedly washed allowed the germination of many herb seeds in comparison to the control. In subsequent work, Lumière (1921) demonstrated the inhibitory potential of leaf litter and he suggested that toxins may act in the soil by competing with seeds for available oxygen. Lumière's work forced Petit (1922) to claim that in 1909 he had also discovered that washing soil would increase the growth of pot plants such as *Calceolaria* and *Heliotropium* (Petit 1910). Chemin (1921) addressed the issue of the existence of soil toxins through experiments involving the heating of soil. Soil collected from a wheat field was found to support the growth of wheat and oats when heated at 70°C for 1 hour; however, untreated soil yielded a poor growth of wheat but a reasonable growth of oats. Chemin, and subsequently Demolon (1922), interpreted these results as supporting the root excretion theories of de Candolle and Whitney. Piettre (1923), whose interest in soil compounds was to span many years, suggested that pyridine was a preferred solvent in the extraction from soil of compounds, including soil toxins. Other original French work stems from the Comoro Islands, west of Madagascar, where it was observed by Advisse-Deruisseaux (1910) that vanilla plants often failed when growing next to certain tree species, such as *Ficus* spp., *Artocarpus* spp., *Mangifera indica*, *Spondias dulcis*, *Anacardium occidentale* and *Acacia lebbek*. He suggested that the antagonism was caused by some toxic effect when vanilla roots came into contact with parts of these trees. In the Philippines, de Peralta and Estioka (1923), in a soil leachate experiment, obtained the remarkable result that soil drainage water from soil cultures of both *Cyperus* sp. and the water-lily *Monochoria hastata*, when applied to rice plants, roughly doubled the growth and yield of the rice. The authors attributed the effect to root excretions, and suggested that these plants should form part of a crop rotation where rice was grown. On the other hand, the soil leachate of a common forage plant, cutgrass, known locally as zacate (*Leersia hexandra*), and that of rice itself slight decreased the growth and yield of rice, which indicated that the former should not be rotated with rice, and rice crops should not be grown in succession.

In Russia, Pryanishnikov[15] (1914), Professor of the Petrovski Agronomic Institute near Moscow, was critical of the Bureau of Soils' work for its lack of objectivity. Pryanishnikov, particularly after meeting A.D. Hall at Rothamsted in 1909, became sceptical of the findings of Whitney and his colleagues, which were unsupported by practical experiments in Europe. Prianishnikov noted that wheat and barley had been grown on the same ground for over 50 years at Rothamsted, and as long as fertiliser was adequately supplied, the yields showed no tendency to decline. Subsequent experimental work at Petrovski Agronomic Institute conducted by Periturin attempted to clarify the reports of the USDA Bureau of Soils. Periturin (1913) initially grew three successive crops of wheat and oats in the same water cultures, but found no evidence of suppressed growth. However, further experiments with repeated crops of plants grown in the same sand showed in many instances reduced growth, of which *Camelina*, flax and graminaceous plants was most marked. Experiments also showed that one crop, for example oats could inhibit the growth of a subsequent different crop such as wheat. As in the work of the USDA Bureau of Soils, inhibittory effects could be alleviated by the application of charcoal. Periturin (1913) and Pryanishnikov (1914) were obliged to conclude that inhibitory substances could develop in soil, but that they were not root excretions, but were more likely microbial products.

Lastly, the work of the Bureau of Soils stimulated interest in India, chiefly through the work of Fletcher, who worked initially in Egypt, and then at Dharwar (known now as Hubli-Dharwad) and Surat, which were in the Bombay Presidency. Fletcher became an ardent supporter of the toxin theory, which led to clashes with E.J. Russell over bioassay methodology and the interpretation of field and soil heating experiments (Fletcher 1913, Russell 1908). Fletcher (1908b) found that *Sorghum vulgare* inhibited the growth of adjacent crops of gingelli (*Sesamum indicum*), cotton, cajanus, as well as sorghum itself, and that a toxic alkaloid-like substance was involved. Also, at this

---

[15] The spelling varies: e.g. Prianichnikov, Prjanishnikov, etc.

time, although apparently independently, there were reports of an organic acid toxin in soils from highland areas near Jorhat, in Assam (Meggitt and Birt 1912, Meggitt 1914). Later experiments demonstrated that maize similarly inhibited the growth of *Sesamum*, despite adequate fertilisation and watering (Fletcher 1912). Experiments with various water cultures led to the isolation of a toxin which had characteristics similar to those of dihydroxystearic acid (Fletcher 1908b, 1910). Furthermore, he demonstrated that soil leachates of some crops, in particular *Cicer arietinum*, were toxic to seedling growth, but the identity of the substances was not determined, although they could be neutralised by tannic substances. Fletcher's work generated considerable interest and was discussed at length in the German agricultural press (Ollech-Steglitz 1912). The work of Fletcher was summarised for the French press by De Wildeman (1909a, 1909b). Fletcher's experiments concerning root excretions (Fletcher 1908a) were later re-examined by Mukerji (1920) who, with more stringent methodology, concluded that that any inhibition was linked to the quality of salts used in the nutrient solutions. Fletcher's work with sorghum and *Sesamum* was also repeated considerably later in the United States by Shull (1932) who could find no evidence of toxic effects. Interest in root excretions came also from the Dutch East Indies (present-day Indonesia), where W.G. Leembruggen, Director of the Bogor Gardens, was of the opinion that root excretions of plants such as sugar cane affected the growth of other plants (Leembruggen 1909).

AFTERMATH

B.E. Livingston, who worked at the Bureau of Soils briefly during 1905, was an early protagonist in the toxin work, but in self-effacing retrospection, he admitted later that there was little evidence to support excretion of toxins by roots, although toxins of unknown origin could be shown to exist in certain soils, particularly if anaerobic (Livingston 1923, Palladin 1918). Henry Cowles, one of the seminal figures in American plant ecology, who had encouraged Livingston's early work on bog toxins at the University of Chicago, lent support to root excretion theory and the work of Schreiner and his colleagues (Cowles 1911, Coulter *et al*. 1911). However, a cogent summary of the lore of soil toxins was given by the other highly influential American plant ecologist of the time, Frederic Clements (1921), and he concluded that there was no evidence to date to support the notion that root excretion under ordinary circumstances contributes to soil toxicity or plant succession; but, he did admit that soil toxins may accumulate in soils due to root excretion or microbial activity under anaerobic conditions (e.g. Harrison and Aiyer 1913):

> Soil toxins are probably to be definitely related to deficient aeration and to anaerobic conditions, as has been indicated by Schreiner, Hall, Russell, and others. This is also shown by the fact that they are readily oxidized, and soon disappear under proper tillage. Hence, they appear to be due to essentially the same conditions and processes as obtain in bogs, the relationship being especially well exhibited by muck soils. In both, the primary causes of toxicity are the direct lack of oxygen and its indirect effect in permitting the accumulation of carbon dioxid in harmful amounts and in producing injurious organic acids and other compounds. (p. 162)

By the 1920s, the Bureau of Soils' views on soil fertility were seen with hindsight as extreme (Usher 1923) and were accepted by few (Weir 1920), although both Schreiner (1923) and Whitney (1921, 1925) continued to maintain that soil toxins do exist. In 1923 the American Society of Agronomy hosted a symposium on "Soil Toxicity" which dealt with both organic and inorganic aspects. This occasion was essentially the last at which Schreiner (1923) presented his views, and Livingston (1923) was also present. R.W. Thatcher, who was Director of the New York State Agricultural Experiment Station, which had research in progress investigating the claims of Pickering and Bedford, gave a critical overview of work relating to the effect of one plant on another. Thatcher (1923) did not dismiss the idea of toxins, but he was forced to conclude that:

> There is as yet no positive proof of the nature of the causative agent or agencies for either the beneficial or the injurious effect of one crop upon another. It may vary widely in different cases, and may be chemical, or bacterial in character. Definite proof that observed injurious effects on a second crop are due to toxic chemical substances in the soil produced by or in association with the first crop has not yet been established.

Sadly, the era is one that has been nearly been erased from the academic memory, as clearly the pursuit of de Candollean theory and the promulgation of misguided ideas concerning fertilisation were ultimately regarded with embarrassment by most of the those concerned. It is noteworthy that among the obituaries or memoirs of the some of the key figures such as Whitney, Livingston and Reed, there is little record of involvement in the soil fertility work (e.g. Anonymous 1927, Shull 1948, Reed 1943). In spite of this, the work of Schreiner and his colleagues, in particular, explored many new frontiers in the realm of the science of soil organic matter, and deserves better recognition.

REFERENCES

Anonymous. 1903a. Chemistry of soils as related to the yield of crops. *Scientific American* **89**: 207.
Anonymous. 1903b. [Review of] The Chemistry of the Soil as related to Crop Production. *Forestry Quarterly* **2**(1): 30-31.
Anonymous. 1911. Revolutionary discovery of the relation of soils to crops. *Current Literature* **51**: 525-526.
Anonymous. 1912. Investigations on "sickness in soil sterilization and some practical applications. *Bulletin of the Bureau of Agricultural Intelligence* **4**: 27-32.
Anonymous. 1922. *In Memoriam Cyril George Hopkins*. University of Illinois Press.
Anonymous. 1927. Prof. Milton Whitney, soil scientist, dead. *U.S. Department of Agriculture Official Record* **6**(46): 4-5.
Advisse-Desruisseaux, P. 1910. De l'influence exercée par quelques plantes sur le vanillier. *L'Agriculture Pratique des Pays Chaude* **10**(2): 33-42.
Anderson, M.S. 1965. Oswald Schreiner (1875-1965). *Journal of the Association of Official Analtical Chemists* **48**: 1270-1271.
André, G. 1922. *Propriétés Générales des Sols en Agriculture*. Librairie Armand Colin, Paris.
Bailey, L.H. 1922. *Cyclopedia of Farm Crops*. MacMillan, New York.
Baker, G.I., W.D. Rasmussen, V. Wiser and J.M. Porter. 1963. *Century of Service: the first 100 years of the United States Department of Agriculture*. USDA, Washington, D.C.
Beal, W.H. 1911. The new science of the soil: bacteriology and chemistry on the farm. *Scientific American* **104**: 168-169; 186-187.
Bottomley, W.B. 1911a. Some effects of bacteriotoxins on the germination and growth of plants. *Report of the British Association for the Advancement of Science* **81**: 584-585.

Bottomley, W.B. 1911b. Some effects of bacteriotoxins on soil organisms. *Report of the British Association for the Advancement of Science* **81**: 806.
Bottomley, W.B. 1914a. The significance of certain food substances for plant growth. *Annals of Botany* **28**: 531-540.
Bottomley, W.B. 1914b. Some accessory factors in plant growth and nutrition. *Proceedings of the Royal Society of London, Series B* **88**: 237-247.
Bottomley, W.B. 1915. A bacterial test for plant food accessories. *Proceedings of the Royal Society of London, Series B* **89**: 102-108.
Bottomley, W.B. 1916a. Bacterised peat. *Science Progress* **11**: 298-303.
Bottomley, W.B. 1916b. Bacterised peat. The problem in relation to plant nutrition. *Journal of the Society of Chemical Industry* **35**: 871.
Bottomley, W.B. 1917. Some effects of organic growth promoting substances (auximones) on plant growth of Lemna minor in mineral culture solutions. *Proceedings of the Royal Society of London, Series B* **89**: 481-507.
Bottomley, W.B. 1920. The effect of organic matter on the growth of various water plants in culture solution. *Annals of Botany* **34**: 353-365.
Breazeale, J.F. 1905. Effect of the concentration of the nutrient solution upon wheat culture. *Science* **22**: 146-149.
Breazeale, J.F. 1906. Effect of certain solids upon the growth of seedlings in water culture. *Botanical Gazette* **41**: 54-63.
Bruère, R.W. 1915. The control of soil fertility. *Harper's Magazine* **126**: 696-704.
Cameron, F.K. 1902. The soil as an economic and social factor. *Popular Science Monthly* **60**: 539-550.
Cameron, F.K. 1904. Reply to an address: present state of soil investigations. *Science* **19**: 343-347.
Cameron, F.K. 1910. An introduction to the study of the soil solution. *Journal of Physical Chemistry* **14**: 320-372; 393-451.
Cameron, F.K. 1911. *The Soil Solution: the Nutrient Medium for Plant Growth*. Chemical Publishing Company, Easton, Pa.
Cameron, F.K. 1912. Concentration of the soil solution. *Original Communications of the 8th International Congress of Applied Chemistry* **15**: 43-48.
Cameron, F.K. and Bell, J.M. 1905. The mineral constituents of the soil solution *U.S. Department of Agriculture, Bureau of Soils Bulletin* **30**: 1-.
Cameron, F.K. and Bell, J.M. 1907. *Les Constituents Minéraux et les Solutions du Sol*. Translated by H. Fabré. Coulet et fils, Montpellier.
Chemin, E. 1921. l'Intoxication du sol par les plantes. *Comptes Rendus de l'Association Français pour l'Avancement de Science* **44**: 202-208.
Clements, F.E. 1921. *Aeration and Air Content: The Role of Oxygen in Root Activity. Carnegie Institution of Washington Publication No. 315*. Carnegie Institution, Washington, D.C.
Coates, L.R. 1910. Process of making soil antitoxin. *Journal of the Society of Chemical Industry* **29**: 442.
Coulter, J.M., C.R. Barnes and H.C. Cowles. 1911. *A Textbook of Botany for Colleges and Universities. Vol. II. Ecology*. American Book Company, New York.
Coville, F.V. 1910. Experiments in blueberry culture. *U.S. Department of Agriculture, Bureau of Plant Industry Bulletin* **193**.
Cowles, H.C. 1911. The causes of vegetative cycles. *Botanical Gazette* **51**: 161-183. Also published in *Annals of the Association of American Geographers* **1**: 3-20.
Dachnowski, A. 1905. Physiological properties of bog water. *Botanical Gazette* **39**: 348-355.
Dachnowski, A. 1908a. Is toxicity a factor in soil problems? Report of the Michigan Academy of Science **10**: 209-210.
Dachnowski, A. 1908b. Toxic properties of bog water and bog soil. *Botanical Gazette* **46**: 130-143.
Dachnowski, A. 1909. Bog toxins and their effect upon soils. *Botanical Gazette* **47**: 389-405.
Dachnowski, A. 1912a. The relation of Ohio bog vegetation to the chemical nature of peat soils. *Journal of Biological Chemistry* **11**: xxxviii.
Dachnowski, A. 1912b. Peat deposits of Ohio, their origin, formation and uses. *Geological Survey of Ohio, 4th Series Bulletin* **16**: 1-424.
Dandeno, J.B. 1904. The relation of mass action and physical affinity to toxicity with incidental discussion as to how far electrolytic dissociation may be involved. *American Journal of Science* **167**: 437-458.
Dandeno, J.B. 1905. Soil fertility. *Popular Science Monthly* **67**: 622-625.

Dandeno, J.B. 1908. Mutual interaction of plant roots. *Report of the Michigan Academy of Science* **10**: 32-36.
Dandeno, J.B. 1909a. Mutual interaction of plants. *Report of the Michigan Academy of Science* **11**: 24-25.
Dandeno, J.B. 1909b. Mutual interaction of plant roots. *Experiment Station Record* **15**: 780.
Dandeno, J.B. 1910. Plant excretion. *Report of the Michigan Academy of Science* **12**: 86-90.
Davenport, E. 1910. The status of soil fertility investigations. *University of Illinois Agricultural Experiment Station Circular* **123**: 1-56.
Davidson, J. 1915. A comparative study of the effect of cumarin and vanillin on wheat growth in soil, sand and water cultures. *Journal of the American Society of Agronomy* **7**: 145-158; 221-238.
Demolon, A. 1922. Fatigue et intoxication du sol. *Journal d'Agriculture Pratique* **86**: 248-249.
Dietrich, T., C. Schnaetzlein and Stift. 1911. Boden und Dungen. *Jahresbericht über die Fortschritte der Agriculturchemie* **12**: 38-175.
Fletcher, F. 1908a. Crop rotation and soil exhaustion. *Cairo Science Journal* **2**(19).
Fletcher, F. 1908b. Note on a toxic substance excreted by the roots of plants. *Memoirs of the Department of Agriculture in India, Botany Series* **2**(3): 1-16.
Fletcher, F. 1910. Effect of previous heating of the soil on the growth of of plants and the germination of seeds. *Cairo Science Journal* **4**: 81-86.
Fletcher, F. 1912. Toxic excreta of plants. *Journal of Agricultural Science* **4**: 245-247.
Fletcher, F. 1913. The bacterial theory of soil fertility. *Nature (London)* **90**: 541-542.
Foussat, J. 1911. 'Methods of crop rotation' *Engrais* **25**: 629-630
Fraps, G.S. 1915. The effect of organic compounds in pot experiments. *Texas Agricultural Experiment Station Bulletin* **174**: 5-13.
Fred, E.B. 1912. Uber die Beschleunigung der Lebenstätigkeit hoherer und niederer Pflanzen durch klein Giftmengen. *Centralblatt für Bakteriologie, Parasitkunde, Infektionskrankheiten (und Hygeine), Abteilung II* **31**: 185-245.
Funchess, M.J. 1916. The effects of certain organic compounds on plant growth. *Alabama Agricultural Experiment Station Bulletin* **191**: 103-132.
Gardner, F.D. 1905. The wire-basket method for determining the manorial requirements of soils. *U.S. Department of Agriculture, Bureau of Soils Circular* **18**: 1-6.
Gardner, F.D. 1916. *Successful Farming*. The Smithsonian Company, Oakland.
Gardner, W.A. 1924. The decomposition of salicylic aldehyde by soil organisms. *Science (Washington, D.C.)* **60**: 503.
Gardner, W.A. 1926. The decomposition of toxins by soil organisms. *Agricultural Experiment Station of the Alabama Polytechnic Institute Bulletin* **225**: 1-38.
Greig-Smith, R. 1910. Contributions to our knowledge of soil fertility. I. The action of wax-solvents and the presence of thermolabile bacteriotoxins in soil. *Proceedings of the Linnean Society of New South Wales* **35**: 808-822.
Greig-Smith, R. 1911a. The bacteriotoxins and the agricere of soils. *Centralblatt für Bakteriologie, Parasitkunde, Infektionskrankheiten (und Hygeine), Abteilung II* **30**: 154-156.
Greig-Smith, R. 1911b. Contributions to our knowledge of soil fertility. IV. The agricere and bacteriotoxins of soil. *Proceedings of the Linnean Society of New South Wales* **36**: 679-699.
Greig-Smith, R. 1912a. The agricere and the bacteriotoxins of the soil. *Centralblatt für Bakteriologie, Parasitkunde, Infektionskrankheiten (und Hygeine), Abteilung II* **34**: 224-226.
Greig-Smith, R. 1912b. Contributions to our knowledge of soil fertility. V. The action of fat-solvents upon sewage-sick soils. *Proceedings of the Linnean Society of New South Wales* **37**: 238-243.
Greig-Smith, R. 1912c. Contributions to our knowledge of soil fertility. VI. The inactivity of the soil-protozoa. *Proceedings of the Linnean Society of New South Wales* **37**: 655-672.
Greig-Smith, R. 1913a. Contributions to our knowledge of soil fertility. VII. The combined action of disinfectants and heat upon soils. *Proceedings of the Linnean Society of New South Wales* **38**: 725-728.
Greig-Smith, R. 1913b. Contributions to our knowledge of soil fertility. VIII. The toxins of soil. *Proceedings of the Linnean Society of New South Wales* **38**: 728-737.
Greig-Smith, R. 1913c. Contributions to our knowledge of soil fertility. IX. The formation of toxins in the soil. *Proceedings of the Linnean Society of New South Wales* **38**: 737-740.
Greig-Smith, R. 1914. Note on the bacteriotoxic action of water. *Proceedings of the Linnean Society of New South Wales* **39**: 533-537.
Greig-Smith, R. 1915. Contribution to our knowledge of soil fertility. XIII. The toxicity of soils. *Proceedings of the Linnean Society of New South Wales* **40**: 631-645.

Greig-Smith, R. 1918a. Contribution to our knowledge of soil fertility. XVI. The search for toxin-producers. *Proceedings of the Linnean Society of New South Wales* **43**: 142-190.

Greig-Smith, R. 1918b. Bacterial toxins in soils. *Journal of the Department of Agriculture of Victoria* **16**: 119-120.

Guthrie, F.B. 1910. Injurious substances in the soil. Bare patches, etc. *Agricultural Gazette of New South Wales* **21**: 434-441.

Guthrie, F.B. 1913. The relation of fertilisers to soil fertility. A short survey of present views on the subject. *Agricultural Gazette of New South Wales* **24**: 221-233.

H. [Hall], A.D. 1904. A new theory of soil. *Nature* **69**: 58-59.

Hall, A.D. 1907. Agricultural chemistry and vegetable physiology. *Annual Report on Progress in Chemistry* **4**: 261-278.

Hall, A.D. 1908a. Agricultural chemistry and vegetable physiology. *Annual Report on Progress in Chemistry* **5**: 242-257.

Hall, A.D. 1908b. Theories of manure and fertiliser action. *Science* **28**: 617-628.

Hall, A.D. 1909. *Fertilisers and Manures*. John Murray, London.

Hall, A.D. 1910. Opening address. Sub-section of B. Agricultural sub-section. The British Association at Sheffield. *Nature* **84**: 309-312. Also published as: "The fertility of the soil" in *Science* **32**: 363-371.

Harrison, W.H. and P.A.S. Aiyer. 1913. The gases of swamp rice soils and their composition and relationship to the crop. *Memoirs of the Department of Agriculture in India, Botany Series* **3**: 65-106.

Harshberger, J.W. 1911. The soil: a living thing. *Science (Washington, D.C.)* **33**: 741-744.

Hartwell, B.L. and S.C. Damon. 1918. The influence of crop plants on those which follow. I. *Agricultural Experiment Station of the Rhode Island State College Bulletin* **175**: 1-29.

Hartwell, B.L., F.R. Pember and G.E. Merkle. 1919. The influence of crop plants on those which follow. II. *Agricultural Experiment Station of the Rhode Island State College Bulletin* **176**: 1-47.

Hartwell, B.L., J.B. Smith and S.C. Damon. 1927. The influence of crop plants on those which follow. III. *Agricultural Experiment Station of the Rhode Island State College Bulletin* **210**: 1-23.

Hays, W.M., A. Boss and A.D. Wilson. 1908. The rotation of crops. 1. Report of 10 years on 44 rotation plots. *Minnesota Agricultural Experiment Station Bulletin* **109**: 281-327.

Hays, W.M. and M. Whitney. 1907. Report on statements of Dr. Cyril G. Hopkins relative to Bureau of soils. *U.S. Department of Agriculture Circular* **22**: 1-12.

Helms, D., A.B.W. Effland and S.E. Phillips. 2002. Founding the USDA's Division of Agricultural Soils: Charles Dabney, Milton Whitney, and the State Experiment Stations. In: D. Helms, A.B.W. Effland and P.J. Durana (eds.), *Profiles in the History of the U.S. Soil Survey*, pp. 1-18. Iowa State Press, Ames.

Hilgard, E.W. 1903. Chemistry of soils as related to crop production. *Science* **18**: 755-760.

Hilgard, E.W. 1904. Chemistry of soils as related to crop production. *U.S. Department of Agriculture, Office of Experiment Stations Bulletin* **142**: 117-121.

Hilgard, E.W. 1904a. Soil work in the United States. *Science* **19**: 233-234.

Hilgard, E.W. 1904b. Soil management. *Science* **20**: 605-608

Hilgard, E.W. 1906. *Soils: their Formation, Properties, Composition, and Relations to Climate and Plant Growth in the Humid and Arid Regions*. MacMillan, New York.

Hopkins, C.G. 1903. The present status of soil investigation. *University of Illinois Agricultural Experiment Station Circular* **72**: 1-21.

Hopkins, C.G. 1904a. The present status of soil investigation. *Science* **19**: 626-629.

Hopkins, C.G. 1904b. The present status of soil investigation. *U.S. Department of Agriculture, Office of Experiment Stations Bulletin* **142**: 95-104.

Hopkins, C.G. 1906. The duty of chemistry to agriculture. *University of Illinois Agricultural Experiment Station Circular* **105**: 1-27.

Hopkins, C.G. 1908. Chemical principles of soil fertility. *University of Illinois Agricultural Experiment Station Circular* **124**: 1-16.

Hopkins, C.G. 1910a. European practice and American theory concerning soil fertility. *University of Illinois Agricultural Experiment Station Circular* **142**: 1-31.

Hopkins, C.G. 1910b. *Soil Fertility and Permanent Agriculture*. Ginn and Co., Boston.

Hopkins, C.G. 1911. *The Story of the Soil*. Richard G. Badger: The Gorham Press, Boston. [reprinted 2004 Kessinger Pub.]]

Hopkins, C.G. 1912. Plant food in relation to soil fertility. *Science* **36**: 616-622.

Hopkins, C.G. 1913. *The Farm That Won't Wear Out*. The Author, Champaign. [reprinted 2004 Kessinger Pub.]

Immendorff, H. 1910. Altes und Neues aus dem Gebiete der Düngerlehre. *Mitteilungen der Ökonomischen Gesellschaft im Königreiche Sachsen* **1909-1910**: 89-105.
Jamieson, T. 1913. Hurtful factors in cultivated plants. *Aberdeenshire Agricultural Research Association Annual Report* **1913**: 15-37.
Jenny, H. 1961. *E.W. Hilgard and the Birth of Modern Soil Science*. Agrochimica, Pisa.
Jodidi, S.I. 1913. The nature of humus and its relationship to plant life. *Biochemical Bulletin* **3**: 17-22.
Johnson, J. 1916. Preliminary studies on heated soils. *Science (Washington, D.C.)* **43**: 434-435.
Johnson, J. 1919. The influence of heated soils on seed germination. *Soil Science* **7**: 1-87.
King, F.H. 1904. *Investigations in Soil Management, Being Three of Six Papers on the Influence of Soil Management Upon the Water-Soluble Salts in Soils and the Yield of Crops*. The Author, Madison.
King, F.H. 1905. Investigations in soil management. *U.S. Department of Agriculture, Bureau of Soils Bulletin* **26**: 1-205.
King, F.H. 1906. *The Soil*. MacMillan, New York.
King, F.H. 1908. Toxicity as a factor in the productive capacity of soils. *Science* **27**: 626-635.
Leather, J.W. 1907. Schreiner and Reed on deleterious excretions by roots. *Torreya* **7**: 220-221.
Leembruggen, W.G. 1909. De gezonheid eenen cultuur-bodem, in verbond met wortelsecreties. *Tydschrift voor Nyverheid en Landbouw in Nederlandsch-Indie* **80**:
Lemmermann, O. 1909. Die Whitneysche Theorie des Wesen der Bodenfruchtbarkeit. *Mittheilungen der Deutsche Landwirtschaftsgesellschaft* **1909**: 739-742.
Lipman, C.B. 1918. On theories concerning soils as media for plant growth. *School Science and Mathematics* **18**: 686-697; 780-791.
Livingston, B.E. 1905a. The relation of soils to natural vegetation in Roscommon and Crawford counties. *Report of the State Board of Geological Survey of Michigan* **5**: 2-14.
Livingston, B.E. 1905b. Physiological properties of bog water. *Botanical Gazette* **39**: 348-353.
Livingston, B.E. 1905c. Relation of transpiration to growth in wheat. *Botanical Gazette* **40**: 178-195.
Livingston, B.E. 1906. A simple method for experiments with water cultures. *Plant World* **9**: 13-16.
Livingston, B.E., J.C. Britton and H.S. Reed. 1905. Studies on the properties of an unproductive soil. *U.S. Department of Agriculture, Bureau of Soils Bulletin* **28**: 1-39.
Livingston, B.E. 1923. Some physiological aspects of soil toxicity. *Journal of the American Society of Agronomy* **15**: 313-323.
Livingston, B.E., J.C. Britton and F.R. Reid. 1905. Studies on the properties of an unproductive soil. *U.S. Department of Agriculture, Bureau of Soils Bulletin* **28**: 1-39.
Livingston, B.E., C.A. Jensen, J.F. Breazeale, F.R. Pember and J.J. Skinner, J.J. 1907. Further studies on the properties of unproductive soils. *U.S. Department of Agriculture, Bureau of Soils Bulletin* **36**: 1-71.
Lumière, A. 1920. Le reveil de la terre arable. *Comptes Rendus hebdomadaire des Séances de l'Academie des Sciences* **171**: 868-871.
Lumière, A. 1921. Action nocive des feuilles mortes sur la germination. *Comptes Rendus hebdomadaire des Séances de l'Academie des Sciences* **172**: 232-234.
Lyon, T.L., E.O. Fippin and H.O. Buckman,. 1916. *Soils: Their Properties and Management*. MacMillan, New York.
Lyon, T.L. and J.K. Wilson. 1921. Liberation of organic matter by roots of growing plants. *Cornell University Agriculture Experiment Station Memoir* **40**: 1-44.
Massart, J. 1912 Le rôle de l'expérimentation en géographie botanique. *Receuil de l'Institut Botanique Léo Errera* **9**: 68-80.
Meggitt, A.A. 1914. Studies on an acid soil in Assam. *Memoir of the Department of Agriculture, India* **3**: 235-269.
Meggitt, A.A. and A.G. Birt. 1912. Preliminary note on the occurrence of acidity in highland soil. *Agricultural Journal of India* **8**: 68-73.
Merrill, M.C. 1915. Electrolytic determination of exosmosis from the roots of plants subjected to the action of various agents. *Annals of the Missouri Botanical Garden* **2**: 507-572.
Miller, N.J.H. 1913. Agricultural chemistry and vegetable physiology. *Annual Report on Progress in Chemistry* **10**: 211-232.
Molliard, M. 1913. Sur la sécrétion par les raciness de substances toxiques pour la plante (Note préliminaire). *Bulletin de la Société Botanique de France* **60**: 442-446.
Molliard, M. 1915. Sécrétion par les racines de substances toxiques pour la plante. *Revue Générale de Botanique* **27**: 289-296.

Mukerji, J.N. 1920. The excretion of toxins from the roots of plants. *Agricultural Journal of India* **15**: 502-507.
Myrick, H. 1904. Editorial. *Orange Judd Farmer*, January 23, 1904.
Newcombe, F.C. 1902. The rheotropism of roots. VI. Location of the sensitive area. *Botanical Gazette* **33**: 341-362.
Odland, T.E., J.B. Smith and S.C. Damon. 1934. The influence of crop plants on those which follow. IV. *Agricultural Experiment Station of the Rhode Island State College Bulletin* **243**: 1-33.
Palladin, V.I. 1918. *Plant Physiology. Authorized English Edition, based on the German Translation of the Sixth Edition and on the Seventh Russian Edition (1914)*, edited by B.E. Livingston. P. Blakiston's Son & Co., Philadelphia.
Parisot, F. 1911. *Rotations et Assolements*. Librairie Larousse, Paris.
Peralta, F. de, and R.P. Estioka.. 1923. A tentative study of the effect of root excretion of common paddy weeds upon crop production of lowland rice. *Philippine Agriculturalist* **11**: 205-216.
Periturin, F.T. 1913. 'Soil fatigue' [Russian]. *Izvestiya Moskovskii Sel'skokhozyaistvennyi Institut* **19**(4): 1-141.
Peter, A.M. and F.K. Cameron. 1909. Soil fertility. *Journal of Industrial and Engineering Chemistry* **1**: 263-267.
Petit, A. 1910. *Principes Généraux de la Culture des Plantes en Pots*. Hachette, Paris
Petit, A. 1922. À propos du "reveil de la terre arable". *Comptes Rendus hedbomadaire des Séances de l'Académie des Sciences* **174**: 1033-1034.
Piettre, M. 1923. Recherche au moyen de la pyridine, des matières humiques et des matières grasses du sol. *Comptes Rendus hebdomadaires des Séances de l'Academie des Sciences* **176**: 1329-1331.
Pouget, I. and D. Chouchak. 1915. Sur la fatigue de terre. *Comptes Rendus hebdomadaire des Séances de l'Académie des Sciences* **145**: 1200-1203.
Pratolongo, U. 1915. Sul probleme della fertilita. *Le Stazioni Sperimentali Agrarie Italiane* **48**: 491-507.
Pryanishnikov, D.N. 1914. Sur la question des excrétions nuisible des racines. *Revue Générale de Botanique* **25**: 563-582.
Reed, H.S. 1907a. The malignant effect of certain trees upon surrounding plants. *Plant World* **10**: 279-282.
Reed, H.S. 1907b. The stimulation of plant growth by means of weak poisons. In: Cyclopedia of American Agriculture, Volume II. (ed. L.H. Bailey) [also in Bailey (1922), pp. 28-29]
Reed, H.S. 1908a. Modern and early work upon the question of root excretions. *Popular Science Monthly* **73**: 257-266.
Reed, H.S. 1908b. Review of *Essai sur la Valeur Antitoxique de l'Aliment Complet et Incomplet*. By A. LeRenard. *Science (Washington, D.C.) N.S.* **28**: 236-238.
Reed, H.S. 1943. *Memories & Vagaries*. Published by the Author, Berkeley.
Reed, H.S. and B. Williams. 1915. The effect of some organic soil constituents upon nitrogen fixation by *Azotobacter*. *Centralblatt für Bakteriologie, Parasitkunde, Infektionskrankheiten (und Hygiene), Abteilung II* **43**: 166-176.
Rigg, G.B. 1913. The effect of some Puget Sound bog waters on the root hairs of *Tradescantia*. *Botanical Gazette* **55**: 314-326.
Rigg, G.B. 1916a. A summary of bog theories. *Plant World* **10**: 310-325.
Rigg, G.B. 1916b. Decay and soil toxins. *Botanical Gazette* **61**: 295-310.
Rigg, G.B., H.I. Turnbull and M. Lincoln. 1916. Physical properties of some toxic substances. *Botanical Gazette* **61**: 408-416.
Robinson, W.J. 1909. Experiments on the effects of the soil of the Hemlock Grove of the New York Botanical Garden upon seedlings. *Journal of the New York Botanical Garden* **10**: 81-87.
Rousset, H. 1908. La fertilité de la terre. *Revue Générale de Chimie Pure et Appliquée* **11**: 152-160. [also in *Science et Nature*]
Russell, E.J. 1905. *Journal of Agricultural Science* **1**:
Russell, E.J. 1908. An alleged excretion of toxic substance by plant roots. *Nature (London)* **78**: 402-403.
Russell, E.J. 1910. Factors which determine fertility in soils. *Science Progress* **4**: 353-365.
Russell, E.J. 1911a. Recent investigations on soil fertility. *Nature (London)* **88**: 486-487.
Russell, E.J. 1911b. The soil and the plant: a review of some recent American hypotheses. *Science Progress* **6**: 135-152.
Russell, E.J. and J. Golding. 1911. Sewage sickness in soils, and its amelioration by partial sterilisation. *Journal of Chemistry in Society and Industry* **30**: 471-474.

Russell, E.J. and H.B. Hutchinson. 1909. The effect of partial sterilisation on the production of plant food. *Journal of Agricultural Science* **3**: 111-144.

Russell, E.J. and F.R. Petheridge. 1912. Investigations on "sickness" in soil. II. Glasshouse soils. *Journal of Agricultural Science* **5**: 86-111.

Schreiner, O. 1911. Symptoms shown by plants under the influence of different toxic compounds. *Journal of Biological Chemistry* **9**: xiii-xiv.

Schreiner, O. 1912. Organic constituents of soils. *Science* **36**: 577-587.

Schreiner, O. 1913. The organic constituents of soils. *U.S. Department of Agriculture, Bureau of Soils Circular* **74**: 1-18.

Schreiner, O. 1913a. Elimination and neutralization of toxic soil substances. *Proceedings of the American Philosophical Society* **52**: 470-480.

Schreiner, O. 1923. Toxic organic soil constituents and the influence of oxidation. *Journal of the American Society of Agronomy* **15**: 270-276.

Schreiner, O. and E.C. Lathrop. 1911a. Dihydroxystearic acid in good and poor soils. *Journal of the American Chemical Society* **33**: 1412-1417.

Schreiner, O. and E.C. Lathrop. 1911b. The distribution of organic constituents in soils. *Journal of the Franklin Institute* **172**: 145-151.

Schreiner, O. and E.C. Lathrop. 1911c. Examination of soils for organic constituents, especially dihydroxystearic acid. *U.S. Department of Agriculture, Bureau of Soils Bulletin* **80**: 1-33.

Schreiner, O. and E.C. Lathrop. 1912a. The chemistry of steam-heated soils. *U.S. Department of Agriculture, Bureau of Soils Bulletin* **89**: 1-37.

Schreiner, O. and E.C. Lathrop. 1912b. The chemistry of steam-heated soils. *Journal of the American Chemical Society* **34**: 1242-1259.

Schreiner, O. and H.S. Reed. 1907a. The production of deleterious excretion by roots. *Bulletin of the Torrey Botanical Club* **34**: 279-303.

Schreiner, O. and H.S. Reed. 1907b. The role of the oxidizing power of roots in soil fertility. *Journal of Biological Chemistry* **3**: xxiv-xxv.

Schreiner, O. and H.S. Reed. 1907c. Some factors influencing soil fertility. *U.S. Department of Agriculture, Bureau of Soils Bulletin* **40**: 1-40.

Schreiner, O. and H.S. Reed. 1908a. The power of sodium nitrate and calcium carbonate to decrease toxicity in conjunction with plants growing in solution culture. *Journal of the American Chemical Society* **30**: 85-97.

Schreiner, O. and H.S. Reed. 1908b. The toxic action of certain organic plant constituents. *Botanical Gazette* **45**: 73-102.

Schreiner, O. and Reed, H.S. 1909. The role of oxidation in soil fertility. *U.S. Department of Agriculture, Bureau of Soils Bulletin* **56**: 1-52.

Schreiner, O., H.S. Reed and J.J. Skinner. 1907. Certain organic constituents of soil in relation to soil fertility. *U.S. Department of Agriculture, Bureau of Soils Bulletin* **47**: 1-52.

Schreiner, O. and E.C. Shorey. 1907. The presence of secondary decomposition products of proteids in soils. *Journal of Biological Chemistry* **3**: xxxviii-xxxix.

Schreiner, O. and E.C. Shorey. 1908a. The isolation and toxic properties of an organic soil constituent. *Journal of Biological Chemistry* **4**: xxvi.

Schreiner, O. and E.C. Shorey. 1908aa. Toxic substances arising during plant metabolism. *Journal of Biological Chemistry* **4**: xxvi-xxvii.

Schreiner, O. and E.C. Shorey. 1908b. The isolation of picoline carboxylic acid from soils and its relation to soil fertility. *Journal of the American Chemical Society* **30**: 1295-1307.

Schreiner, O. and E.C. Shorey. 1908c. The isolation of dihydroxystearic acid from soils. *Journal of the American Chemical Society* **30**: 1599-1607.

Schreiner, O. and E.C. Shorey. 1909a. The isolation of harmful organic substances from soils. *U.S. Department of Agriculture, Bureau of Soils Bulletin* **53**: 1-33.

Schreiner, O. and E.C. Shorey. 1909b. The presence of cholesterol substances in soil-agrosterol. *Journal of the American Chemical Society* **31**: 116-118.

Schreiner, O. and E.C. Shorey. 1909c. Soil fatigue caused by organic matter. *Journal of Biological Chemistry* **6**: 39-50.

Schreiner, O. and E.C. Shorey. 1910a. Chemical nature of soil organic matter. *U.S. Department of Agriculture, Bureau of Soils Bulletin* **74**: 1-48.

Schreiner, O. and E.C. Shorey. 1910b. The presence of arginine and histidine in soils. *Journal of Biological Chemistry* **8**: 381-384.
Schreiner, O. and E.C. Shorey. 1910c. Pyrimidine derivatives and purine bases in soils. *Journal of Biological Chemistry* **8**: 385-393.
Schreiner, O. and E.C. Shorey. 1910d. Some acid constituents of soil humus. *Journal of the American Chemical Society* **32**: 1674-1680.
Schreiner, O. and E.C. Shorey. 1911a. Cholesterol bodies in soils: phytosterol. *Journal of Biological Chemistry* **9**: 9-11.
Schreiner, O. and E.C. Shorey. 1911b, Glycerides of fatty acids in soils. *Journal of the American Chemical Society* **33**: 78-80.
Schreiner, O. and E.C. Shorey. 1911c. Paraffin hydrocarbons in soils. Journal of the American Chemical Society **33**: 81-83.
Schreiner, O. and E.C. Shorey. 1911d. Soil organic matter as material for biochemical investigation. *Journal of the Franklin Institute* **171**: 295-300.
Schreiner, O. and J.J. Skinner. 1909. Ratio of plant nutrients as affected by harmful soil compounds. *Journal of Biological Chemistry* **7**: xxxiii-xxxiv.
Schreiner, O. and J.J. Skinner. 1911. Lawn soils. *U.S. Department of Agriculture, Bureau of Soils Bulletin* **75**: 1-55.
Schreiner, O., J.J. Skinner, L.C. Corbett and F.L. Mulford. 1912. Lawn soils and lawns. *USDA Farmers' Bulletin* **494**: 1-48.
Schreiner, O. and J.J. Skinner. 1910. Some effects of a harmful organic soil constituent. *Botanical Gazette* **50**: 161-181.
Schreiner, O. and J.J. Skinner. 1912a. Nitrogenous soil constituents and their bearing on soil fertility. *U.S. Department of Agriculture, Bureau of Soils Bulletin* **87**: 1-84.
Schreiner, O. and J.J. Skinner. (1912b). The toxic action of organic compounds as modified by fertilizer salts. *Botanical Gazette* **54**: 31-48.
Schreiner, O. and J.J. Skinner. 1914a. Harmful effects of aldehydes in soils. *U.S. Department of Agriculture, Bureau of Soils Bulletin* **108**: 1-26.
Schreiner, O. and J.J. Skinner. 1914b. Occurrence of aldehydes in garden and field soils. *Journal of the Franklin Institute* **178**: 329-343.
Schreiner, O. and J.J. Skinner. 1914c. Field tests with a toxic soil constituent: salicyclic aldehyde. *Journal of the American Society of Agronomy* **6**: 108.
Schreiner, O. and M.X. Sullivan. 1907. The products of germination affecting soil fertility. *Journal of Biological Chemistry* **3**: xxv-xxvi.
Schreiner, O. and M.X. Sullivan. 1908a. Toxic substances arising during plant metabolism. *Journal of Biological Chemistry* **4**: xxvi-xxvii.
Schreiner, O. and M.X. Sullivan. 1908b. Toxic substances arising during plant metabolism. *Science* (Washington, D.C.) **27**: 329.
Schreiner, O. and M.X. Sullivan. 1909a. Soil fatigue caused by organic compounds. *Journal of Biological Chemistry* **6**: 39-50.
Schreiner, O. and M.X. Sullivan. 1909b. Concurrent oxidizing and reducing power of roots. *Journal of Biological Chemistry* **7**: xxxii-xxxiii.
Schreiner, O. and M.X. Sullivan. 1910. Studies in soil oxidation. *U.S. Department of Agriculture, Bureau of Soils Bulletin* **73**: 1-57.
Schreiner, O. and M.X. Sullivan. 1911a. Biological analogies in soil oxidation. *Journal of Biological Chemistry* **9**: xvii.
Schreiner, O. and M.X. Sullivan. 1911b. Reduction by roots. *Botanical Gazette* **51**: 121-130.
Seaver, F.J. and Clark, E.D. 1910. Studies in pyrophilous fungi. II. Changes brought about by the heating of soils and their relation to the growth of *Pyronema* and other fungi. *Mycologia* **2**: 109-124.
Seaver, F.J. and Clark, E.D. 1912. Biochemical studies on soils subjected to dry heat. *Biochemical Bulletin* **1**: 413-427.
Shorey, E.C. (1906). *Report of the Hawaii Agricultural Experiment Station* **1906**: 37.
Shorey, E.C. (1912). The isolation of creatinine from soils. *Journal of the American Chemical Society* **34**: 99-107.
Shorey, E.C. (1913). Some organic soil constituents. *U.S. Department of Agriculture, Bureau of Soils Bulletin* **88**: 1-41.

Shorey, E.C. (1914). The presence of some benzene derivatives in soils. *U.S. Department of Agriculture, Journal of Agricultural Research* **1**: 357-363.
Shull, A.E. 1932. Toxicity of root excretions. *Plant Physiology* **7**: 339-341.
Shull, C.A. 1948. Burton Edward Livingston 1875-1948. *Science (Washington, D.C.)* **107**: 558-560.
Skinner, J.J. 1913. Illustration of the effect of previous vegetation on a following crop: cabbage after sesame. *Plant World* **16**: 342-6.
Skinner, J.J. 1914. Effect of salicyclic aldehyde on plants in soil and solution cultures. *Biochemical Bulletin* **3**: 390-402.
Skinner, J.J. 1915a. The antizymotic action of a harmful soil constituent: salicylic aldehyde and mannitol. *Plant World* **18**: 162-168.
Skinner, J.J. 1915b. Effect of vanillin as a soil constituent. *Plant World* **18**: 321-330.
Skinner, J.J. 1915c. Field tests with a toxic soil constituent: vanillin. *U.S. Department of Agriculture Journal of Agricultural Research* **164**.
Skinner, J.J. 1918a. Soil aldehydes, a scientific study of a new class of soil constituents unfavourable to crops, their occurrence, properties and elimination in practical agriculture. 1. *Journal of the Franklin Institute* **186**: 165-186.
Skinner, J.J. 1918b. Soil aldehydes, a scientific study of a new class of soil constituents unfavourable to crops, their occurrence, properties and elimination in practical agriculture. 2. *Journal of the Franklin Institute* **186**: 289-316.
Skinner, J.J. 1918c. Soil aldehydes, a scientific study of a new class of soil constituents unfavourable to crops, their occurrence, properties and elimination in practical agriculture. 3. *Journal of the Franklin Institute* **186**: 449-480.
Skinner, J.J. 1918d. Soil aldehydes, a scientific study of a new class of soil constituents unfavourable to crops, their occurrence, properties and elimination in practical agriculture. 4. *Journal of the Franklin Institute* **186**: 547-584.
Skinner, J.J. 1918e. Soil aldehydes, a scientific study of a new class of soil constituents unfavourable to crops, their occurrence, properties and elimination in practical agriculture. 5. *Journal of the Franklin Institute* **186**: 723-741.
Skinner, J.J. and J.H. Beattie. 1916. A study of the action of carbon black and similar absorbing materials in soils. *Soil Science* **2**: 93-101.
Skinner, J.J. and C.F. Noll. 1916. Field tests of fertiliser action on soil aldehydes. *Journal of the American Society of Agronomy* **8**: 273-298.
Skinner, J.J. and F.R. Reid. 1919. The influence of phosphates on the action of alpha-crotonic acid on plants. *American Journal of Botany* **6**: 167-80.
Smith, S. 1849. *A Word in Season;or, how the corn-grower may yet grow rich, and his labourer happy.* James Ridgway, London.
Spillman, W.J. 1906. Renovation of worn-out soils. *U.S. Department of Agriculture, Farmers' Bulletin* **245**: 1-16.
Stead, A. 1918. Plant toxins, a cause of infertility in soils: a South African observation. *South African Journal of Science* **14**: 439-442.
Stubbs, W.C. 1913. Organic matter in soils. *The Louisiana Planter and Sugar Manufacturer* **50**(24): 379-382.
Sullivan, M.X. 1912. Biochemical factors in soil. *Original Communications of the 8th International Congress on Applied Chemistry* **15**: 305-312.
Thatcher, R.W. 1923. The effect of one crop on another. *Journal of the American Society of Agronomy* **15**: 331-338.
Thomas, A.W. 1914. A review of methods for the isolation and identification of the organic constituents of soil. *Biochemical Bulletin* **3**: 210-221.
Transeau, E.N. 1914. On the development of palisade tissue and resinous deposits in leaves. *Science (Washington, D.C.)* **19**: 866-867.
True, A.C., W.H. Beal and H.C. White. 1904. Proceedings of the Seventeenth Annual Convention of the Association of American Agricultural Colleges and Experiment Stations, held at Washington, D.C., November 17-19, 1903. *U.S. Department of Agriculture, Office of Experiment Stations Bulletin* **142**. See pp. 117-121.
Ulpiani, C, 1910. La chimica fisica e l'agricoltura. *Conferencia al IV Congresso della Societa Italiana per il Progresso delle Scienze*. Naples.

United States House of Representatives. 1908. *Hearings Before the Committee on Agriculture, of the Honorable Secretary of Agriculture and Chiefs of Bureaus and Divisions of the Department of Agriculture, on the Estimates of Appropriations for the Fiscal Year Ending June 30, 1909*. Government Printing Office, Washington, D.C.

Upson, F.W. and A.R. Powell. 1915. Effect of certain organic compounds on wheat plants in the soil. Preliminary report. *Journal of Industrial and Engineering Chemistry* **7**: 420-422.

Usher, A.P. 1923. Soil fertility, soil exhaustion, and their historical significance. *Quarterly Journal of Economics* **37**: 385-411.

Voelcker, J.A. 1904. Agricultural chemistry and vegetable physiology. *Annual Report on the Progress of Chemistry* **1**: 192-221.

Walters, E.H. and L.E. Wise. 1916. α-Crotonic acid, a soil constituent. *U.S. Department of Agriculture, Journal of Agricultural Research* **6**: 1043-1045.

Weber, G.A. 1928. *The Bureau of Chemistry and Soils: its History, Activities and Organization*. Johns Hopkins Press, Baltimore.

Weir, W.W. 1920. *Productive Soils: the Fundamentals of Successful Soil Management and Profitable Crop Production*. J.B. Lippincott Co., Philadelphia.

Westerhoff, R. 1859. *Verhandling over de kol- of hekselringen, ool wel tooverkringen genaamd; voorgelzen in de vergadering van het Genootschap ter Beförderung der Natuurundige Wetenschappen te Groningen, op Woensdag den 3 Februarij 1859, door R. Westerhoff*. C.M. van Hoitsema, Groningen.

Wheeler, H.J, 1911. Concerning the action of pyrogallol on unproductive soil. *Proceedings of the Society for the Promulgation of Agricultural Science* **30**: 43-54.

Whitney, M. 1892. Some physical properties of soils in their relation to crop production. *U.S. Department of Agriculture, Weather Bureau Bulletin* **4**: 1-90.

Whitney, M. 1901. Exhaustion and abandonment of soils. *U.S. Department of Agriculture Report* **70**: 1-48.

Whitney, M. 1906. Soil fertility. An address delivered before the Rich Neck Farmers' Club of Queen Anne County, Maryland. *U.S. Department of Agriculture, Farmers' Bulletin* **257**: 1-35.

Whitney, M. 1907. *La Fertilité du Sol. Conférence à l'association des fermiers de Rich Neck de comté de Queen Anne (Maryland)*. Translated by Henri Fabré. Coulet et fils, Montpellier.

Whitney, M. 1909. Soils of the United States, based upon the work of the Bureau of Soils to January 1, 1908. *U.S. Department of Agriculture, Bureau of Soils Bulletin* **55**: 1-243.

Whitney, M. 1921. Fundamental principles established by recent soil investigations. *Science (Washington, D.C.) N.S.* **54**: 348-351.

Whitney, M. 1925. *Soil and Civilization*. D. Van Nostrand, New York.

Whitney, M. and Cameron, F.K. 1903. The chemistry of the soil, as related to crop production. *U.S. Department of Agriculture, Bureau of Soils Bulletin* **22**: 1-71.

Whitney, M. and Cameron, F.K. 1904. Investigations in soil fertility. *U.S. Department of Agriculture, Bureau of Soils Bulletin* **23**: 1-48.

Whitney, M. and F.K. Cameron. 1907. *Invesigaciones Acerca de la Fertilidad del Terreno*. Translated by D. Fernando Flores E Iñiguez. Biblioteça Agraria Solariana, Seville.

De Wildeman, E. 1908a. Les racines des plantes excrètent-elles des poisons? *L'Agronomie Tropicale (Bruxelles)* **1**(7): 109-112.

De Wildeman, E. 1908b. Les racines des plantes excrètent-elles des poisons? *L'Agronomie Tropicale (Bruxelles)* **1**(10): 154-161.

Willis, R.J. 1996. Pioneers of allelopathy. V. Oswald Schreiner 1875-1965. *Allelopathy Journal* **3**: 1-8.

Willis, R.J. 1997. The history of allelopathy. 2. The second phase (1900-1920): The era of S.U. Pickering and the U.S.D.A. Bureau of Soils. *Allelopathy Journal* **4**: 7-56.

Woll, F.W. 1908. The Proceedings of the Association of Official Agricultural Chemists. *Science (Washington, D.C.)* (Oct. 26)

Zolla, D. 1907. Revue annuelle d'agronomie. *Revue Générale des Sciences* **18**:

Zolla, D. 1908. Revue annuelle d'agronomie. *Revue Générale des Sciences* **19**: 577-585.

CHAPTER 11

APPROACHING THE MODERN ERA

> I saw the white waves beat
> upon the shore
> by blows of the wind,
> and I was struck
> by the clean and clear view.
>
> Anonymous from *The Manyoshu* c. 759 A.D.

There was a general hiatus in interest in the chemical interaction of plants for at least a decade following about 1910. The reasons for this were twofold: 1) the work of Pickering at the Woburn Experimental Fruit Farm, and the relevant work of Whitney, Schreiner and various associates at United States Department of Agriculture, Bureau of Soils were seen as either unconvincing, unrepeatable, uneven, misinterpreted or biased, and were essentially regarded as best forgotten; and 2) events in Europe, including World War I (1914-1918), and the Russian Revolution of 1917 had far-reaching effects on resources, and consequently directions and funding for agricultural and botanical research.

As Livingston (1923) noted, the intervening years allowed healing following the bitter disputes of Whitney, Hilgard, King, Hopkins and others, and encouraged a more balanced outlook of the dynamics of plant interactions, particularly in the light of new discoveries in soil chemistry, microbiology and plant physiology. Furthermore, the advent of ecology as a science encouraged greater acceptance of the notion of complexity in nature. It also must be remembered that the concept of bacteria in the environment only gained currency toward the end of the nineteenth century, and interest was greatly accelerated by the discovery of microbial antibiosis. Another important discovery that gave substance to the well-known positive effects of legumes was that of rhizobial bacteria in the root nodules of leguminous plants.

The problems of infertile soils and declining crop yields were still ever-present; however, new information and new technology allowed manifold explanations, and researchers were more willing to accept a multiplicity and/or simultaneity of causes. For example, causes of soil infertility were acknowledged to include nutrient deficiency, acidification and changes in nutrient availability and or toxicity (such as aluminium), changes in soil microbial populations leading to nutrient immobilization and/or pathogenesis, and finally the accumulation of harmful substances in soil either due to root excretion or the microbial breakdown of plant organic matter.

The accumulation of data from a variety of scattered sources, including soil microbiology, agriculture, and plant ecology left the door open for further enquiries. Notably the new discoveries of the antibiotic interactions from various microorganisms served, at least in some circles, as a guiding light for kindred discoveries amongst higher organisms. The German biologist Ernst Küster is regarded as something of a founding father in chemical ecology, as in 1909 he published one of the earliest reviews on the topic, entitled "On the chemical influence of organisms on one another", which was largely concerned with the new information emerging from the nascent field of microbiology. He did include mention of the interactions of higher plants, and wrote:

> We come to an extension of the term of poisonous plant, after we have seen that the materials produced by a plant are not only able to be poisonous for humans and animals, but also for any other plants. If the different plant species can be mutually exclusive or co-existent by chemical effects, then it seems no longer impossible that some features of the ecology, observations on those so-called underplants, etc. are able to be explained on new, chemico-physiological bases. An abundance of new tasks, which must not lag in importance behind aspects supplied by the newly blossoming discipline of the soil bacteriology, is placed for agricultural research: the viewpoints suggested here in the theory of crop rotation and of soil sickness and with the study of the relevant observations must not be ignored. Some features, which have been attributed to the effectiveness of soil-inhabiting micro-organisms, will perhaps find better explanation with the study of the macro-organisms effective in the soil and their chemico-physiological peculiarities. I draw attention to the investigations of A. Koch on "soil sickness"; he succeeded in showing that vine-sick soil is improved through heating; it is suggested that heating may act less due a killing of the germs contained in the soil than due to a destruction of thermolabile materials of the soil. (p. 18)

Similarly, the Russian Nobel laureate, I.I. Mechnikov[1] used the paradigm of autotoxicity in microorganisms as a possible explanation of the asynchronous senescence of plant structures such as flowers, and the death of annual plants after fruiting (Metchnikoff 1907).

While the period prior to about 1925, as discussed in the previous chapters, was rich in anecdotes, commentaries, simple experiments, theories and polemic regarding the chemical interactions of plants, it must be said that little, if any of it, was utterly compelling in convincing anyone of the scientific basis of allelopathy, with the possible exception of the work of Pickering, which remains essentially unexplained to this day. It was the thirty years following 1925 that paved the way for the development of allelopathy as a science. As discussed initially above, this was for a whole variety of reasons: economic, sociological and technical. Firstly, there was a massive increase in knowledge that had continually gained momentum in the sciences, especially in chemistry, physics, and in entirely new disciplines such as microbiology and ecology. There developed widespread recognition of the economic and

---

[1] In the early twentieth century, I.I. Mechnikov was known in translation as E. Metchnikoff.

social value of studying plants, especially in order to develop improved methods in agriculture, forestry and phytopathology. This was manifested in the establishment of hundreds of research institutions devoted to agriculture and forestry in developed countries and their dependencies around the globe during the letter parts of the nineteenth century and the early decades of the twentieth century. There developed an increased awareness of the value of management and conservation of native natural resources, that was caused, at least in part, by the gradual disintregation of entrenched imperial regimes, a changing order in world politics, and the spectre of two global wars. The study of biology, or natural history as it formerly was known, had been the domain primarily of the more privileged classes, but it became increasingly accessible to anyone who was interested, with a will to learn . The first decades of the twentieth century witnessed the birth and proliferation of many of the scientific and professional societies and their associated journals, that have served to cement the cornerstones of allelopathic research. Some journals of importance to allelopathy that originated during this period include the *Journal of Ecology* (British Ecological Society and journal founded 1913), *Ecology* (Ecological Society of America and journal founded 1915), *Bulletin of the Torrey Botanical Club* (journal founded 1870), *Proceedings of the American Society of Agronomy*, continued by the *Journal of Agronomy* (American Society of Agronomy and journal founded 1907), *American Midland Naturalist* (journal founded at University of Notre Dame 1900), *American Journal of Botany* (Botanical Society of America founded 1893; journal founded 1914) and so forth. The technological revolution of the twentieth century provided an unprecedented array of analytical tools, including spectrophotometry, chromatography and statistics.

Despite the stigma that had come to accompany research involving root exudates or soil toxins, there still persisted a steady flow of articles that indicated investigators had found evidence of plant interaction via toxic substances, or a strong suspicion that such factors were at work, although there tended to be a markedly increased interest in foliar secondary metabolites and their release and effect in the environment. Furthermore, in company with the growing disciplines of both plant physiology and plant ecology, there was a distinct increase in interest in the role of toxic substances in natural plant communities (Evenari 1949, Bonner 1950), whereas previously interest had centred on agriculture, and sometimes forestry. As a result, many species hitherto ignored were added to the list of allelopathic candidates. The discovery of plant hormones during the 1920s and the consequent synthesis of indole acetic acid analogs such as the herbicide 2,4-D (2,4-dichlorophenoxyacetic acid) gave further impetus to understanding phytotoxicity in plants (Went 1950). The domain of allelopathy was widened eventually to embrace phenomena such as chemical interactions between pollen grains (e.g. Branscheidt 1930). Material from the years preceding 1955 has been well reviewed by Grümmer (1955), and I will here focus on personalities and studies from the era c. 1925-1955 that have served to form the bases of the modern concepts of allelopathy.

GERMAN BOTANISTS c. 1937

The year 1937 is crucial in the study of allelopathy, as a number of publications significant to the development of allelopathy appeared, notably those of Hans Molisch and Gerhard Madaus. The German-American botanist W.F. Loehwing (1936) had presented a brief review concerning the injurious effects of one crop on another to the Sixth International Botanical Congress in Amsterdam, and then in 1937 he comprehensively reviewed the world literature on the root interactions of plants, including the realm of root excretions (Loehwing 1937). He concluded that "most of the older data ascribing soil sickness and plant injury to toxic root excretions now have been re-interpreted as the results of disturbed nitrogen nutrition rather than direct injury by toxins." However, he did foreshadow the modern definition of allelopathy (Rice 1984) in writing that "the distinction between toxic and beneficial root secretions, though objective in a practical sense, is actually arbitrary."

Toward 1937, particularly in Germany, there seemed to be a growing awareness of the importance of metabolites in the ecology of plants, including their interactions with herbivores. While this latter topic is beyond this scope of the present discussion, it is interesting that in 1937 the Liebig Company[2], known for its meat extract product Oxo, issued among its many sets of popular trading cards, a set of six rather sophisticated cards devoted to the defences of plants against animals, that covered nettles, raphides, spines, latex, and even described the defensive symbiosis between the Central American ant *Azteca* and the tree *Cecropia* (Figure 11.1) that was described subsequently as an extreme form of allelopathy by Janzen (1969).

Figure 11.1. A Liebig Co. trading card from the series "Plant Defenses" (c. 1937) illustrating the relationship between the ant Azteca *and the tree* Cecropia.

---

[2] Justus von Liebig invented the process for making concentrated meat extracts in about 1840 and founded the companies that originally made these products commercially.

Hans Molisch

Hans Molisch (1856-1937; see Chapter 1, Figure 1.1) is sometimes referred to as "the father of allelopathy" (Narwal and Jain 1994, Lovett 2005), but as the preceding chapters have revealed, this attribution seems overly generous. Molisch was born into a middle-class German family in Brunn[3]. The family business was gardening, and thus Molisch was raised in an environment in which he learned about the practical aspects of plants. Molisch earned his Ph.D. in 1879, and then served at a variety of academic institutions in Vienna, Graz, and Prague, and this culminated in 1908 with his appointment as Director of the University of Vienna. He was the author of 24 different books, which spanned the breadth of plant biology, and his last book, published in 1937, was *Der Einfluss einer Pflanze auf die andere – Allelopathie* (Molisch 1937a), which has been only recently translated into English (Molisch 2001).

For the purposes of our discussion here, Hans Molisch is remembered as the originator of the word "allelopathy", which he coined, of course, in German in his book on allelopathy[4] (see Chapter 1). In 1936, Molisch, aged 80, began a series of fairly simple experiments in which plants of one species were exposed to the volatile emanations of another species. Advance notice of his findings were read to the Akademie der Wissenschaften in Vienna in April 1937 in a paper entitled "On the influence of one plant on another when separated spatially", but there was no apparent mention of allelopathy as yet (Molisch 1937b). Molisch's book, published later in the year, gave details of his findings, and consideration of other work, notably observations and experiments concerning the effects of air polluted by lighting gas on plants. These experiments, many of which involved the gas ethylene, led Molisch to speculate that such interactions must be more general than previously thought. However, the chemical phenomena that had molded Molisch's thinking were intimate relationships such as plant parasitism, grafting, and effects of microorganisms. Thus to Molisch, allelopathy was the realm of the immediate effects of plant volatiles, a topic which had been broached in a far less convincing fashion fifty years earlier by Gustav Jaeger.

Molisch's view of allelopathy in natural plant interaction did not sit entirely comfortably, and his view did not take into account the subtlety that allelopathy can assist in the maintenance of plant diversity[5], but he saw allelopathy as a factor associated with dominance:

> Therefore, gaseous inhibitory substances from the root (excluding carbonic acid) can exert such detrimental effect on roots of other plants, that it would be difficult for different kinds of plants to exist together at the same place. Perhaps plant sociology, to which this question is still completely strange, will in the future provide us more precise information about it.

---

[3] At the time Brunn was within the Austro-Hungarian Empire. Today this city is known as Brno, and is within the Czech Republic. Brno is best known in science as the home of Gregor Mendel, who knew the Molisch family.

[4] The *Macquarie Dictionary*, the definitive dictionary of English as used in Australia, erroneously gives the origins as French. Ironically, the French until recently have avoided using the term *allélopathie*, and have often preferred the endemic term *télétoxie* (Bertrand 1945, Wildeman 1946).

[5] See Chapter 1.

Thus, it is somewhat ironic that the name Molisch has become synonymous with allelopathy; indeed, for many writers, Molisch's work represents the starting point of the discipline. The book had little impact at the time, for a number of reasons: the book had limited accessibility as it was written in German[6], Molisch died a couple of months after his book was published, and finally, the political climate in Europe was rapidly changing and by 1939 Europe was embroiled in war. The work of Molisch, Gerhard Madaus, and others had a moderate impact in Germany, but little elsewhere. Molisch's work with volatile substances was emulated in Spain by the biologist Abilio Rodriguez Rosillo (1944). It is little realised that one of the earliest works, at least in English, to cite Molisch's monograph was the Swiss holistic agriculturalist, Ehrenfried Pfeiffer (1943). Pfeiffer was an adherent of the controversial farming principles of Rudolf Steiner, who championed a system of farming which he termed "Bio-dynamic Agriculture", the principles of which today are embodied in the various systems of "alternative" agriculture.

Gerhard Madaus

One of the most underrated figures in the development of allelopathy was Gerhard Madaus (1890-1942; Figure 11.2). His work on plant interactions was commenced before that of Molisch, had much more ecological relevance, and in many respects was much closer to what we regard as allelopathic research today.

Madaus' mother was an authority on homeopathy, and not surprisingly, Gerhard Madaus trained as a doctor. Due largely to the influence of Hugo Schulz at the University of Greifswald, Madaus developed what was to become a life-long interest in pharmacology and phytotherapeutic agents (Dietrichkeit 1991). In 1919 Gerhard Madaus and his two brothers established the pharmaceutical company Dr. Madaus & Co. By 1936 the company had numerous facilities throughout Germany and over 500 employees. The company became a household name, known especially for preparations of *Echinacea purpurea*[7], and the company is still thriving today as Madaus AG.

Gerhard Madaus had a strong and vested interest in the effects of plant chemicals on biological processes, and considerable sums were spent on research. He employed a number of research scientists, and maintained publication either through recognised journals or company periodical publications, such as the artistically ambitious *Jahrbuch Dr. Madaus* (1926-1938), which was followed by the more mundane and technical *Madaus Jahresbericht* (1937-1954). In 1936, the research activities became centralised with the founding of the Madaus Biological Institute in Radebeul (near Dresden), under the directorship of F.E. Koch. During the 1930's Madaus gained an interest in

---

[6] The book was reviewed in a number of German periodicals, but I could locate only two reviews in English (Anonymous 1938, Weiss 1938) and one in Russian (Razdorskii 1941.

[7] It is part of the Madaus folklore that when Gerhard Madaus went to the United States in search of herbal remedies, he gained an interest in *Echinacea angustifolia*, known since the nineteenth century as a native American remedy. He returned to Germany with *Echinacea angustifolia* seed; however, the seed had been misidentified and proved to be the untried *E. purpurea*.

Figure 11.2. Photograph of Gerhard Madaus (reproduced from Dietrichkeit 1991).

the effects of plant volatile substances and exudates on the growth of other plants. This likely occurred as Dr. Madaus & Co. had invested heavily in the local production of medicinal plants, and Madaus was naturally concerned about the effects of a previous crop on a successive crop, but also became aware of the possibility of increasing yield through companion planting. According to Dietrichkeit (1991), he was much influenced by E. J. Russell's book, *Soil Conditions and Plant Growth*, available in German as *Boden und Pflanze* (Russell 1936), and he became acquainted with the ideas of de Candolle. An inkling of this influence is given in a paper outlining an experiment in which wormwood (*Artemisia absinthium*) grew markedly more poorly in bitter lupin (*Lupinus albus*)[8] soil than soft lupin (*Lupinus* sp.) soil. Madaus (1937) concluded:

> A plant forming bitter substances alters the soil such that a second successive bitter plant finds conditions less favourable for the formation of these substances, and that consequently its bitter principle content declines. (p. 30)

His findings on the effects of one plant on another were published as early as 1934 in the *Jahrbuch Dr. Madaus*. His experiments included work very similar to

---

[8] Since ancient times lupins have featured in agriculture (see Chapter 2); the curious generic name derives from wolf, and this is in reference to the reputation of lupins in destroying soil fertility.

that of Molisch (1937), as he demonstrated the effects of volatiles, including ethylene, on other plants, and those of *Conium* on wheat (Madaus 1937a, 1937c; Madaus 1938b), and indeed he is among the very first authors to cite Molisch's book on allelopathy. In 1938 he began publication of his monumental *Lehrbuch der biologischen Heilmittel*, of which volume 1 contained a review of the chemical interactions of plants (Madaus 1938a). Madaus developed an interest in companion planting (*Freundschaft und Feindschaft*), and he strongly suspected that root exudates were involved in many plant interactions. He cited anecdotal examples from German culture: plant pairs that were "friendly" included the rose and lily, rue and figwort[9], Scotch pine and birch, Scotch pine and alder; antagonistic were peppermint and parsley, cabbage species and marjoram, borage, and cyclamen, oak and walnut, cornelian cherry (*Cornus mas*) and dog rose (*Rosa canina*). Experiments showed, for example, that *Viola tricolor*[10], which commonly appeared among rye stubble, germinated and grew better when there were neighbouring plants of rye (Schindler 1936, Madaus 1937a, Madaus 1938b). In another experiment, *Nasturtium officinale* appeared to unduly inhibit the growth of *Veronica beccabunga*. In a set of plot experiments, Madaus showed that the growth of *Atropa belladonna* was enhanced when grown with *Galega* or *Artemisia*, but was markedly depressed when grown with *Sinapis* (Madaus 1935, Madaus 1937a, Madaus 1938b). Madaus set up an ambitious experiment in which numerous species of herbaceous plants were grown either alone or in mixture with another species; altogether there 3225 field plots (Madaus 1937d). No attempt was made to establish whether the results were allelopathic or not; however, there was, for example, a striking reduction of growth of *Cannabis sativa*, *Papaver somniferum*, or *Galega officinalis* when grown with rye (*Secale cereale*). While Madaus was foremost a phytotherapist, he had a strong ecological bent, and he was convinced that the chemical components of plants had much to do with the maintenance of the integrity of plant communities; conversely he was aware that the artificial conditions generated in monocultures were likely to cause problems, as in so-called "soil sickness", which he noted was a problem with cultures of peppermint (*Mentha piperita*), marshmallow (*Althaea officinalis*), and madder (*Rubia tinctorum*). Perhaps because Madaus was essentially known for his phytotherapeutic work, his plant ecology studies are seldom cited; however, brief summaries did appear in 1938 (Anonymous 1938) and 1948 (Bässler 1948).

In addition to Schindler, other personnel at the Madaus Biological Institute also contributed to the allelopathic research. Bässler (1938) investigated the effect of various essential oils, e.g. garlic, peppermint and eucalyptus oil, on the germination and seedling growth of sweet pea (*Lathyrus odoratus*). Gerhard Madaus also collaborated with R. Thren (1940). The period 1939-1945 was not kind to Dr. Madaus & Co. Gerhard Madaus died in 1942, and aerial bombing by Allied foreces during World War II caused the loss of much of the company's buildings and factories.

---

[9] See Chapter 8, p.183.

[10] *Viola tricolor* in the wild is a small-flowered European plant, often a weed of fields; however, most people are familiar with this plant through the common pansy or heart's ease, which is a large-flowered and often odiferous cultivar.

Finally, the post-war occupation of the eastern part of Germany in 1945-1946 resulted in the dismantling of the company's headquarters in Radebeul, and its appropriation by the Soviet forces. Dr. Madaus & Co. re-established its headquarters in Bonn, which also became the new home for the Biological Institute. In 1950 the company decided to develop a second research facility, the Botanical Institute, in Cologne. This paved the way for another important associate of the firm Dr. Madaus & Co. - Arrien.G. Winter (1910-1960). Winter studied in microbiology and began his academic career at the University of Bonn. He became interested in soil bacteria, and wrote numerous papers about the occurrence and effects of bacterial phytotoxins in straw and in soil (Winter and Bublitz 1953a, Winter and Schönbeck 1953a, 1953b, 1954, Winter and Sievers 1954). In 1952 he was appointed Head of the new Botanical Institute of Dr. Madaus & Co., in Cologne, and also became Professor of Soil Microbiology at the University of Cologne. Winter published results on the antibiotic effects of nasturtium (*Tropaelium majus*) (Winter 1953a, 1953b, 1953c) and beech (*Fagus sylvatica*) litter (Winter and Bublitz 1953b). Another recruit to the Madaus Botanical Institute was Wolfgang Bublitz who investigated the inhibitory effects of spruce (*Picea* sp.) litter extracts on spruce seed germination (Bublitz 1953a, 1953b). The medicinal plant cultures at Dr. Madaus & Co., Cologne were made available to other researchers on allelopathy (e.g. Knapp and Thyssen 1952), and the phytosociologist Rüdiger Knapp, firstly at the University of Cologne and then at the University of Giessen, has subsequently maintained a life-long interest in allelopathy.

It is interesting that so much of the work on allelopathy from the middle decades from the twentieth century was published in Germany, and one can speculate that the German interest in homeopathy, as founded by Samuel Hahnemann, may have had something to do with this (see Chapter 8). It is little appreciated that another nation that has been very receptive to both homeopathy and allelopathy is India. Homeopathy was introduced into India through German missionaries in the early nineteenth century, and the holistic nature of Indian culture made it highly predisposed to Hahnemann's ideas. Perhaps this is parallelled by allelopathy, and in the past twenty-five years, concentrated interest in allelopathy has come from India, and more recently from China, another nation with a substantial interest in holistic medicine. India has shown leadership in allelopathic research, and the majority of monographs on allelopathy published during the past two decades have been authored or edited by Indian authors, e.g. S.S. Narwal, S.J.V. Rizvi and V. Rizvi, R.K.S. Kohli, and Inderjit. The only journal devoted to allelopathic research, the *Allelopathy Journal*, now in its fourteenth year, originates from India.

Friedrich Boas

Friedrich Boas was another German plant physiologist who had a strong interest in plant growth substances in relation to plant ecology. He was Professor of Botany at the Technischen Hochschule in Munich, and he was an associate of Karl Rippel (q.v.), who had an interest in soil sickness. Boas (1949) recognised that Molisch's term "allelopathy" tended to indicate only harmful effects and suggested, evidently without impact, the term "allelobiology". He noted that Koegel had also suggested a replacement term: allelergy. Boas had a longstanding interest in the chemical interactions of organisms, and curiously, he erroneously cited Molisch as the originator of the term "allelobiology" in a supposed posthumous work dated 1938 (Boas 1939).

SOIL MICROORGANISMS AND PLANT INTERACTIONS

The appreciation of the extent and importance of microorganisms in soil, in the twentieth century, led to numerous new lines of investigation concerning problems of low or declining soil fertility. It became increasingly apparent that organic substances, released into the soil either through exudation or the decomposition of plant parts, had the capacity to affect the abundance and relative balance of microorganisms including rhizospheric microorganisms, nitrifying bacteria, mycorrhizal fungi, pathogenic bacteria, fungi and nematodes (Waksman 1936). Research notably in 1950s confirmed that rhizospheric micro-organisms were commonly different from free-living soil micro-organisms, and were frequently dependent upon substances, such as certain amino acids, released from plant roots. Similarly, it was shown by investigators, such as Metz (1955) that root material from plants such as *Chelidonium majus*, *Crepis virens*, *Hieracium pilosella*, *Hypericum perforatum*, and *Viola tricolor* could be inhibitory to certain bacteria. It was also recognised that the plant roots of different species, or even different varieties of the same species, achieved their individual mixture of rhizospheric microorganism species that was a reflection of a mixture of factors including root exudation, antibiosis or stimulation among the microorganisms themselves, and altered nutritive conditions caused by the microorganisms. However, during this formative investigative period, there was comparatively little interdiciplinary research, and there was only gradual realisation of the potential role of microorganisms in allelopathy, for example, as in the decomposition of grass residues as investigated by A.G. Winter and his associates, and the decomposition of peach roots in the peach replant problem.

An important study was conducted over several years in Forestry Commission plots at Wareham Heath, Dorset, England. It had been seen that trial plantings of *Pinus* spp. at Wareham frequently resulted in poor seedling growth. William Neilson-Jones (1934) determined that a major cause of the checked growth was the poor establishment of mycorrhizal fungi on the pine roots, which was required for the satisfactory growth of the seedlings on the infertile heath soils. Subsequent work (Neilson-Jones 1940, 1941, Rayner and Neilson-Jones 1944) demonstrated that there existed a chemical factor in the soil, that was inimical to mycorrhizal fungi, such as *Boletus*. Neilson-Jones (1947) acknowledged that plants were capable of excreting materials into the soil, but indicated that there was still comparatively little evidence indicating them to be of any great significance in plant-plant interactions; however, there was a growing mass of information about the chemical interactions of microorganisms and the realm of antibiotics. Neilson-Jones, with his extensive experience in dealing with mycorrhizae, was well aware of the subtleties of the interactions that could occur in the soil.

The matter was significantly advanced by the work of Percy Brian, who is perhaps best remembered as the discoverer of the antibiotic griseofulvin. Brian *et al.*

(1945) found that the Wareham Heath soils contained an unusually large proportion of *Penicillium* species, and they showed that the soil containing an antibiotic substance later identified as gliotoxin. While it is not within the scope of this review, it is worth noting that this work, and other observations involving fungistasis, stimulated greater interest in the phenomenon of antibiosis in soil, particularly with the control of plant diesase in mind (Stevenson 1954, Jeffreys and Brian 1954). Work in the Soviet Union showed that rhizospheric bacteria could also suppress the establishment of mycorrhizae (Sideri and Zolotun 1952).

Another line of allelopathic research which had its origins during this time, and which eventually reached its full expression under the aegis of Elroy Rice during the 1970s, was that concerning the inhibition of nitrifying bacteria by higher plants. In the nineteenth century it had been noted that nitrate was often scarce in soil from certain vegetation types such as forests (Grandeau 1907). Early work by Yves Dommergues showed that forest soils in Madagascar exhibited much lower rates of nitrification than neighbouring agricultural soils (Dommergues 1954). It was subsequently suggested that in certain envornments plant roots may inhibit the growth of nitrifying bacteria, and there was a burst of activity in relation to grasses (Theron 1951, Theron and Haylett 1953). Stiven (1952) suggested that the low rates of nitrification from veldt soils in South Africa dominated by *Trachypogon plumosus* were due to antibiotic substances produced by the grass roots, although he only demonstrated antibiotic potential using non-nitrifying bacteria. Perhaps more importantly, Stiven noted that the same soil seemed inhibitory to the germination and seedling growth of various weeds. The matter was progressed by Roux (1953) who grew grass plants from seed in sand culture, and found that the root leachate was severely inhibitory to the germination of *Tagetes minuta*. Roux (1954) subsequently presaged the controversial theory that nitrogen availability may play a role in successional dynamics.

AGRICULTURE AND HORTICULTURE

Interest in the chemical interactions of plants largely originated as a result of trying to understand the causes of crop failure and apparent associated soil infertility. While interest in allelopathic phenomena in natural plant communites began to grow in the twenthieth century, it is also became increasingly apparent that many problems in agriculture remained unresolved, and possibly involved substances released by plants or microorganisms. There were many unexplained results that came from countless experiments involving crops in mixed monoculture or monoculture. Ahlgren and Aamodt (1939) found that when various grasses were grown in mixture, grasses such as redtop (*Agrostis* sp.) and Kentucky bluegrass (*Poa pratensis*) appeared to inhibit other species through root exudations. Schuphan (1948) found that peas and carrots grown in mixed culture or monoculture had no differences in overall yield per species, but pea plants in monoculture gave a higher yield of peas, and carrots from monoculture were markedly better quality and richer in carotene. Particulerly in Europe, there was considerable interest in the age-old problem of "soil sickness", whereas in the United States, there developed a more mechanistic understanding of the effects of individual species, including weeds, on the capacity of the soil to support crop growth.

Soil Sickness

In the first half of the twentieth century there was substantial revived interest in what is variously known as soil sickness or soil fatigue (German: *Bodenmüdigkeit*; French: *fatigue de sol*; Italian: *stanchezza del terreno*). This topic has been known to farmers since antiquity, and as much as anything has fuelled the idea that the land can become "poisoned". Soil sickness was often discussed under the name of the crop that had affected the soil; thus farmers often spoke of "clover-sick", "pea-sick" or "flax-sick" soil, and so forth. The problem was well known with many herbaceous crops and in horticulture, especially concerning fruit trees including apple, peach and various citrus species (Table 11.1).

One may define "soil sickness" in the broad sense as the phenomenon in which an agricultural soil shows a reduced yield in crop growth for reasons other than a lack of normal nutrients. The development of soil microbiology in the latter part of

Table 11.1. Some species reported as demonstrating "soil sickness".

Common Name	Species	References
Alfalfa	*Medicago sativa*	Demolon 1936
Apple	*Malus* spp.	Fastabend 1955, Schander 1956
Beans	*Phaseolus vulgaris*	Pantanelli 1926
Cabbage	*Brassica oleraceus*	Froehlich 1957
Coffee	*Coffea arabica*	Canargo 1945, Chevalier 1929, Piettre 1950
Cucumber	*Cucumis sativa*	Rienhold 1935
Ficus	*Ficus carica*	Hirai & Hirano 1949a, 1949b, Hirai & Nishitani 1951
Flax	*Linum usitatissimum*	Becquerel and Rousseau 1941, Papadakis 1941
Oat	*Avena sativa*	Pantanelli 1926
Orange	*Citrus aurantium*	Martin 1950, Martin and Batchelor 1952
Peach	*Prunus persica*	Proebsting and Gilmore 1941
Peas	*Pisum sativum*	Rippel 1936
Pineapple	*Ananas comosus*	Stead 1929
Plum	*Prunus* .spp.	Montard 1926
Sunflower	*Helianthus* spp.	Pantanelli 1926, Curtis and Cottam 1950
Tomato	*Lycopersicum esculentum*	Hirano 1940

the nineteenth century and early twentieth century helped elucidate the causes of many cases of soil sickness. For example, Bolley (1901) identified the cause of flax-sick soils in North Dakota as the fungal pathogen *Fusarium lini*. The reasons for soil sickness in various crops were often site specific, and have been variously attributed to increased populations of soil pathogenic organisms, or parasites such as nematodes, changes in soil structure, changes in soil microbiology, micro-nutrient depletion, or the accumulation of toxic substances in the soil (Reisinger 1947, Winter 1952). Rippel (1936) attempted to address the confusion regarding soil sickness, and believed that decreased yields due to known causes such as pathogens should be recognised as such, that is as a consequence of disease, whereas the term "soil sickness" should be reserved for situations where a particular type of plant alters the biological equilibrium of the soil essentially through its root excretions. In recent years, "soil sickness" has been viewed mainly as an autotoxic form of allelopathy, or autotoxicity (Gupta 2005). There is a substantial body of work which has looked at the issues of soil sicknesss, especially in regard to fruit tree orchards and plant nurseries in Germany (e.g. Vogel 1929, Kaven 1930, Vogel and Weber 1931, von Bronsart 1933, Klaus 1940, Kobernuss 1952, Schander 1956). There was also substantial interest in Japan in soil sickness in many crops, including tomato, figs, peach, and the topic of soil sickness was reviewed by Asami (1947).

Sorghum

Sorghum (*Sorghum bicolor*), of which some varieties are known as kaffir, is a grain crop that is grown world-wide, and is valued both for its grain and its utility as a forage crop. It was introduced into the United States in the early twentieth century, and was found valuable as a forage crop. Breazeale (1924) reported that sorghum, notably in the western states, was notorious for causing diminished growth to a crop planted immediately afterwards. Breazeale believed that there was a toxic effect that was related to the decomposition of the sorghum stubble in the soil. This toxic effect usually lasted only a few weeks, which suggested that the toxin either decomposed or volatilised. Breazeale suggested that the toxin was cyanide, but he failed to find a toxin in his pot experiments. Hawkins (1925) confirmed the damaging effects, but added that sorghum stalks plowed in were more detrimental than roots. The issue of sorghum toxicity was controversial and both Conrad (1927) and Wilson and Wilson (1928) demonstrated that the injurious effects of sorghum could be satisfactorily explained by the relatively high concentration of sugars in the residues. The microbial breakdown of the residues was associated with the immobilisation of significant amounts of nitrogen which made the soil temporarily unsuitable for new plant growth. McKinley (1931) reported that sorghum residues added to the soil in moderate amounts (up 0.6 tonnes per hectare) had no deleterious effects on crop production.

Peach Replant Disease

One of the more enduring cases of soil sickness concerns peach trees. Soil sickness is relatively common amongst Rosaceae species, and has been described for roses, peach, plum, cherry and apple. Fruit trees such as the peach and apple are comparatively short-lived, and frequently decline substantially in yield after about 20-30 years, after which they need to be replaced. It has been known, at least since the eighteenth century that fruit trees such as peach and apple often failed when replanted on the same ground.

It had been reported by Upshall and Ruhnke (1933) that old peach orchards were difficult to re-establish with various types of fruit trees, and it was thought that the problem was associated with a long-term nitrogen deficiency. Study of the peach replant problem from a different point of view had begun in 1922, somewhat fortuitously, when a covercrop experiment was started at the University of California at Davis, with eight species of fruit trees (Proebsting 1950). In 1942, the apple and peach trees were removed, and their rows were subsequently replanted with peach trees. Within a year it was noted that the peach trees, replanted in the rows that had previously grown peaches, grew more poorly, and soil tests failed to show pathogens or parasites. Proebsting and Gilmore (1941) had already begun consideration of the peach replant problem in 1937, in response to reports from California peach growers. It was found that the addition of roots of peach, as well as other stone fruit trees, such as cherry, apricot and myrobalan, to the soil in which peach seedlings were then planted, caused substantial growth reduction. It was hypothesised that the inhibition may have been caused by the presence of amygdalin, through its decomposition products. The addition of amygdalin plus emulsin to trees in sand culture caused severe inhibition, assumedly through the release of benzaldehyde. Hildebrand (1945) also found that the addition of peach root material to soil caused inhibition. However, these results were not supported by the findings of Havis and Gilkeson (1947), who found no evidence of peach inhibition in sand culture amended with various treatments based on peach bark or peach roots. The issue of the toxicity of peach orchard soil was taken by others including Hewetson (1953) in Pennsylvania, and Koch (1955) and Patrick (1955) in Ontario, Canada. Patrick found that it was possible to demonstrate that peach roots residues contained relatively large amounts of cyanogenic glycoside amygdalin, and that this was readily broken down to the phytotoxic products cyanide and benzaldehyde by soil microorganisms. Living roots were also found to contain amygdalin and the hydrolysing enzyme emulsin, which could also cause the formation of thedegradation products. As Patrick's results were found under laboratory conditions, he was cautious in extrapolating his findings to field conditions, but he emphasised the point that the field situation was undoubtedly complex, and likely involved several factors, including soil nutrients, pathogens including parasitic nematodes and phytotoxins.

It is little known that there was substantial work concentrating on soil sickness problems in Japan, that was led by Juzo Hirai and Satoru Hirano[11] at the Horticultural Institute of Kyoto University, and later at the Chiba Horticultural Experiment Station. Beginning in 1947, this group began looking at soil sickness among fruit

---

[11] The names of Japanese workers on soil sickness are confusing. There was a T. Hirano (1940) who published his work with tomatoes. The early articles by Hirano (e.g. Hirai and Hirano 1947a, Hirano 1951) gave his name as Akira Hirano, but it appears that he later preferred the name Satoru Hirano (e.g. Hirano 1955).

trees, such as, but especially peach, and the research continued for two decades. Initially Hirano (1951, 1955) found, as had American and Canadian workers, that toxicity was associated mainly with peach roots.

Analagous work was initiated concerned the problem of replanting apple orchards. For example, Polishchuk and Snezhko (1954) found that seven different apple varieties planted with fertiliser in old apple soil all showed sustained poor growth over a period of four years. Fastabend (1955) indicated that breakdown of the phenolic glycoside phloridzin to toxic products was likely involved in the apple replant problem. The citrus replant problem was worked on extensively in California by James P. Martin and his associates (Martin 1950, Martin and Batchelor 1952), but while Martin suspected that soil toxins were involved, only relatively unconvincing evidence was found: soil treated with strong acid or alkali was found to be less inhibittory, and Martin suggested that accumulated toxins were destroyed.

Juglans spp.

While the subject of the toxic effects of walnut trees, particularly black walnut (*Juglans nigra*) on apple trees had become a much discussed topic in the latter part of nineteenth century (see Chapter 8), the subject then languished until about 1920. Most reports of the allelopathic effects of walnuts have concerned *J. nigra*, and occasionally *J. regia* and *J. cinerea*, although the California native *J. hindsii* has also been claimed as allelopathic (Pratt and Dufrenoy 1949), and the matter has been made difficult where grafted stock has been used. There had been a couple of minor reports at the beginning of the twentieth century; Jones and Morse (1903) reported observations by A.H. Gilbert that cinquefoil (*Potentilla fruticosa*), a common weed was generally found dead within a circular area often greater than the that of the canopy of butternut (*Juglans cinerea*, also known as white walnut) trees. The effect seemed related to the intermingling of cinquefoil roots with butternut roots, and the extent of the effect generally increased with the age. Many years later, it was reported that *J. cinerea* was associated with the death of the introduced *Pinus mugo*[12] (Smith 1941). In 1905, the prolific American horticultural authority U.P. Hedrick stated that grape vines were harmed by nearby black walnut trees.

However, the subject of allelopathy in black walnut, or black walnut toxicity, as it was then called, became a controversial subject once more in the 1920's. On this occasion, the reports centred largely on the toxic effects to tomatoes and similar annual crops. In 1921 Cook described the wilting of tomato plants, evidently due to the effect of nearby black walnut trees. The subject achieved prominence with the appearance of the first substantial paper on the topic (Massey 1925) in which the author, A.B. Massey (1925, 1928) collated many observations of his colleagues on the effects of black walnut trees on tomato plants. In one experiment, the region of a group of small square plots of tomato plants that happened to be bordered by two black walnut trees was sprayed with water, and all the tomato plants within 12-15 m of the walnut trees were injured. Massey noted that in an alfalfa field containing a

---

[12] *Pinus mugo* is an old name for *Pinus montana* var. *mughus*, a variety of Swiss mountain pine.

solitary black walnut tree, alfalfa was generally absent in the vicinity of the walnut tree, and replaced by grass. The response of different crop plant to nearby walnut trees seemed differential; generally tomato plants and to a lesser extent potato plants fared poorly, whereas beets, snap beans and corn were comparatively unaffected. Massey demonstrated that a toxic effect was linked to root bark, and he concluded that in black walnut there must be some toxic principle in the walnut root which is initially insoluble in water, but may become altered once it exits the root. He guessed that the toxin was likely the orangeish hydroquinone known as juglone.

Haasis (1930) in North Carolina reported that tree seedlings seemed to suffer when they were planted near black walnut trees. Perry (1932) at the Mount Alto State Forest Nursery in Pennsylvania, also found that conifer seedlings became severely damaged in beds that happened to adjoin a plantation of black walnut trees. Subsequent trenching and severing of the walnut roots eliminated further injury to the conifer seedlings. Similar observations of injury to a plantation of mixed pines by black walnut were reported by Schreiner (1949, 1950).

Another major contributor to the argument was Schneiderhan, who reported on the demise of apple trees growing in the vicinity of black walnut trees (Schneiderhan 1926, 1927a, 1927b), which was a common problem in West Virginia. From his observations, only *J. nigra* and occasionally the butternut (*J. cinerea*) were injurious; the other American walnuts occasionally used in plantations, *J. hindsii* and *J. californica* could support intercropping, and the Persian walnut *J. regia* seemed not to cause any significant effect. In a cursory survey of effects in Frederick County, Virginia, Schneiderhan found eighteen instances of black walnut injury to apple trees. Dead apple trees (48) were found to be an average 11.9 m distant from a walnut tree, and injured apple trees (14) were an average 14.3 m distant from a walnut tree. Schneider speculated that the relatively common occurrence of walnut toxicity in Frederick County, Virginia was due to the high degree of intermingling roots associated with shallow soils. He also suggested that there may be variation in sensitivity among different varieties of apple rootstock, and that the variety Stayman may be more resistant than others.

In 1928, E.F. Davis, a researcher at the Virginia Agricultural Experiment Station, isolated juglone (5-hydroxy-1,4-naphthoquinone; Figure 11.3) from both the fruit hulls and the roots of black walnut. Furthermore, he demonstrated dramatically the toxic effects of juglone, both extracted from plant material and that prepared by synthesis, by injecting it into the stems of tomato and potato plants. Davis' findings were picked up by the newspaper press including the *New York Times* (Anonymous 1929b) and agricultural press (Anonymous 1929a), and the cause of walnut toxicity was regarded as solved.

The triumph of science in apparently solving black walnut toxicity led to some curious consequences. Firstly, just as the upas tree, a century earlier, had become a metaphor for moral corruption, the same occurred with the black walnut (Naylor 1930). Secondly, there were concerns that the bad press labeling the black walnut as a "poison tree" would lead to its demise as an important part of the American landscape and as a valuable tree crop (Anonymous 1926, 1929c). The villainous reputation ascribed to black walnut was not unanimously agreed, and was treated by some

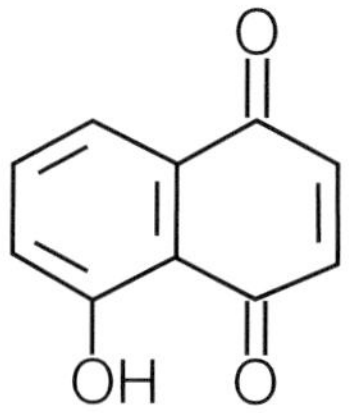

Figure 11.3. Structure of juglone (5-hydroxy-1,4,-naphthoquinone)

with ridicule (Hershey 1929). Miller (1926) objected to Schneiderhan's findings, and stated that in one orchard bordering walnut trees, dead apple trees could also be found distant from walnut trees. Bixby (1926) argued that the injurious effects of walnut were due to competition and were no different from those due to any other large tree. Finally, in 1948 the USDA published an extraordinary press release that was designed to reassure the public that the black walnut tree was harmless, even to tomato plants (USDA 1948).

The controversy concerning the allelopathic nature of black walnut lasted, more or less unabated, from the 1920's until the outbreak of the World War II, and was largely debated within the annual meetings of the Northern Nut Growers Association. In 1939 Davidson, highlighted the confusion surrounding the reputation of the black walnut. Subsequently in 1940, MacDaniels[13] and Muenscher (1940) at Cornell University reported some experiments in which tomato and alfalfa plants had been grown in black walnut soil. There seemed little evidence of inhibition, and in another trial, walnut hulls placed around grape plants caused increased growth. Tomato plants planted near a black walnut tree showed little inhibition until late in the growing season. MacDaniels and Muenscher (1940) reiterated the contradictory nature of much of the evidence concerning the alleged toxicity of black walnut. Brown (1942), a student of Muenscher's, was able to show that the germination and growth of alfalfa and tomato were inhibited when seeds or seedlings were in contact with walnut root bark. Tomato seedlings grown in nutrient solution with added walnut root bark were significantly inhibited, especially when nitrogen was deficient. Gries (1943) was among the first to realise that the discrepancies reported for walnut toxicity was related to the nature of the root system of neighbouring plants and the age of the walnut root tissue. In fresh walnut tissue, relatively non-toxic hydrojuglone is oxidised to juglone, but on exposure to air and/or with age juglone is further oxidised, often to a black non-toxic product. In any case there were several other reports citing instances of the toxic effects of walnut on other plants, notably *Rhododendron* spp. (Pirone 1938), apple (Smith 1941, Orton 1943, Orton and Jenny 1948, Wilkinson 1948), alfalfa (Smith 1941), tomato (Reinking 1943, 1947; Orton 1943, Strong 1944) and cabbage (Reinking 1943, 1947; see Figure 11.4), potato (Orton 1944), grasses (Smith 1942);

---

[13] Laurence H. MacDaniels (1888-1986) had a remarkable record in publication on the effects of black walnut, and his work in this area alone spanned almost half a century, from 1940 to 1986..

Figure 11.4. The effect of black walnut trees on cabbages (reproduced from Reinking 1943).

on the other hand, numerous reports claimed that the plants grew remarkably well under black walnut, or not differently than that under any other shade tree (Greene 1929, Mattoon 1944). MacDaniels and Muenscher (1940) had realised that there was substantial disquiet from commercial walnut growers, especially in California where most of the walnuts originated from plantations of the apparently more benign *J. regia*, either with their own roots or on grafted stock; in any case, the growers reported no toxic effects with these walnuts.

In 1951, Maurice Brooks attempted to redress the anecdotal nature of much of the black walnut information, by surviving the vegetation in the vicinity of 300 mature black walnut trees in various situations, mainly in West Virginia. On the basis of his data and observations, he was able to confirm that black walnut did appear to exert an antagonistic effect on certain, but not all, plants. He found that black walnut appeared harmful to apple trees, potatoes, tomatoes alfalfa, blackberry, and Ericaceous plants, as had been reported by others. Generally the antagonism was most apparent where there was demonstrable contact between black walnut roots and those of the afflicted plants. Where exceptions to these patterns occurred, there was usually a lack of root contact, enforced by a soil barrier, either natural or artificial. Brooks noted that the effect of black walnut was selective, and that certain plants, especially grasses such as Kentucky bluegrass (*Poa pratensis*) thrived within the vicinity of black walnut. Brooks did not attempt any detailed soil studies, but he did find that soil within the vicinity of black walnut trees was generally less acidic than soil unaffected by black walnut trees. Brooks (1952) also reported that black walnut had little effect on fern species.

Weeds

Perhaps the best researched weedy species during these formative years was *Agropyron repens*, known commonly as either couchgrass or quackgrass. At the Seventh International Botanical Congress in Stockholm in 1950, the Swedish botanist Hugo Osvald related that he had a long-standing suspicion that plants, especially grasses, could have a deleterious effect on other plants, but that the failure of earlier excretion theories had dampened his inclination to investigate the matter, although he had been impressed by the the experiments of Hartwell *et al.* (see Chapter 10), which he had seen at the Rhode Island Agricultural Experiment Station (Osvald 1953).

Osvald's latent interest in allelopathy was rekindled through new discoveries concerning the potent effects of antibiotic substances, which were found to be effective at very low concentrations, fresh results from the United States, notably that concerning desert shrubs, and the work of Benedict (1941) who found the decaying roots of *Agropyron repens* to be inhibitory to the growth of brome grass (*Bromus* sp.). Ahlgren and Aamodt (1939) had also noted red clover (*Trifolium pratense*) and white clover (*T. repens*) rarely occurred in *Agropyron* sod, whereas alsike clover (*T. hybridum*) did. In 1945 Osvald had found that rape seemed to grow poorly on soil which had patches of couch grass (*Agropyron repens*), and previously he questioned why certain grasses were less likely to be invaded by weeds, or why clover disappeared more quickly when sown with a grass such as cocksfoot (*Dactylis glomerata*). Preliminary experiments with *Agropyron repens* showed that extracts of stolons and roots were inhibitory to germination and seedling growth of both rape and oats, although at very low concentrations, the extracts were stimulatory (Osvald 1947, 1948). Osvald's results were criticised for the artificial nature of the extracts, and he subsequently attempted work which better demonstrated the field situation. Subsequent observation of a meadow with dense red-fescue (*Festuca rubra*), seemingly without any incursions by other species, led Osvald to bioassay an extract of the red fescue soil, which was found to be inhibitory to germination of several plants. In another experiment, various grasses were grown for three weeks on filter papers, and the filter papers were then bioassayed with rape seedlings, which showed abnormal growth, but normal germination, evidently due to some factor left behind by the grasses, including oats, barley and perennial ryegrass (Osvald 1949). A leachate was collected and concentrated from containers which had been growing rye grass plants in sterilised sand. The extract, which was attributed to root exudation, was found to slightly inhibit germination, and to have a very marked effect on seedling growth, but other ecologists such as John Harper questioned whether the effects were mainly osmotic (Osvald 1953). Hamilton and Buchholtz (1955), working in Wisconsin, investigated the effect of quackgrass rhizomes on the growth of weed species by means of small field plots containing soil with or without quackgrass rhizomes. The authors found that the presence of living quackgrass rhizomes decreased the abundance of species such as *Veronica peregrina, Polygonum persicaria, Oxalis stricta* and *Setaria lutescens*, but favoured *Taraxacum officinale*. Revived interest in the ecological effects of root exudates, especially from grasses, caused Selander (1950) to speculate that certain plants that are usually confined to bare mineral soils and almost never found in association with meadow grasses, were essentially non-commensal possibly due to their sensitivity to the root exudates of other plants, especially grasses.

The work of W.C. J. Kooper (1927) in eastern Java has been largely ignored, as it many respects it was simply too challenging. Kooper meticulously investigated the weed infestations of the plantations, rotated with sugar cane, rice and maize, at Pasuruan near Surabaya in eastern Java, and he found that there were distinct weed communities that recurred on the same ground, despite variation in cultivation treatment or even climate. It seemed that the pre-existing weed community had a profound influence on the success of the next weed community and less to do with seed availability or competition as one might expect. This was illustrated in one set of trials where experimental plots usually infested with a post-harvest weed community dominated by *Polanisia viscosa* (Capparidaceae) were seeded with a variety of species typical of other weed communities. Despite favourable conditions and a lack of competition, there was almost no germination, and any germinants that did appear soon died. The results suggested that that there was some soil factor, either chemical or microbiological, that differentially affected the germination and growth of species, and that originated with the pre-existing weed flora.

Another grass that attracted attention was bromegrass (*Bromus* spp.). It had been noted by Robbins *et al*. (1942) that certain plants, including bromegrass, were effective as "smother crops", that is as a crop useful in suppressing weed growth. Evidence of toxicity in bromegrass rhizomes was provided by Myers and Anderson (1942) and subsequently by Went *et al*. (1952). The competition experiments of Mann and Barnes (1947, 1952) provided further results that tended to indicate that bromegrass produced substances that affected neighbouring plants. Froeschel and Funke (1939, 1941) investigated the effects of various plants extracts on the germination and growth of other plants, and suggested that the apparent absence of corncockle (*Agrostemma githago*) from beet fields was due to beet exudations. Similar findings were reported by Hurtig (1953) who found that inhibitory properties were associated with the fruits. Rademacher (1941) demonstated that the growth of winter rye (*Secale cereale*) was effective in reducing the growth of several weed species, especially *Matricaria maritima*. In Australia Greenham (1943) reported that the noxious weed, skeleton weed (*Chondrilla juncea*) was phytotoxic to wheat. Helgeson and Konzak (1950), working with various noxious weeds in North Dakota, found that extracts of the shoots of field bindweed (*Convolvulus arvensis*) and the roots of Canada thistle (*Cirsium arvense*) were inhibitory to the root growth of the seedlings of flax and wheat. The phytotoxicity became more pronounced when the crop seedlings were subject to a diurnal temperature regime with a 12° C minimum and 25° maximum.

Medicinal Plants

While the subject of allelopathy among medicinal plants was discussed in some detail in consideration of the work of Gerhard Madaus, there was other work, notably that of Hans Bode on the effect of wormwood (*Artemisia absinthium*) on other plants. Bode became interested in the problem when he observed, at a medicinal plant farm in Giesenheim, Germany, that many types of herbs, for example, fennel (*Foeniculum vulgare*), lovage (*Levisticum officinale*), carraway (*Carum carvi*), basil (*Ocimum basilicum*), lemon balm (*Melissa officinale*), catnip (*Nepeta cataria*), and sage (*Salvia sclarea*), all showed stunted growth when growing in proximity to wormwood. Bode (1939, 1940) investigated in detail the effects of wormwood on fennel, and he showed there was an inverse relationship between the growth of fennel plants and their proximity to wormwood. Leachates collected from intact foliage of wormwood inhibited fennel seed germination and reduced the growth of fennel plants. Relatively large amounts of organic material, including absinthin, were released from the abundant T-shaped trichomes that covered the surface of wormwood leaves. Bode's results were confirmed by Funke (1943a) who also noted the selective nature of inhibition; in his study, *Senecio* was particularly affected, whereas *Stellaria* and *Datura* were not. George Lodewijk Funke (1896-1946) was a Dutch-Belgian plant ecologist whose writings on allelopathy (e.g. Funke 1943b) have been largely overlooked. Several other investigators subsequently found inhibitory effects associated with wormwood (Golomyedova 1952, Shenderetzkii 1952, and Knapp and Thiessen 1952).

Guayule

Interest in the root interactions of plants received a boost, largely as a benefit of the Emergency Rubber Program, which had been established in 1942, shortly after the United States became directly embroiled in the Second World War, when Japanese aircraft attacked the American Navy at Pearl Harbour, Hawaii[14]. Americans had realised early in the war that the supply of natural rubber was under threat, as almost all natural rubber then originated from plantations of *Hevea brasiliensis* in Malaya, which was subsequently invaded by Japanese forces in late 1941. Guayule (*Parthenium argentatum*), a shrub native to Mexico and Central America, and the only viable Western alternative to *Hevea*, in earlier decades had served as an important source of American rubber, but its cultivation languished, although guayule latex is regarded as less allergenic than that of *Hevea*. Research at the California Institute of Technology began in early 1940 to focus on selection and breeding, cultivation, and methods of improving yield of guayule. Led by the plant physiologist James Bonner (1910-1996), the Special Guayule Research Project, U.S. Bureau of Plant Industry, Soils and Agricultural Engineering became part of the Emergency Rubber Program (Figure 11.5).

---

[14] Ironically, both in the U.S. and Australia, interned Japanese were used extensively in wartime guayule cultivation.

Figure 11.5. *Guayule plantings being inspected by Doctor Robert Emerson (third from right), a biochemist and botanist from the California Institute of Technology and Director of the Guayule Rubber Experiment. He is shown conferring with several of his staff of young evacuee scientists at Plot 4. There are plants in this plot that have been in the ground for ten days. Photo dated 6/28/42.*

Another young plant physiologist, A.W. Galston, joined Bonner in the research during much of 1944. The project was a qualified success; however, the synchronous development of means of producing synthetic rubber ultimately made guayule rubber production redundant.

It had been observed during field trials of guayule at Salinas, California that in closely planted nurseries, the outermost plants were much larger despite ample water and fertilisation, which suggested a root interaction problem. Initial experiments demonstrated that leachates from guayule gravel culture were inhibitory to both tomato plants and guayule seedlings (Bonner and Galston 1944). In another experiment, guayule seedlings were planted under mature guayule plants, and seedlings, which were shielded from root contact, grew better. Further bioassays of seedlings in aqueous culture, using guayule culture leachate, demonstrated toxic effects. Extraction of about 227.4 l (60 gallons) of guayule culture medium with ether yielded 1.8 g of physiologically active organic substances, or about 7.5 mg per plant. Two inhibitory compounds were subsequently isolated, of which one was identified as trans-cinnamic acid, which in pure form caused 50% reduction in guayule height growth at a concentration of 30 ppm. Further studies (Bonner 1946) demonstrated that trans-cinnamic acid added to soil could cause growth inhibition of guayule seedlings. However, Bonner was disappointed to find that field soils which had sustained guayule for several years had no detectable amounts of trans-cinnamic acid. Soil toxicity was found only in certain pot cultures, where it was assumed the root density was unusually high. Bonner was forced to conclude that whatever toxins were released by guayule roots, they were rapidly decomposed or neutralised in the soil.

AQUATIC SYSTEMS

The term "allelopathy" has been used in connection with aquatic systems only in comparatively recent years. Even more recently, it has also gained some currency among zoologists in relation to the chemical interactions of aquatic sessile invertebrates. For most of the twentieth century, the chemical interactions of planktonic organisms seemed to have been regarded as more similar to the antibioses of bacteria and fungi, although an integrative approach was broached by Lucas (1938, 1944). A separate terminology was introduced, and biologically active substances released into aquatic environments were recognised as ectocrine substances (Lucas 1947, 1949), as opposed to internal or endocrine substances, and inhibitory effects were sometimes separated as either hetero-antagonistic or auto-antagonistic (Lefevre *et al.* 1952).

Maestrini and Bonin (1981) have provided a scheme that recognised the basic steps in the progress of understanding chemical interactions among phytoplankton: 1) recognition of dissolved organic matter in water, 2) discovery of inhibitory processes in culture media, 3) concept of ectocrine substances in natural systems, 4) isolation of inhibitors, e.g. chlorellin, and 5) quantitative and physiological studies.

In the nineteenth century it was realised that aquatic environments often contain relatively high amounts of diverse organic substances, which are collectively known as the dissolved organic matter (DOM). The more generalised concept that organic substances releases by phytoplankton may be important in their natural dynamics was first suggested by Pütter (1907a, 1907b), and eventually achieved wide recognition with the work of Lucas (1947).

The laboratory growth of cyanobacteria and green algae in pure culture, or mixtures of two competitive species, as became popular in early population studies, for example by Gause, led to observations of reduced growth in old cultures, or inhibition of one species by another, which in turn led to the hypothesis that growth-inhibiting substances were involved. Harder (1917) found growth-inhibiting substances in cultures of *Nostoc punctiforme*, and both Hoyt (1913) and Sakamura (1922) reported similar phenomena from cultures of *Spirogyra*. Flint and Moreland (1946) reported that the exudates of blue-green algae were inhibitory to other algal species. Schreiter (1928) suggested that aquatic plants such as *Elodea canadenisis* might affect phytoplankton via inhibitory substances. Similarly, it was suggested that the substances released by phytoplankton may affect zooplankton species (Allen 1934, Harding 1935).

Nielsen (1934) hypothesised that metabolites produced by phytoplankton may play a role in regulating the abundance of the dinoflagellate *Ceratium* in warmer waters. Akehurst (1931) who was aware of the work by Pickering, developed a set of hypotheses, reminiscent of those of de Candolle, to explain the succession of algal species in freshwater ponds. He suggested that algae, such as dinoflagellates and diatoms, produce substances that favour members of the Chlorophyta and vice versa. Dominant species such as *Asterionella* ultimately become inhibited by their own metabolites, and thus algal diversity was maintained. Reich and Aschner (1947) found that *Prymnesium parvum* growing at high densities became inhibited by its own metabolites. Proctor (1957a, 1957b) suggested that the short-term occurrence of *Haematococcus pluvialis*, a species found primarily in temporary water bodies, was in part linked to its sensitivity to inhibitors eventually produced by other slower-growing green algae such as *Chlamydomonas reinhardii* and *Scenedesmus quadricauda*. In marine environments, species that can attain high densities, sometimes known as blooms, such as found with the diatoms and dinoflagellates, attracted interest (Levring 1945, von Denffer 1948, Talling 1957, Graham and Bronikovsky 1944).

Chlorellin

The outbreak of World War II brought some allelopathic work prematurely to a conclusion, yet was responsible for the initiation of others. The most comprehensive studies in relation to potential allelopathy within an algal species were those of the microbiologist Roberston Pratt (1909-1976), who had done some experimental work on cell division in *Chlorella vulgaris* (now *C. kessleri*), while a postgraduate student at Columbia University. He joined the staff of the University of California College of Pharmacognosy, San Francisco in 1938, and resumed his studies there on *Chlorella*. From 1940-1948, he published a series of papers that detailed his experimental work on the behaviour of *Chlorella vulgaris* in cultures at differing densities. Firstly, he found that cultures with initial larger inocula showed a decreased growth rate, which suggested that an autoinhibitor was a factor (Pratt 1940). Cells transferred to "used" growth medium showed a rapid reduction in growth, and more detailed experiments demonstrated the inhibitory effect of culture filtrates (Pratt and Fong 1940). Such "staling" substances were known from bacterial and fungal cultures, but were unknown in algae. By 1942, the United States was directly involved in the World War II, and Pratt's research with *Chlorella* was regarded as of significance as it provided a model for understanding the dynamics of micro-organisms under mass culture as required in the production of antibiotics, and secondly it ultimately led to the isolation of the first known antibiotic from an alga.

Pratt (1942) was able to determine that the inhibitor produced by *Chlorella vulgaris* was obviously water-soluble, a relatively small molecule, and heat labile. It had limited solubility in organic solvents, and was slightly basic. Further work showed it to be an inhibitor of photosynthesis, in particular, carbon fixation, but it had no apparent effect on respiration (Pratt 1943). By 1944 the project had become very large, as it was now funded by industry, and had engaged twelve researchers at two campuses. Pratt *et al.* (1944) named the autoinhibitor chlorellin, and showed that its effects could be negated through dilution or adsorption (Pratt 1944). A bioassay for chlorellin using *Staphylococcus aureus* showed that the production of chlorellin was irregular, with a significant lag from days 3-5 (Pratt *et al.* 1945). Spoehr (1945) attempted to duplicate some of Pratt's work, but used the species *Chlorella pyrenoidosa*, which Pratt himself had used briefly. Spoehr believed that chlorellin in *C. pyrenoidosa* was a mixture of unsaturated fatty acids. This seemed to run counter to Pratt's earlier findings, and Pratt further showed that there was no obvious negative correlation between chlorellin concentration (as gauged with *Staphylococcus aureus*) of surface tension of the culture filtrates, which suggested that his substance was not similar to that described by Spoehr (Pratt 1948). In the end, chlorellin proved impractical as an antibiotic for both medical and commercial reasons, but its discovery did pave the wave for ecological research decades later that showed other algal genera, such as *Scenedesmus*, are also capable of producing potent inhibitors.

Subsequent work with other algal species, notably those capable of forming blooms, led to observations that supported the idea of inhibitors. Water collected from a lake in which *Asterionella* had bloomed was found to be unsuitable for preparation of growth medium for *Asterionella* cultures (Worthington 1943). Rodhe (1948) found that *Asterionella formosa* cultured with *Chlorella* had a lower rate of division than when cultured alone.

The French phycologist Maurice Lefevre and his colleagues published extensively on the effects of culture filtrates of various freshwater and soil algae on the development and morphology of other algae, but the results were mainly qualitative (Lefevre and Jakob 1949, Lefevre 1951, Lefevre *et al.* 1948, 1952, Jakob 1954). Lefevre *et al.* (1948) found that the culture media in which freshwater green algae, such as *Scenedesmus* or *Pandorina*, had previously grown had an inhibitory effect on the growth of subsequent or the same species; autoxicity in colonial forms was often indicated by distortions in growth.

Theodore Rice investigated the interactions of the diatom *Nitzschia frustulum* and the green algae *Chlorella vulgaris* using agar culture (Rice 1954). The two species were found to have reciprocal negative effects that seemed due to the release of inhibitory substances that could be neutralised with carbon or heating. Pond water that had supported a bloom of *Pandorina* was also found to be inhibitory to the growth of both *Nitzschia* and *Chlorella*.

ALLELOPATHY IN NATURAL TERRESTRIAL SYSTEMS

"Fairy Rings"

The distinctive patterning, known as "fairy-rings", well known in fungi (see Chapter 9) was revisited in relation to flowering plants. It had been noted by several researchers that fairy-ring patterns were occasionally found in higher plants, especially among colonising species, for example, *Juncus squarrosus* (Farrow 1915) and *Andropogon scoparius* (Olmsted 1937). In 1931, Cooper and Stoesz reported that the rhizomatous plant, *Helianthus rigidus*, demonstrated distinct rings of inhibited growth which seemed associated with compounds left in the soil. Some years later, Curtis and Cottam (1950) reported fairy-rings in *H. rigidus* and *H. occidentalis*, as well as inhibition of *Poa pratensis* and *Monarda fistulosa* at the centre of the colonies. Soil manipulation experiments demonstrated that the inhibition was likely associated with relatively short-lived inhibitors produced by young rhizomes. The authors speculated that the fairy-rings sometimes seen in other Asteraceae, such as *H. tuberosus*, *Antennaria fallax*, *Aster macrophyllus*, and *Erigeron pulchellus*, occurred for similar reasons. F.W. Went (1956), having visited Australia in 1955, suggested that the distinctive clonal rings of the arid zone tussock grasses *Triodia* spp. and *Plechtrachne* spp., that sometimes attain diameters of 5 m or more, may be due to inhibitors formed from the decomposing roots of the older plants.

American Desert Plants

In large measure, the revival of American interest in allelopathy was due to discoveries made concerning the chemical interactions of plants from arid environments in the western United States. Pioneering research by Frits Went, James Bonner and Arthur Galston, based at the California Institute of Technology, led to interest in similar phenomena by Cornelius (Neil) H. Muller, who subsequently in the 1960's and 1970's became the leading figure in allelopathic research in the United States (Willis 1995, Halsey 2004). Reviews by Bonner (1949, 1950) provided an overview of the contemporary work.

Frits W. Went (1903-1990; Figure 11.6) was the son of Dutch botanist F.A.F.C. Went. After working in Netherlands and Java, he moved to California in 1933, and his colleagues at the California Institute of Technology (CalTech) included some of the best known names in plant hormone research: James Bonner, Kenneth Thimann, Folke Skoog and Johannes van Overbeek. Went developed an enduring fascination with the American desert flora, which was combined with a long-term interest in the causes of plant associations, notably those in the tropics. Went, during his time in Java, had witnessed and affirmed the controversial findings of Kooper (1927), and Went (1940) had a personal interest in the specificity and chemical relationship of rainforest epiphytes and their hosts. In 1941 Went began a phytosociological study of the shrubs and associated annual plants in the Colorado and Mojave deserts in southern California. Went noted that many annuals were found preferentially associated with certain shrubs; for example *Rafinesquia neomexicana* was often found primarily with the shrub burro-weed (*Ambrosia dumosa*[15]), and *Phacelia tanacetifolia* with sagebush (*Larrea tridentata*); however, the degree of specificity was often dependent on locality. In contrast, some annuals seemed negatively correlated with certain shrubs, *Malacothrix* with brittlebush (*Encelia farinosa*), especially if the shrubs were live plants. As the United States became involved in the Second World War in 1941, it is likely that Went was unable to spend further time pursuing these problems at the time. He could only conclude that many factors were likely involved in the positive and negative interactions of desert annuals with desert shrubs, among which must be considered were chemical factors emanating from the shrubs themselves. In 1952 Went *et al.* investigated the poor seedling recruitment of native species in burnt and unburnt chaparral near Pasadena. Some of the burnt areas had been seeded subsequently with *Brassica nigra*, and there was comparatively poor growth of certain species, such as *Salvia mellifera* and *Rhus laurina*. Studies showed that germination was affected by seed exudates and litter of mustard. *Salvia mellifera* prevented the establishment of *Adenostoma fasciculatum* seedlings. In other studies, barley, tobacco and beets were found to inhibit seedling growth.

---

[15] Formerly known as *Franseria dumosa*.

Figure 11.6. Photograph of Frits Warmolt Went

Another important botanist who was had served with the Guayule Special Research Project was Cornelius H. Muller (1909-1997). Muller became well familiar with the guayule root problem studied by Bonner and Galston, and wrote a monograph on the root ecology of guayule plants (Muller 1946). Following the end of the Second World War, Muller was appointed as plant ecologist at the University of California, Santa Barbara. Muller, encouraged by Went, revisited the problem of shrub-herb associations in the California desert (Muller 1953). He was able to corroborate most of Went's observations, and in the light of the discovery and identification of inhibitory substances in the foliage of *Encelia farinosa* by Reed Gray under the supervision of Bonner (Gray and Bonner 1948, 1948a), he focussed on trying to establish if the shrubs *Encelia farinosa* and *Ambrosia dumosa* had any inhibitory effect on the growth of underlying herbs. Aqueous extracts from dried leaves of the two shrubs proved relatively phytotoxic to tomato seedlings in aqueous culture. While the paucity of the shrub-dependent annuals beneath *Encelia* and *Ambrosia* was reputedly due to inhibitors, the data were contradictory. The widespread occurrence of other annuals beneath the canopies of both *Encelia* and *Ambrosia* troubled Muller, and he ignored the possibility that there could species-specific responses to a phytotoxin. Muller was forced to consider that any toxins were generally ineffective due to lack of a transport mechanism, decomposition by soil microorganisms or due to inactivation by soil colloids. Muller did note that where *Encelia* was found growing in sandy or gravelly soils, poor in organic matter, there was a marked absence of herbs, which was possibly caused by allelopathy. In 1956, Muller collaborated with his unrelated namesake and colleague Walter.H. Muller (Muller and Muller 1956), and found that foliar extracts of *Encelia*, *Ambrosia* and *Thamnosma* were all inhibitory in bioassay to tomato seedlings; however, ultimately the authors regarded their findings of little ecological relevance, as there was little evidence of toxicity under field conditions. In New Mexico, Shields (1956) also observed bare zones around bushes of *Larrea tridenta*, which she suggested were possibly due to toxins. As a result of his work, Cornelius Muller developed a healthy scepticism concerning allelopathy, and it was this cautious approach that was to greatly enhance the credibility of his studies on allelopathy in later years (see Chapter 1). Muller and Muller (1956) concluded their paper with an eloquent summary of the ecological problem:

> It should be emphasised that the natural habitat is a complex of physical and biological factors that influence growth. Even though plant distributions may give the impression of an antibiotic effect by some of the individuals, careful investigation may indicate that the situation cannot be explained in such a simple fashion. Environmental influences and the metabolic activities of organisms are complex factors which are variously intermingled, and in most cases it is doubtful whether any one factor would be distinguishable as the primary causative influence.

Secondary Succession - Catherine Keever

Another indirect effect of the Second World War was that a shortage of male applicants increased opportunities for women in postgraduate study. Catherine Keever (1908-2003) completed her doctoral study of the dynamics of early old-field succession in 1949, under the supervision of H.J. Oosting[16] at Duke University, North Carolina. Another Duke alumna of 1949 and colleague was Elsie Quarterman[17]. Keever, who is remembered as one of the pioneering women in American ecology, attempted to understand the factors causing the succession of plant species in abandoned old-field on the Piedmont of North Carolina (Keever 1950). A common sequence was horseweed (*Erigeron canadensis*) and crabgrass (*Digitaria sanguinalis*) becoming dominant in the first year, *Aster pilosus* the dominant daisy during the second year, and in the third year broomsedge (*Andropogon virginicus*) dominant. Keever was aware of Pickering's work with grass leachates, discoveries concerning the effects of juglone from black walnut, and the recent work in California concerning desert shrubs. Consequently, as part of her study, she investigated whether leachates and decomposition products of the dominant plants played any role in growth inhibition of promotion of other species. Firstly soil leachates from pots containing either a plant of horseweed, aster or broomsedge were fed to pots containing young plants of the same three species. After five months, Keever could find no significant difference in plant growth attributable to leachates from living plants. A second pot experiment was conducted in which either roots or the aboveground parts of each of the above three species were incorporated into soil, at rates approximating field conditions. The pots were sown with seed of the same three species, and were subsequently weeded and thinned until there were two substantial plants of the requisite species. Plants were either watered with distilled water or nutrient solution, in order to assess any nutrient depletion, and similarly the horseweed was grown in both sand and field soil to assess any effects due to soil microorganisms. Keever found that decomposing horseweed roots inhibited horseweed and aster seedlings, and that horseweed tops, aster tops and broomsedge tops all inhibit the growth of

---

[16] Oosting was a student of W.S. Cooper who in turn was a student of H. Cowles (see Chapter 9).

[17] The Southeast Chapter of the Ecological Society of America has honoured these two women by creating the Quarterman-Keever Award. Quarterman also has published on allelopathy, in the Tennessee cedar glade communities, in the 1970's.

asters. However, interpretation of these findings was confounded as the addition of extra nutrients eliminated any inhibitory effects, which suggested that microbial immobilisation of nutrients played a substantial role in seedling suppression. Nonetheless, Keever (1950) concluded that the decomposing plant parts, but not the living parts, of species such as horseweed does play a small role in inhibiting the successive growth of horseweed seedlings, and thus in determining the successional sequence in abandoned fields. Keever's study is esteemed in the ecological literature, not because of any outstanding finding concerning allelopathy, but because of the meticulous consideration of a host of factors in assessing an ecological question. Keever (1983) had the opportunity to look at her classic study with the hindsight of thirty-five years of further information, and she concluded that her findings had essentially stood the test of time.

France - Plant Communities Dominated by Perennial Species

During the middle of the last century, there was concentrated interest by French ecologists regarding the unusually low percentage of annuals (therophytes), in certain plant communities in France. Molinier (1934), who had studied a type of Mediterranean shrubland dominated by *Rosmarinus officinalis* and *Erica arborea* in western Provence known as "garigue", had suggested that the root excretions of the dominants or inhibition due to microorganisms or cryptogams may have been the cause of the paucity of therophytes. The subject of the chemical interaction of species had also been raised by Bertrand (1945), who suggested the term "*télétoxie*" to denote the concept of toxicity at a distance, in both the plant and animal worlds. Generally, French plant ecologists have avoided adopting the term "allelopathy", and have used other terms such as "*télétoxie*" (Bertrand 1950, Deleuil 1950), and "antibiosis" in the broad sense as suggested originally by Vuellemin (Guyot 1954). French researchers had generally shown a favourable reaction to the work of the USDA Bureau of Soils, and had sustained interest in matters relating to "soil sickness". According to Guyot (1951), the respected French soil scientist Albert Demolon had found that charcoal added to soil, especially to that which had supported a second crop of the same species, caused increased growth. In the early 1950's, the question of the lack of therophytes in several vegetation types, chalk soil communities in Normandy, Champagne and Picardie, garigue, and evergreen oak shrublands near the Mediterranean coast, was investigated in detail by Lucien Guyot and Yvette Becker and their colleagues at the École Nationale d'Agriculture at Grignon, and by Gabriel Deleuil at the Université de Provence.

Deleuil (1950) found that in the garigue, despite an apparent lack of annual plants, there was substantial seed germination of annual species after good rainfall. However, the seedlings were short-lived, and seedling mortality seemed related to soil toxicity, whereas two-year perennial plants grown in the same soil were little affected. Deleuil found that annuals, grown in various soils preparations that had grown garigue plants, died in conjunction with excessive permeability to iron. Root extracts of dominant garigue plants were found to inhibit germination and growth of therophytes (Deleuil 1951a). In order of toxicity from highest to lowest were the following: *Erica multiflora, Lithospermum fruticosum, Helianthemum lavandulaefolium, Andropogon pubescens, Globularia alypum, Staehelina dubia, Linum glandulosum, Rosmarinus officinalis*. Deleuil, also found that the effective toxicity was mitigated by the amount of carbonate material in the soil, and that the toxicity was destroyed by heating the soil to 50° C. Where therophytes were found in the garigue, Deleuil (1951b) noticed that they often had unusual life histories; such as nodulation (legumes such as *Hippocrepis* spp. and *Ervum gracile*), hemiparasitism (*Odontites lutea*), or early spring growth (*Draba verna, Hutchinsia petraea, Linum strictum*). He found that root nodules and parasitised roots were capable of producing substances that neutralised the toxins produced by shrubs of the garigue. In another report, Deleuil (1954) described how occasionally one finds a micro-association of plants in coastal grassland in Provence: *Allium chamaemoly, Hyoseris scabra* and *Bellis annua*, but never the former two species on their own. Deleuil believed that his experimental results indicated an extraordinary interaction of three species: *Allium chameamoly* produced phytotoxins that could inhibit the seedling growth of species such as *Hyoseris scabra*; however, *Bellis annua* in the presence of *A. chamaemoly* produced an antitoxin that allowed *Hyoseris scabra* to survive, in a manner analogous to immunisation.

Guyot, Becker and their co-workers concentrated their research on grasslands, in northern France, dominated by the perennial grasses *Brachypodium pinnatum* and *Bromus erectus* (Becker 1950; Becker and Guyot 1951a, 1951b; Becker *et al.* 1950a, 1950b, 1951, 1954; Guyot and Massenot 1950; Guyot *et al.* 1951, 1955). The group found that extracts prepared from roots or aboveground parts of certain plants inhibited the germination and growth of therophytes. From 41 species tested, five species, the composites *Hieracium murorum, H. umbellatum, H. vulgatum, Hypochoeris radicata, Solidago virgaurea* showed the greatest phytotoxicity. *Anatractylis cancellata*, a thistle, was later added to the list of very toxic species (Guyot 1951). The relative biological activity of soil and root extracts of various plants was found to be dependent upon season and concentration (Becker and Guyot 1951b). It was hypothesised that the replacement of perennial plants such as *Hieracium pilosella* by grasses may be due to a combination of autotoxicity in colonies of *H. pilosella* accompanied by competition from the grasses (Becker *et al.* 1951a). Furthermore, the authors suggested that parasitic plants such as *Melampyrum arvense* and *Loroglossum hircinum* had developed seed germination strategies that utilised the root excretions of their hosts. The researchers were aware that interactions mediated in the soil were not necessarily dictated simply by root excretions or leaf leachates, but could be due to complex microbiological problems (Becker *et al.* 1951b). These could be initiated by the breakdown of plant parts in the soil, such as were well known in crops in certain crops including rapeseed, alfalfa, barley and flax cultivation. In particular, it was found that extracts of plant parts of *Helleborus foetidus*, and soil in contact with roots of plants such as *Centaurea cyanus* and *Sinapis arvensis*, exerted a selective effect on the growth of fungi and bacteria (Guillemat *et al.* 1954). A very useful and illustrated overview of the French and related work was provided by Guyot (1951). He also drew attention to the fact that in many different locales dominance by various weedy composites, including *Solidago* spp., *Erigeron* spp., *Cirsium* spp., *Hieracium* spp., *Helianthus* spp., and *Inula* spp. may lead to localised degradation of the vegetation.

Other Studies

Portères (1948) found that the small seeds of *Cinchona* sp. displayed density-dependent germination, which was possibly mediated by root excretions of the early germinated seedlings. Seed sown at low density (1500-3000 m^{-2}) completed germination with 80-85% success, and no further seeds germinated, even when space was created by transplanting some of the seedlings. On the other hand, when the initial seed density was high (12000-18000 m^{-2}), the comparable initial germination rate was only 50–55%, and if seedlings were pricked out, further germination then occurred, for up to two years, despite the fact stored seed generally loses half of its viability within six months. Portères suggested that *Cinchona* offered a mechanism of avoiding excessive intraspecific competition and optimising germination success.

There have been many studies that have suggested that various species in natural communities were inhibitory to other species. This is a marked departure from the past, where the principal rationale for understanding plant interactions was in increasing yield. In some respects, this reflected the changing approaches to the study of the ecology of plants, that is, from the nineteenth century natural history viewpoint, to the early twentieth century development of community ecology, to a more individualistic approach in understanding the occurrence and growth plant species, sometimes captured in the phrase "pattern and process". The pioneering studies of Went, Bonner and Galston are prime examples of the letter. Plant ecologists sought explanations for why a particular species seemed excluded from one position yet may be favoured in another, and biotic factors, including biologically active substances became part of the general model. For example, in Australia, Costin (1954) having read the work of Bonner, suggested that the lack of certain herbs near the shrub *Pimelea pauciflora*, and near the trees *Eucalyptus dives* and *E. radiata* may be due to inhibitors released by them. There emerged many reports of various plant extracts being toxic to the germination or seedling growth of other plants, and these were regarded as offering evidence of allelopathic interactions. Examples are the effects of *Robinia pseudoacacia* (Perry 1932, Waks 1936), *Sorbus aucuparia* (Kuhn *et al*. 1943), that of *Castanopsis sempervirens* on *Ribes roezli* (Offord 1952), that of *Acer campestre* on *Caragana arborescens* (Tribunskaya 1953), and the inhibitory effects of substances in *Pinus densiflora*, *Cryptomeria japonica*, and *Chamaecyparis obtusa* litter (Ooyama 1954), and *Alnus glutinosa* litter (McVean 1955). Revived interest in the ecological effects of root exudates, especially from grasses, caused Selander (1950) to speculate that certain rare montane plants that are usually confined to bare mineral soils and almost never found in association with meadow grasses, were essentially non-commensal possibly due to their sensitivity to the root exudates of other plants, especially grasses.

ALLELOPATHY COMES OF AGE

As illustrated above, there occurred, especially during the early 1950s an unprecedented wealth of disparate publications that endorsed and reinforced the ideas that chemicals were involved in the interactions of plants. However, what gave greater coherence to the subject of allelopathy in the botanical literature, was the appearance of four books devoted to allelopathy within the short period 1955-1957: 1) *Die gegenseitige Beeinflussing höherer Pflanzen – Allelopathie* by Gerhard Grümmer (1955); 2) a book in Russian by S.I. Chernobrivenko (1956) with the translated title, "The Biological Role of Plant Excretions and Interspecies Interactions in Mixed Crops"; 3) a German translation of a previously little known Russian work about "phytoncides", *Phytonzide*, by B.P. Tokin (1956); and finally 4) *Chemical Aspects of Ecology in Relation to Agriculture* by Hubert Martin (1957).

Gerhard Grümmer

Revival of interest in allelopathy, particularly in Europe, following World War II was largely due to Gerhard Grümmer (1926-1995; Figure 11.7). Grümmer acquired an interest in the chemical interactions of plants during his doctoral studies on *Papaver* (Willis 1997). He spent most of his academic career at the University of

Figure 11.7. Photograph of Gerhard Grümmer c. 1957 *(Courtesy of Harald Grümmer)*

Greifswald, in the northeast of the former Deutsche Demokratik Republik, or East Germany. In 1953 he published his first review on allelopathy (Grümmer 1953), and this was greatly expanded, and published in 1955 in book form as the *Die gegenseitige Beeinflussing höherer Pflanzen – Allelopathie*, the first monograph on allelopathy since that of Molisch (1937).

Grümmer's contribution was considerable as he organised the disparate literature involving the chemical interactions of plants into a coherent framework, and he raised the profile of an aspect of ecology that had been substantially neglected. Grümmer recognised that there was terminology concerning the chemical interaction of organisms that had emanated from different disciplines, and he presented and preserved some of these terms in his scheme of chemical interactions (Figure 11.8).

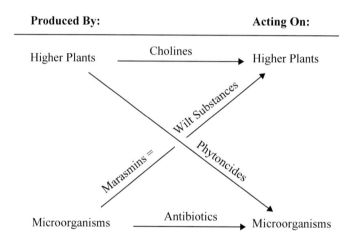

Figure 11.8. Grümmer's scheme of allelopathic interactions (Grümmer 1955)

In this scheme, he adopted the term "antibiotic", used originally by the Frenchman Vuillemin in 1889 to denote a harmful substance produced notably by microorganisms, and which had come gradually into use in the late nineteenth century. He adopted the term "phytoncide" coined by the Russian Academician B.P. Tokin (1942) to describe substances produced by higher plants that affect microorganisms. The term "choline" or "blastocholine" was rather less satisfactory, because of the use as the former as the name of a specific compound in physiology, but was borrowed from the work of Kockemann (1934). The term "marasmin" was coined by Gäumann and Jaag (1946).

Grümmer (1955) addressed the topic of allelopathy in seven main chapters dealing with:

    Ethylene
    Other volatile substances (e.g. terpenes, allyl sulphides, cyanide) released by plants
    Other exudates from leaves
    Root exudates
    The significance of inhibitory substances in agriculture and forestry

Semi-parasitic and parasitic plants
Pollen interactions

Despite its seminal importance, Grümmer's book has never been translated into any other western European language[18]; however, it has been translated into Russian (Grümmer 1957). Grümmer (1956-1957) also published a supplement to his book.

In the light of the progress that has been made in recent years, it is instructive to look at Grümmer's closing remarks, made over fifty years ago, regarding priorities for allelopathic research (Grümmer 1955):

> The future tasks in research on areas of allelopathy are suggested here accordingly in a few sentences:
>
> 1. Further investigations on the physiological effectiveness of ethylene and clarification of its interaction with the growth and inhibitory substances. Determination of its origin in the context of the total metabolism. Investigations on whether there is any significance connected to the release of ethylene by leaves and flowering organs in free nature.
> 2. Analysis of the plant excreted volatile substances with consideration of the natural communities of plants (above all dwarf bush associations), with which such exudations are possibly of importance in the composition of the communities.
> 3. Increase in the number of examples, with which the leaf and root exudations can be proven with certainty to have a noticeable effect on neighbouring plants. The study of the relations between weeds and cultivated plants is of special importance here.
> 4. Clarification of the questions concerning the term "soil sickness". It must be decided for each individual case whether enrichment of cholines or lack of trace elements is responsible for the soil sickness (in the strict sense).
> 5. Elucidation of the composition of the compounds, which are responsible for the relations between the parasitic and semiparasitic higher plants and their hosts. Thus are created hypotheses for the further manipulation of mistletoe problems and the influence of germination in root parasites.
> 6. Extension of the number of examples of the mutual interaction of pollen, as well as of pollen and stigma, with the goal, of approaching the questions of "selective fertilisation ", the self-sterility and the capacity to hybridise or not to hybridise of related forms. (p. 137)

Grümmer maintained an interest in allelopathy over the next few years; however, in later life, particularly with the impact of the Cold War on divided Germany, he became better known in the West for his unflinching criticism of American policy of using napalm and herbicides in the Vietnam war, which likely hindered recognition of his work on allelopathy, at least in the United States.

S.I. Chernobrivenko

The work of Sergei Ivanovich Chernobrivenko (1899-1967; Figure 11.9) is virtually unknown outside of the former Soviet Union and the Ukraine, both because of the apparent rarity of his book on allelopathy, and inaccessibility due to it being unavailable in translation. Chernobrivenko was a Ukrainian who worked at numerous agricultural

---

[18] A English translation is in preparation.

Figure 11.9. Photograph of Sergei Ivanovich Chernobrivenko (1899-1967), (courtesy of Dr. Vladimir Grakhov, Central Botanical Garden, Kiev).

research stations in the Soviet Union, and became well known for his expertise in plant breeding. During the latter part of his career, he was working at the All-Union Scientific-Investigative Maize Institute, Dnipropetrovsk, where he was Head of the Leguminous Plants Breeding Program.

Chernobrivenko's experimental work in the nascent field of allelopathy began in 1948, and at the Sinel'nikov Selection-Experiment Station Institute he investigated the interactions between various combinations of 30 different crops, where factors concerning competition for water and nutrients were excluded. His work was published in 1956 in Russian under the title "The Biological Role of Plant Excretions and Interspecies Interactions in Mixed Crops". Oddly, Chinese students of allelopathy are familiar with Chernobrivenko's book as a major portion was translated into the Chinese language and published in China in 1961. Chernobrivenko recognised that allelopathic interactions encompassed any of the possible interactions between a pair of species (Figure 11.10).

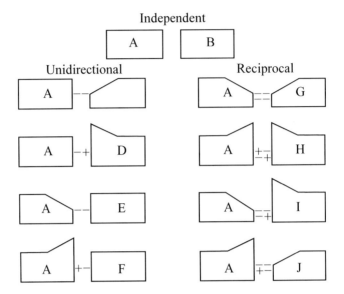

Figure 11.10. Chernobrivenko's scheme of interactions between species (Chernobrivenko 1956)

As Chernobrivenko's book is so inaccessible, it is reasonable to include here a list of the chapter headings:

> Chapter 1. Some general ideas on the organic exudates and biochemical interactions of plants
> Chapter 2. The effect of plant exudates on plants of one species
> Chapter 3. The effect of plant exudates on animals
> Chapter 4. On interspecific interactions in mixed crops
> Chapter 5. Experiments on the study of interspecific relations with crops by field cultures
> Chapter 6. Conclusion

Boris Petrovich Tokin

Boris Petrovich (1900-1984; Figure 11.11) was an influential biologist who became head of the prestigious Timirayezev Biological Institute in Moscow in 1931, and from 1949-1984 was head of the Department of Embryology at Leningrad State University (Roshchina and Narwal 1998). Tokin became widely known for his work in developmental zoology, but for our point here, he was the person that popularised the idea that plants produce biologically active, notably volatile substances, that can affect all sorts of living organisms. This concept originated with his work within the Soviet Union in 1928, where he demonstrated that plant products could affect the growth of microorganisms. In 1942 Tokin published a monograph in Russian entitled "Bactericides of Plant Origin (Phytoncides)". The original bactericidal function of phytoncides was gradually expanded into a more general framework in which these plant products were seen as capable of influencing everything from microorganisms to the growth of other plants in ecosystems to animal behaviour. Tokin has received less credit as a pioneer in allelopathy than is due, in part because his early books on

Figure 11.11. Photograph of Boris Petrovich Tokin (1900-1984). Reproduced from Tokin (1974).

the subject were published in Russian and were produced in the Soviet Union during a period when relations between the Soviet Union and the West were poor. Furthermore, Tokin was politically astute and he aligned himself with T.D. Lysenko[19] and Marxist views on scientific work. Lysenko (1948) himself gave some insight in the dichotomy of Marxist and Darwinian views of an allelopathic phenomenon. He believed that an interpretation of the effects of inhibitory root exudates on seed germination as a phenomenon relating to "survival of the fittest" was a capitalist/bourgeois view of the natural world; on the other hand, he adopted a "socialist" interpretation, in that inhibitors relieved the seed from the stresses of competition, until such time as the environment was favourable for germination and success of the species.

Tokin enjoyed some recognition for his ideas on phytoncides for two reasons. Firstly, Grümmer (1955) adopted Tokin's term "phytoncide" within his scheme of allelopathic interactions to denote substances produced by plants that affect microorganisms. Secondly, the 1951 edition of Tokin's Russian book on phytoncides was augmented and translated into German, and published in 1956 as *Phytonzide*. Again, it is instructive to provide an overview of the contents:

Chapter 1 The phenomenon of phytoncides
Chapters 2 and 3. The effect of phytoncides on protozoa

---

[19] Trofim Denisovich Lysenko (1898-1976) was a Soviet agronomist who attained political influence and power during the eras of Stalin and Krushchev. He rejected Mendelian genetics in favour of Lamarckian principles of crop improvement. He was regarded outside the Soviet Union as a charlatan, and was ultimately blamed for the relative failure of Russian and Chinese agriculture from 1940-1970, as well as the demise and death of many Mendelian geneticists, including Nikolai Vavilov, in the U.S.S.R.

Chapters 4 and 5. The bactericidal characteristics of phytoncides
Chapter 6 Phytoncides and lower fungi; On the effect of phytoncides on *Phytophthora infestans*
Chapter 7 The effect of phytoncides on the eggs of freshwater animals. Some considerations on the chemical interrelations between plants and animals
Chapter 8 Some explanations on the chemistry of phytoncides in connection with the question about their role in nature
Chapter 9 The role of phytoncides in nature
Chapter 10 Supplementary views on the phytoncides of lower plants
Conclusion
Supplement to the German edition:
    The phenomenon of phytoncides is not unique to any group of plants, but belongs to the entire plant world
    The production of volatile phytoncides occurs not only *in vitro* with damaged plants, but phytoncides and biozonoses[20] are demonstrated under natural conditions
    Remarks on the chemistry of phytoncides
    The classification of phenomena with phytoncides

It is worth adding that while the term "phytoncide" is now rarely used in relation to ecology, and has been replaced by terms such as antibiotic, fungistatic or allelopathic substance, the term still enjoys some popularity, especially in countries such as Japan, Taiwan and Korea, in relation to the therapeutic effects of volatile plant substances for humans.

Hubert Martin

The fourth member of this eclectic vanguard of authors was Hubert Martin (1899-1988; Figure 11.12). Martin was an Englishman who was educated in chemistry at

Figure 11.12. Photograph of Hubert Martin (Courtesy of Dr Gilles Saindon, Agriculture and Agri-Food Canada)

---

[20] The term "biozonose", sometimes biocenosis, refers to an ecological community.

the Royal College of Science, University of London, and served at the University of Bristol until 1950. He then migrated to Canada to assume the post of Director of the newly created Science Service Laboratory of the Canada Department of Agriculture, which was affiliated with the University of Western Ontario, London, Ontario. Martin served in this position until 1960. Martin had a distinguished career in the area of plant protection through chemical means, and is well known for books such as *Scientific Principles of Crop Protection, with Special Reference to Chemical Control* which was first published in 1928, and has been republished in numerous editions.

The work that concerns us here, however, is not well known, and is a thin monograph of 96 pages, entitled *Chemical Aspects of Ecology in Relation to Agriculture*. It was published in 1957 by the Canada Department of Agriculture, and it reviews the literature relating to chemical ecology, with special reference to plants, fungi, bacteria and insects. Again, as this publication is scarce, it may help the reader to have knowledge of its contents:

  Chapter 1. Introduction
  Chapter 2. The production of phytotoxins by higher plants
  Chapter 3. Higher plants vs. insects: being mainly concerned with the chemical defenses of plants against insects
  Chapter 4. Higher plants vs. fungi: the chemical defenses of plants against fungal attack
  Chapter 5. Higher plants vs. bacteria: being mainly concerned with the rhizosphere
  Chapter 6. The reaction of higher plants to insect attack
  Chapter 7. Fungi vs. higher plants: the role of phytotoxins in plant pathology
  Chapter 8. Fungi vs. fungi: the chemical basis of biological control
  Chapter 9. The production of insecticides and insect attractants by insects and microorganisms
  Chapter 10. Bacteria vs. higher plants
  Chapter 11. The ecological chemistry of bacteria

This book is of considerable interest as it is really the first book devoted to chemical ecology, at least in English, and it is believed to be one of the earliest books to use "allelopathy" as an English word (p. 11)[21].

REFERENCES

Anonymous 1926. Is the walnut poison? *Washington Star* **1926** (August 2), p. 8.
Anonymous 1929a. Black walnut tree effect upon plant life. *American Nut Journal* **31**: 28.
Anonymous 1929b. Walnut poison isolated. *New York Times* 10 February, X, p. 11.
Anonymous 1929c. Would hide criticism of black walnut. *American Nut Journal* **31**: 91.
Anonymous. 1938. Pflanzenfreundschaft und Pflanzenfeindschaft. *Wiener Allgemeine Forst- und Jagd-Zeitung* **56**: 105-106.
Anonymous. 1938. Soil toxicity in the Dorset heaths. Nature
Anonymous. 1948. London Meeting. *Journal of Ecology* **36**: 192-193.
Aamodt, O.S. 1942. War among plants. *Turf Culture* **2**: 240-244.

---

[21] See note 4, Chapter 1. A reference titled "Allelopathy in walnuts" by A.E. Smith (1949) is a fictional reference which has been supplied in an document, available on the Internet and likely elsewhere, to illustrate citation practice, and is unwittingly anachronic; however, A.E. Smith is a real researcher on allelopathy.

Ahlgren, H.L. and O.S. Aamodt. 1939. Harmful root interactions as a possible explanation for effects noted between various species of grasses and legumes. *Journal of the American Society of Agronomy* **31**: 982-985.

Akehurst 1931. Observations on pond life with special reference to the possible causation of swarming of phytoplankton. *Journal of the Royal Microscopy Society* **51**: 237-265.

Asami, Y. 1947. 'Issues concerning the nutrient status of fruit trees. 4. Replanting on the same ground' [Japanese]. *Agriculture and Horticulture (Tokyo)* **22**(9): 485-488.

Bässler, F.A. 1938. Der Einfluss von Duftstoffen auf die keimung von *Lathyrus odoratus*. *Madaus Jahresbericht* **2**: 133-137.

Bässler, F.A. 1948. Vom gegenseitigen Einfluss der Pflanzen. *Urania* **11**: 387-391.

Becker, Y., J. Guillemat, L. Guyot and D. Lelievre. 1951. Sur un aspect phytopathologique du problème des substances racinaires toxiques. *Comptes Rendus des Séances hebdomadaires de l'Academie des Sciences* **233**: 198-199.

Becker, Y. and L. Guyot. 1951a. Sur les toxines racinaires des sols incultes. *Comptes Rendus des Séances hebdomadaires de l'Academie des Sciences* **232**: 105-107.

Becker, Y. and L. Guyot. 1951b. Sur une particularité fonctionelle des exsudats racinaires de certains végétaux. *Comptes Rendus des Séances hebdomadaires de l'Academie des Sciences* **232**: 1585-1587.

Becker, Y. and L. Guyot. 1951c. Sur la présence d'excrétats racinaires toxiques dans le sol de la pelouse herbeuse à Brachypodium pinnatum (faciès du Xero-Brometum erecti) du nord de la France. *Bulletin de la Societé d'Histoire Naturelle de Toulouse* **86**: 7-17.

Becker, Y., L. Guyot. and M. Massenot. 1951. Sur quelques incidences phytosociologiques du problème des excrétions racinaires. *Comptes Rendus des Séances hebdomadaires de l'Académie des Sciences* **232**: 2472-2473.

Becker, Y., L. Guyot, M. Massenot and J. Montegut. 1950a. Sur la présence d'excrétats racinaires toxiques dans le sol de certains groupements végétales spontanés. *Comptes Rendus de l'Académie d'Agriculture de France* **36**: 689-696.

Becker, Y., L. Guyot, M. Massenot and J. Montegut. 1950b. Sur la présence d'excrétats racinaires toxiques dans le sol de la pelouse herbeuse à *Brachypodium pinnatum* du nord de la France. *Comptes Rendus des Séances hebdomadaires. de l'Académie des Sciences* **231**: 165-167.

Becquerel, P. and J. Rousseau. 1941. Sécrétions par les racines du lin d'une substance specifique toxique pour une nouvelle culture de cette plante. *Comptes Rendus des Séances hebdomadaires de l'Academie des Sciences* **213**: 1028-1030.

Benedict, H.M. 1941. The inhibiting effect of dead roots on the growth of bromegrass. *Journal of the American Society of Agronomy* **33**:

Bertrand, G. 1945. Sur un processus de défense du milieu habité par des espèces vivantes à l'aide d'une substance toxique. *Comptes Rendus des Séances hebdomadaires de l'Académie des Sciences* 525-529.

Bertrand, G. 1950. Sur un cas de télétoxie récemment mis à jour. *Comptes Rendus des Séances hebdomadaires de l'Académie des Sciences* **230**: 1990-1991.

Bixby, W.G. 1929. Black walnut tree's effects explained. *American Nut Journal* **1929** (August), p. 51.

Boas, F. 1939. Über einige Ergebnisse der dynamischen Botanik. *Berichte der deutschen botanischen Gesellschaften* **57**: 100-113.

Boas, F. 1949. *Dynamische Botanik*. 3rd ed. Carl Hanfer, München.

Bode, H.R. 1939. Über die Einwirkung von Heilpflanzen auf andere Pflanzen. *Geisenheimer Mitteilungen für den praktischen Obst- und Gartenbau* **54**: 163-164.

Bode, H. R. 1940. Über die Blattausscheidungen der Wermuts und ihre Wirkung auf andere Pflanzen. *Planta* **30**: 567-589.

Bolley, H.L 1901. Flax wilt and flax sick soils. *Bulletin North Dakota Agricultural Experiment Station* **50**: 27-58.

Bonner, J. and A. W. Galston 1944. Toxic substances from the culture media of guayule which may inhibit growth. *Botanical Gazette* **106**: 185-198.

Bonner, J. 1946. Further investigation of toxic substances which arise from guayule plants: relation of toxic substances to growth of guayule in soil. *Botanical Gazette* **107**: 343-351.

Bonner, J. 1949. Chemical sociology among the plants. *Scientific American* (reprinted in *Plant Life* 1957, Simon and Schuster, New York)

Bonner, J. 1950. The role of toxic substances in the interactions of higher plants. *Botanical Review* **16**: 51-65.

Branscheidt, P. 1930. Zur Physiologie der Pollenkeimung und ihrer experimentellen Beeinflussung. *Planta* **11**: 368-456.
Breazeale, J.F. 1924. The injurious after-effects of sorghum. *Journal of the American Society of Agronomy* **16**: 689-700.
Brian, P.W., H.G. Hemming and J.C. McGowan. 1945. Origin of a toxicity to mycorrhiza in Wareham Heath soil. *Nature* **155**: 637-638.
Bronsart, H. von. 1931. *Bodenmüdigkeit ihre Ursachen und Bekämpfung.* Verlag von J. Neumann, Neudamm.
Brooks, M.G. 1951. Effects of black walnut trees and their products on other vegetation. *West Virginia Agricultural Experiment Station Bulletin* **347**: 1-37.
Brooks, M.G. 1952. Ferns asssociated with black walnut trees. *American Fern Journal* **42**: 124-130.
Brown, B.I. 1942. Injurious influence of bark of black walnut on seedlings of tomato and alfalfa. *Northern Nut Growers Association Annual Report* **33**: 97-102.
Bublitz, W. 1953a. Über die keimhemmende Wirkung der Fichtenstreu. *Madaus Jahresbericht* **6**: 147-155.
Bublitz, W. 1953b. Über die keimhemmende Wirkung der Fichtenstreu. *Naturwissenschaften* **40**: 275-276.
Bublitz, W. 1954. Die Bedeutung von Hemmstoffen für die Forstwirtschaft. II. Mitteilung. Über den Nachweis antibiotisch wirksamer Fichtenrohhumussubstanzen und ihren Einfluss auf die Entwicklung von Bodenbakterien. *Madaus Jahresbericht* **7**: 92-106.
Chernobrivenko, S.I. 1956. *Biologicheskaya Rol' Rastitelnykh Vydelenii i Mezhvidovye Vzaimootnosheniya v Smeshannykh Posevakh.* Sovetskya Nauka, Moscow.
Chernobrivenko, S.I. 1961. *Zhi wu fen mi wu di sheng wu xue zuo yong he jian zuo zhong di zhong jian xiang hu guan xi* [Pinyin] Ke xue chu ban she, Beijing.
Conrad, J.P. 1927. Some causes of the injurious after-effects of sorghums and suggested remedies. *Journal of the American Society of Agronomy* **19**: 1091-1111.
Cook, M.T. 1921. Wilting caused by walnut trees. *Phytopathology* **11**: 346.
Cooper, W.S. and A.D. Stoesz. 1931. The subterranean organs of *Helianthus scaberrimus*. *Bulletin of the Torrey Botanical Club* **58**: 67-72.
Costin, A.B. 1954. *A Study of the Ecosystems of the Monaro Region of New South Wales with Special Reference to Soil erosion.* Government Printer, Sydney.
Curtis, J.T. and G. Cottam. 1950. Antibiotic and autotoxic effects in prairie sunflower. *Bulletin of the Torrey Botanical Club* **77**: 187-191.
Davidson, J. 1939. Some unanswered questions. *Northern Nut Growers Annual Report* **30**: 58-66.
Davis, E.F. 1928. The toxic principle of *Juglans nigra* as identified with synthetic juglone, and its effects on tomato and alfalfa plants. *American Journal of Botany* **15**: 620.
Deleuil, G. 1950. Mise en évidence de substances toxiques pour les thérophytes dans les associations du Rosmarino-Ericion. *Comptes Rendus des Séances hebdomadaires. de l'Academie des Sciences* **230**: 1362-1364.
Deleuil, G. 1951a. Origine des substances toxiques du sol des associations sans thérophytes du Rosamarino-Ericion. *Comptes Rendus des Séances hebdomadaires. de l'Académie des Sciences* **232**: 2038-2039.
Deleuil, G. 1951b. Explication de la presence de certains thérophytes rencontres parfois dans les associations du Rosmarino-Ericion. *Comptes Rendus des Séances hebdomadaires. de l'Académie des Sciences* **232**: 2476-2477.
Deleuil, G., 1954. Action réciproque et interspécifique des substances toxiques radiculaires. *Comptes Rendus hebdomaires des Séances de l'Académie des Sciences* **238**: 2185-2186.
Demolon, A. 1936. Fatigue des luzernières. Causes et remèdes. *Comptes Rendues hebdomaires des Séances de l'Académie d'Agriculture de France* **22**: 579-588.
Denffer, D. von. 1948. Über einem Wachstumshemmstoff in alternden Diatomeenkulturen. *Biologich Zentralblatt* **67**: 7-13.
Dietrichkeit, G. 1991. *Gerhard Madaus (1890-1942): Ein Beitrag zu Leben und Werk.* Marburg/Lahn.
Dommergues, Y. 1954. Biology of forest soils of central and eastern Madagascar. *Transactions of the the 5th International Congress of Soil Science* **3**: 23-28.
Evenari, M. 1949. Germination inhibitors. *Botanical Review* **15**: 153-194.
Farrow, E.P. 1915. On the ecology of the vegetation of breckland. V. Observations relating to competition between plants. *Journal of Ecology* **5**: 155-172.
Fastabend, H. 1955. *Über die Ursachen der Bodenmüdigkeit bei Obstgehölzen.* Landwirtschaftsverlag, Hiltrup bei Münster.

Flint, L.H. and C.F. Moreland. 1946. Antibiosis in the blue-green algae. *American Journal of Botany* **33**: 218.
Froeschel, P and Funke, G.L. 1939. Een Pogin tot experimenteele Plantensociologie I. *Natuurwetenschappelijk Tijdschrift* **21**: 348-355.
Froeschel, P. and Funke, G.L. 1941. Ein Versuch zur experimentellen Pflanzensoziologie. II. *Biologisch Jaarboek (Gent)* **7**: 267-273.
Funke, G.L. 1943a. The effect of *Artemisia absinthum* on neighbouring plants. *Blumea* **5**: 281-293.
Funke, G.L. 1943b. *Experimenteele Plantensociologie*. J. Noorduijn en Zoon, Gorinchem.
Gaumann, E. and O. Jaag. 1946 Über das Problem der Welkekrankheiten bei Pflanzen. *Experientia (Basel)* **2**: 215-220.
Gericke, W.F. 1924. Growth-inhibiting and growth-stimulating substances. *Botanical Gazette* **78**: 440-445.
Golomyedova, T.I. 1952. 'On the mutual toxic interaction of plants through their aqueous extracts' [Russian]. *Agrobiologiya* **2**: 132-134.
Graham, H.W. and N. Bronikovsky. 1944. The genus *Ceratium* in the Pacific and North Atlantic Oceans. Scientific Results of Cruise VII Carnegie 5
Grandeau, L. 1907. Les sols forestiers ne nitrifient pas. *Journal d'Agriculture Pratique* **21**: 645-646.
Gray, R. and J. Bonner. (1948a) An inhibitor of plant growth from the leaves of *Encelia farinosa. American Journal of Botany* **34**: 52-57.
Gray, R. and J. Bonner. (1948b) Structure determination and synthesis of a plant growth inhibitor, 3-acetyl-6-methoxybenzaldehyde found in the leaves of *Encelia farinosa Journal of the American Chemical Society* **70**: 1249-1253.
Greene, K.W. 1929. The toxic (?) effect of the black walnut. *Northern Nut Growers Annual Report* **34**: 152-157. Also in: *Nut Grower (Downington, Pennsylvania)* **7**(3): 1-3 (1930).
Gries, G.A. 1943. Juglone (5-hydroxy-1,4-naphthoquinone) – a promising fungicide. *Phytopathology* **33**: 1112.
Gries, G.A. 1943. Juglone – the active agent in walnut toxicity. *Northern Nut Growers Annual Report* **34**: 52-55.
Grümmer, G. 1953. Die gegenseitige Beeinflussung höherer Pflanzen – Allelopathie. *Biologsiches Centralblatt* **72**: 494-518.
Grümmer, G. 1955. *Die gegenseitige Beeinflussung höherer Pflanzen – Allelopathie*. Gustav Fischer, Jena.
Grümmer, G. 1956-1957. Neuer Erkenntnisse über die gegenstige Beeinflussung höherer Pflanzen. Wissenschaftliche *Zeitschrift der Ernst Moritz Arndt-Universität Greifswald, mathematisch-naturwissenschaftliche Reihe* **6**: 245-250.
Grümmer, G. 1957. *Vzaimnoye Vliyaniye Vysshikh Rasteniy – Allelopatiya*. Inostrannoi Literatury, Moscow.
Guillemat, J., L. Guyot and J. Montegut. 1954. De l'action selective exercée par certaines plantes phytotoxiques sur la microflore du sol. In: *Huitième Congrès International de Botanique, Rapports et Communications, Sections 21-27*, pp. 53-54. Paris.
Gupta, U.S. 2005. *Physiology of Stressed Crops. Volume 3. The Stress of Allelochemicals*. Science Publishers, Enfield, New Hampshire.
Guyot, L. 1951. Les excrétions racinaires toxiques chez les végétaux. *Bulletin Technique d'Information, Minstère de l'Agriculture* **59**: 345-359.
Guyot, L. 1953. Sur l'effet fongostatique selectif de l'extrait aqueux de poudre de somites fleuries d'hellebore. *Comptes Rendus des Séances hebdomadaires de l'Académie des Sciences* **237**: 200-202.
Guyot, L. 1954. Effets antibiotiques provoqués par des lichens et des végétaux supérieurs. Repurcussion sur l'équilibre fungique de profondeur et l'équilibre phanérogamique de surface. In: *Huitième Congrès International de Botanique, Rapports et Communications, Sections 21-27*, pp. 47-52. Paris.
Guyot, L., Y. Becker, J. Guillemat, D. Lelievre, M. Massenot and J. Montegut. 1951. Sur un aspect de la determinisme biologique de l'evolution floristique de quelques groupes de plantes. *Compte Rendu sommaire des Séances de la Société de Biogéographie* **239**: 2-14.
Guyot, L., J. Guillemat and J. Montegut. 1955. L'effet selectif biostatique excercé par certaines plantes. *Annales des Epiphyties* **6**: 119-163.
Guyot, L. and M. Massenot. 1950. Sur la persistence prolongée de semences dormantes dans le sol de la pelouse herbeuse à *Brachypodium pinnatum* du Nord de la France. *Comptes Rendus des Séances hebdomadaires de l'Académie des Sciences* **230**: 1894-1896.
Haasis, F.W. 1930. Forest plantations at Biltmore, N.C. *USDA Miscellaneous Publications* **63**: 1-30.

Halsey, R.W 2004. In search of allelopathy: an eco-historical view of the investigation of chemical inhibition in California coastal sage scrub and chamise chaparral. *Journal of the Torrey Botanical Society* 131: 343-367.

Harder, R. 1917. Ernahrungsphysiologische Untersuchungen an Cyanophyceen hauptsachlich dem endophytischen *Nostoc punctiforme*. *Z. Botanik* 9: 145-242.

Havis, L. and A.L. Gilkeson. 1947. Toxicity of peach roots. *Proceedings of the American Society for Horticultural Science* 50: 203-205.

Hawkins, R.S. 1925. The deleterious effect of sorghum on the soil and on the succeeeding crops. *Journal of the American Society of Agronomy* 17: 91.

Hedrick, U.P. 1905. The relationship of plants in the orchard. *Proceedings of the Society for Horticultural Science* 1903: 72-82.

Helgeson, E.-A. and R. Konzak 1950. Phytotoxic effects of aqueous extracts of field bindweed and of Canada thistle – a preliminary report. *North Dakota Agricultural Experiment Station Bimonthly Bulletin* 12(3): 71-76.

Hershey, J.W. 1929. Some interesting data on the toxic effect of walnut species on themselves and each other. *Nut Grower (Downington, Pa.)* 7(4): 3.

Hewetson, F.N. 1953. Re-establishing the peach orchard. *Pennsylvania Agricultural Experiment Station Progress Report* 106: 1-4.

Hildebrand, E.M. 1945. Peach root toxicity in a New York orchard. *Plant Disease Reporter* 45: 462-465.

Hirai, J. and A. [or S.]. Hirano. 1949a. Studies on the fig-sick soil. I. Effect of sick soil upon growth. *Studies Horticultural Institute Kyoto University* 4: 96-102.

Hirai, J. and A. [or S.]. Hirano. 1949b. Studies on the fig-sick soil. II. Effect of bark and trunk powder upon the germination of some vegetable seeds. *Studies Horticultural Institute Kyoto University* 4: 103-110.

Hirai, J. and Y. Nishitani. 1951. Studies on the fig-sick soil. III. Effect of sick soil upon growth. *Studies Horticultural Institute Kyoto University* 6: 11-14.

Hirano, T. 1940. 'Studies on the soil sickness of the tomato' [Japanese]. *Japanese Journal of Soil Science and Plant Nutrition* 14: 521-529.

Hirano, A. [or S.] 1951 Studies on the peach sick soil. *Studies Horticultural Institute Kyoto University* 5: 6-10.

Hirano, S. 1955. Studies on the peach sick soil. II. *Studies Horticultural Institute Kyoto Unierity* 7: 13-17.

Hirano, S. 1957. Studies on the peach sick soil. III. On the toxic substances in peach root. *Studies Horticultural Institute Kyoto University* 8: 27-31.

Hurtig, I. 1953. Über die allelopathische Beeinflussung der Keimfähigkeit und Triebkraft von Samen verschiedener Kulturpflanzen und Unkräuter. *Wissenschaftliche Zeitschrift der Universität Rostock* 2: 145-157.

Jakob, H. 1954. Compatabilités et antagonismes entre algues du sol. *Comptes Rendus des Séances hebdomadaires de l'Académie de Science* 238: 928.

Janzen, D.H. 1969. Allelopathy by myrmecophytes: the ant *Azteca* as an llelopathic agent of *Cecropia*. *Ecology* 50: 147-153.

Jeffreys, E.G. and P.W. Brian. 1954. Are antibiotics produced in soil? In: *Huitième Congrès International de Botanique, Rapports et Communications, Sections 21-27*, pp. 71-72. Paris.

Jones, L.R. and W.J. Morse. 1903. The shrubby cinquefoil as a weed. *Vermont Agricultural Experiment Station Annual Report* 16: 173-179.

Kaven, G. 1930. Bodenmüdigkeit in Obstbau. *Die Kranke Pflanze* 7: 9-11.

Keever, C. 1950. Causes of succession on old fields of the Piedmont, North Carolina. *Ecological Monographs* 20: 229-250.

Keever, C. 1983. A retrospective view of old-field succession after 35 years. *American Midland Naturalist* 110: 397-404.

Klaus, H. 1940. Das Problem der Bodenmüdigkeit unter Berücksichtigung des Obstbaues. *Landwirtschaftliche Jahrbücher* 89: 413-459.

Knapp, R. and Thyssen, P. 1952. Untersuchungen über die gegenseitige Beeinflussung von Heilpflanzen in Mischkulturen. *Berichte der Deutsche botanische Gesselschaften* 65: 60-70.

Kobernuss, E. 1952. *Untersuchungen zur Ursache und Behehung der Bodenmüdigkeit bei Obstgehölzen*. Max Niemeyer Verlag, Halle.

Koch, L.W. 1955. The peach replant problem in Ontario. I. Sypmtomatology and distribution. *Canadian Journal of Botany* 33: 450-460.

Köckemann, A. 1934. Zur Frage der Keimungshemmenden Substanzen in fleischigen Früchten. *Berichte der deutschen botanischen Gesselschaft* **52**: 523-526.
Kooper, W.J.C. 1927. Sociological and ecological studies on the tropical weed-vegetation of Pasuruan (the island of Java). *Recueil de Travaux Botaniques Néerlandais*. **24**: 1-255.
Kubler, M. 1954. Ein Beitrag zur Allelopathie der Weise. *Zeitschrift für Acker- und Pflanzenbau* **97**: 399-422.
Kuhn, R., D. Jerchel, F. Moewus and E.F. Moeller. 1943. Über die chemische Natur der Blastokoline und ihre Einwirkung auf keimende Samen, Pollenkörner, Hefer, Baketerien, Epithelgewebe und Fibroblasten. *Naturwissenschaften* **31**: 468.
Küster, E. 1909. Über chemische Beeinflussung der Organismen durch einander. *Vorträge und Aufsätze über Entwicklungsmechanik der Organismen. Heft VI.* (ed. W. Roux), pp. 1-25. Wilhelm Engelmann, Leipzig.
Lastuvka, Z. 1955a. The influence of leaf extracts and root secretions of couch grass on the germination of wheat and rye. *Cseskoslovenosko Biologie* **4**: 103-108.
Lastuvka, Z. 1955b. The influence of couch grass on the growth of wheat and rye. *Cseskoslovenosko Biologie* **4**: 165-174.
Lastuvka, Z. 1955c. The influence of couch grass on glycide and nitrogen metabolism of wheat and rye. *Cseskoslovenosko Biologie* **4**: 422-428.
Lefevre, M. and H. Jakob. 1949. Sur quelques propriétés des substances actives tirées des cultures. d'Algues d'eau douce. *Comptes Rendus des Séances hebdomadaires de l'Académie des Sciences* **229**: 334-276.
Lefevre, M., H. Jakob. and M. Nisbet. 1948. Action des substances excretées en culture, par certaines espèces d'algues sur le metabolisme d'autres especes d'algues. *Proceedings of the International Assoiation of Theoretical and Applied Limnology* **10**: 259-264.
Lefevre, M., H. Jakob. and M. Nisbet. 1952. Auto et hétéroantagonisme chez les algues d`eau douce *in vitro* et dans les collections d`eaux naturelles. Annales de la Station Centrale de Hydrobiologie Appliquée. **4**: 5-198.
Lefevre, M. and H. Jakob. 1952. Compatibilités et antagonismes entre algues d'eau douce dans les collections d'eau naturelles. *Proceedings of the International Association of Theoretical and Applied Limnology* **11**: 224-229.
Lefevre, M. and M. Nisbet. 1948. Sur la secrétion, par certaines. espèces d'algues de substances inhibitrices d'autre espèces. d'algues. *Comptes Rendus des Séances hebdomadaires de l'Académie des Sciences* **226**: 107-109.
Levring, T. 1945. Some culture experiments with marine plankton diatoms. *Goteborgs Kungl. Vetenskaps-och Vitterhets samhälles Handlingar 6, Folj. Ser. B* **3**(12): 1-18.
Loehwing 1936. Interactions between different plants through their roots. *Proceedings of the Sixth International Botanical Congress* **1**: 128-142.
Loehwing, 1937. Root interactions in plants. *Botanical Review* **3**: 195-239.
Lovett, J. 2005. Hans Molisch' legacy. In: J.D.I. Harper, M. An, H. Wu and J.H. Kent (eds), *Establishing the Scientific Base. Proceedings of the Fourth World Congress on Allelopathy 21-26 August 2005, Charles Sturt University, Wagga Wagga, NSW, Australia*, pp. 22-30. Regional Institute Limited, Gosford. Also in *Allelopathy Journal* **19**: 49-60 (2007).
Lucas, C.E. 1938. Some aspects of integration in plankton communities. *Journal du Conseil (Conseil International pour l'Exploration de la Mer)* **13**: 309 322.
Lucas, C.E. 1944. Excretions, ecology and evolution. *Nature* **153**: 378-379.
Lucas, C.E. 1947. The ecological effects of external metabolites. *Biological Review* **22**: 270-295.
Lucas, C.E. 1949. External metabolites and ecological adaptation. *Symposium Society Exp. Biol.* **3**: 336-356.
Lysenko, T.D. 1948. *Agrobiologiya*. GISkIL, Moscow. Published in English in 1954 as *Agrobiology*, Foreign Languages Publishing House, Moscow.
MacDaniels, LH. and W.C. Muenscher. 1940. Black walnut toxicity. *Northern Nut Growers Annual Report* **31**: 172-179.
Madaus, G. 1934. Unsere Heilpflanzen heilen auch die Erde. *Jahrbuch Dr. Madaus* **1934**: 14-15; 18-19; 22-23
Madaus, G. 1935. Wurzelausscheidungen machen krank, machen gesund. *Jahrbuch Dr. Madaus* **1935**: 32-36.
Madaus, G. 1937. Der Einfluss von Bitterpflanzen auf den Bitterstoffgehalt der Folgepflanzen. *Madaus Jahresbericht* **1**: 29-30.

Madaus, G. 1937a. Pflanzenstudien I. Ein Beitrag zur Frage der Wirkungssteigerung der Heilpflanze. *Madaus Jahresbericht* **1**: 31-36.
Madaus, G. 1937b. Pflanzenstudien II. Ein Beitrag zur Frage der Gesunderhaltung des Bodens durch Pflanzen. *Madaus Jahresbericht* **1**: 37-41.
Madaus, G. 1937c. Experimenteller Nachweis der wachstumfordernden Wirkung von Blattausscheidungen. *Madaus Jahresbericht* **1**: 38-39.
Madaus, G. 1937d. Ein grosszügiger Kombinationsversuch zur Feststellung des Einflusses vershiedener Pflanzenarten aufeinander. *Madaus Jahresbericht* **1**: 42-48.
Madaus, G. 1938a. *Lehrbuch der biologischen Heilmittel*. Band 1. Leipzig.
Madaus, G. 1938b. Pflanzenstudien I. Ein Beitrag zur Frage der Wirkungssteigerung der Heilpflanzen. *Deutsche medizinische Wochenschrift* **26**: 925-928. (This is an expanded version of Madaus 1937a)
Madaus, G. 1938c. Pflanzenstudien II. Ein Beitrag zur Frage der Gesunderhaltung des Bodens durch Pflanzen. *Deutsche medizinische Wochenschrift* **27**: 998-1000. (This is an expanded version of Madaus 1937b)
Madaus, G. 1939. Von Einfluss verschiedener Pflanzenarten aufeinander. *Deutsche Landw. Presse* **66**: 453-454.
Madaus, G. and R. Thren. 1940. Über die gegenseitige Beeinflussung von Nachbarpflanzen. *Madaus Jahresbericht* **4**: 60-65.
Maestrini, S.Y. and Bonin, D.J. 1981. Allelopathic relationships between phytoplankton species. In: T. Platt (ed), *Physiological Bases of Phytoplankton Ecology*, Canadian Bulletin of Fisheries and Aquatic Sciences Bulletin 210, pp. 323-338. Department of Fisheries and Oceans, Ottawa.
Mann, H. and T.W. Barnes. 1947. The competition between barley and certain weeds. II. Competition with Holcus mollis. *Annals of Applied Biology* **34**: 252-266.
Mann, H. and T.W. Barnes. 1952. The competition between barley and certain weeds. V. Competition with clover considered as a weed. *Annals of Applied Biology* **39**: 111-119.
Martin, H. 1957. *Chemical Aspects of Ecology in Relation to Agriculture*. Science Service Laboratory Monograph 1, Canada Department of Agriculture Publication no. 1015. Canada Department of Agriculture,. Ottawa.
Martin, J.P. and Batchelor, L.D. 1952. The difficulties of replanting lands to the same species of orchard trees. *Proceedings of the Annual Rio Grande Valley Horticultural Institute* **6**: 1-10.
Massey, A.B. 1925. Antagonisms of the walnut (*Juglans nigra* L. and *J. cinerea* L.) in certain plant associations. *Phytopathology* **15**: 773-784.
Massey, A.B. 1928. Are nut trees poisonous to other trees and plants? *Flower Grower* **15**: 4.
Mattoon, H.G. 1944. A commercial black walnut venture. *Northern Nut Growers Annual Report* **35**: 79-82.
McKinley, A.D. 1931. Effect of sorghum residues on crop yields. *Journal of the American Society of Agronomy* **23**: 844-849.
McVean, D.N. 1955. Ecology of *Alnus glutinosa* (L.) Gaertn. II. Seed distribution and germination. *Journal of Ecology* **43**: 62-71.
McVeigh, I. and W. Brown. 1954. In vitro growth of *Chlamydomonas chlamydogama* Bold and *Haematococcus pluvialis* Flotow em. Wille in mixed cultures. *Bulletin of the Torrey Botanical Club* **81**: 218-233.
Metchnikoff, E. 1907. *The Prolongation of Life – Optimistic Studies*. W. Heinemann, London.
Miller, A.G. 1926. Walnuts and apples. *Farm Journal* **1926**(July): 17.
Molinier, R. 1934. Études phytosociologiques et écologiques en Provence occidentale. *Annales du Musée d'Histoire Naturelle de Marseille* **27**: 1-274.
Molisch, H. 1937a. *Der Einfluss einer Pflanze auf die andere – Allelopathie*. Gustav Fischer, Jena.
Molisch, H. 1937b. Über den Einfluss einer Pflanze auf eine andere räumlich davon getrennte. *Anzeiger der Akademie der Wissenschaften in Wien* **1937** (8): 53-54.
Molisch, H. 2001. *The Influence of One Plant on Another: Allelopathy*. Edited by S.S. Narwal; translated by L.J. La Fleur and M.A.B. Mallik. Scientific Publishers, Jodhpur.
Muller, C.H. 1953. The association of desert annuals with shrubs. *American Journal of Botany* **40**: 53-60.
Muller, W.H. and C.H. Muller. 1956. Association patterns involving desert plants that contain toxic products. *American Journal of Botany* **43**: 354-361.
Myers, H.E. and K. Anderson. 1942. Bromegrass toxicity vs. nitrogen starvation. *Journal of the American Society of Agronomy* **34**: 770-773.
Narwal, S.S. and S.K. Jain. 1994. Hans Molisch (1856-1937): the father of allelopathy. *Allelopathy Journal* **1**: 1-5.

Naylor, C.W. 1930. *The Secret of the Singing Heart*. Warner Press.
Neilson-Jones, W. 1940. Some biological aspects of soil fertility. *Nature* **145**: 411-412.
Neilson-Jones, W. 1941. Biological aspects of soil fertility. *Journal of Agricultural Science* **31**: 379-411.
Neilson-Jones, W. 1947. *The Growing Plant*. Faber and Faber, London.
Nielsen, E.S. 1934. Untersuchungen über die Verbreitung, Biologie und Variation der Ceratien im südlichrn Ozean. *Dana Reports* **4**: 1-67.
Offord, H.R. 1952. Chemico-ecologic suppression of *Ribes* in forest areas. *Proceedings of the Western Society of Weed Science* **13**: 35-40.
Olmsted, C.E. 1937. Vegetation of certain sand plains of Connecticut. *Botanical Gazette* **99**: 209-300.
Ooyama, 1954. 'The growth inhibiting substances contained in the leaf litter of trees. I. The inhibition effect on germination of coniferous trees' [Japanese]. *Journal of the Japanese Forestry Society* **36**(2): 38-41.
Orton, C.R. and G. Jenny. 1948. Research powers of the farm. *West Virginia Agricultural Experiment Station Bulletin* **334**: 3-35.
Osvald, H. 1947. Växterna vapen i kampen om utrymmet. *Växtodlung* **2**: 288-303.
Osvald, H. 1949. Root exudates and seed germination. *Annals of the Royal Agricultural College Uppsala* **16**:
Osvald, H. 1953. On antagonisms between plants. In: *Proceedings of the Seventh Internation Botanical Congress, Stockholm July 12-20, 1950.* (edited Osvald, H. and Åberg, E.), pp. 167-171. Almqvist & Wiksell, Stockholm.
Pantanelli, E. 1926. Stanchezza del terreno. *Actes IV Conference Internationale de Pedologie* **3**: 560-563.
Parker, R.E. 1949-1950. The inhibitory effects of *Madia glomerata* upon seed germination and plant growth. *Utah Academy of Sciences, Arts and Letters Proceeedings* **27**:74.
Patrick, Z.A. 1955. The peach replant problem in Ontario. II. Toxic substances from microbial decomposition products of peach root residues. *Canadian Journal of Botany* **33**: 461-486.
Perry, G.S. 1932. Some tree antagonisms. *Proceedings of the Pennsylvania Academy of Science* **6**: 136-141.
Pfeiffer, E. 1943. *Bio-Dynamic Farming and Gardening*. Rudolf Steiner Publishing Co., London.
Pirone, P.P. 1938. The detrimental effect of walnut on rhododendrons and other ornamentals. *Nursery Disease Notes (New Jersey Agricultural Experiment Station)* **11**(4): 1-4. Also in: *Northern Nut Growers Annual Report* **30**: 73-74 (1939).
Polishchuk, O.D. and Snezhko, V.L. 1954. *Remont Starikh Sadiv*. Kiev.
Portères, R. 1948. Compétition ou entr'aide au sein de l'espèce et de la race. Le cas des germinations de *Cinchona*. *Comptes Rendus des Séances hebdomadaires de l'Academie des Sciences* **227**: 1114-1115.
Pratt, R. 1940. Influence of the size of the inoculum in the growth of *Chlorella vulgaris* in freshly prepared culture medium. *American Journal of Botany* **27**: 52-56.
Pratt, R. and J. Fong. 1940. Studies in *Chlorella vulgaris*. II. Further evidence that *Chlorella* cells form a growth-inhibiting substance. *American Journal of Botany* **27**: 431-436.
Pratt, R. 1942. Studies in *Chlorella vulgaris*. V. Some properties of the growth-inhibitor formed by *Chlorella* cells. *American Journal of Botany* **29**: 142-148.
Pratt, R. 1943. Studies in *Chlorella vulgaris*. VI. Retardation of photosynthesis by a growth-inhibiting substance from *Chlorella vulgaris*. *American Journal of Botany* **30**: 32-37.
Pratt, R. 1944. Studies in *Chlorella vulgaris*. IX. Influence on growth of *Chlorella* on continuous removal of chlorellin from the culture solution. American Journal of Botany **31**: 418-421.
Pratt, R., J.F. Oneto and J. Pratt. 1945. Studies in *Chlorella vulgaris*. X. Influence of the age of the culture on the accumulation of chlorellin. *American Journal of Botany* **32**: 405-408.
Pratt, R. 1948. Studies in *Chlorella vulgaris*. XI. Relation between surface tension and accumulation of chlorellin. *American Journal of Botany* **35**: 634-637.
Pratt, R., T.C. Daniels, J.J. Eiler, J.B. Gunnison, W.D. Kumler, J.F. Oneto, L.A. Strait, H.A. Spoehr, G.J. Hardin, H.W. Milner, J.H.C. Smith and H.H. Strain. 1944. Chlorellin, an antibacterial substance from *Chlorella*. *Science* **99**: 351-352.
Pratt, R. and J. Dufrenoy. 1949. *Antibiotics*. J.B. Lippincott Company, Philadelphia.
Pratt, R., and H. Mautner, H. 1951. Antibiotic activity of seaweed extracts. *Journal of the American Pharm. Association (Scientific Edition)* **40**: 575.
Proctor, V.W. 1957a. Some controlling factors in the distribution of *Haematococcus pluvialis*. *Ecology* **38**: 457-462.
Proctor, V.W. 1957b. Studies of algal antibiosis using *Haematococcus* and *Chlamydomonas*. *Limnology and Oceanography* **2**: 125-139.

Proebsting, E.L. 1950. A case history of a "peach replant" situation. *Proceedings of the American Society for Horticultural Science* **56**: 46-48.
Proebsting, E.L. and Gilmore, A.E. 1941. The relation of peach root toxicity to the re-establishing of peach orchards. *Proceedings of the American Society for Horticultural Science* **38**: 21-26.
Pütter, A. 1907a. Der Ernährung der Wassertiere. *Zeitschrift für allgemeine Physiologie* **7**: 283-320.
Pütter, A. 1907b. Der Stoffhaushalt des Meeres. *Zeitschrift für allgemeine Physiologie* **7**: 431-368.
Rademacher, B. 1941. Über die antagonistische Einfluss von Roggen und Weizen auf Keimung und Entwicklung mancher Unkrüuter. *Pflanzenbau* **17**: 131-147.
Rayner, M.C. and W. Neilson-Jones. 1944. *Problems in Tree Nutrition*. Faber and Faber, London.
Rayner, M.C. 1934. Mycorrhiza in relation to forestry. I. Researches on the genus *Pinus*, with an account of experimental work in a selected area. *Forestry* **8**: 96-125.
Razdorskii, V. 1941. 'Allelopathy in plants' [Russian]. *Uspekhi Sovremennoi Biologii* **14**: 549-550.
Reich, K. and M. Aschner. 1947. Mass development and control of the phytoflagellate *Prymnesium parvum* in fish ponds in Palestine. *Palestine Journal of Botany* **4**: 14.
Reinhold, J. 1935. *Die Gerkentreiberei in Gewächshäusern*. Stuttgart.
Reinking, O.A. 1943. Possible black walnut toxicity on tomato and cabbage. *Northern Nut Growers Annual Report* 34: 56-58.
Reinking, O.A. 1947. Black walnut is deadly to tomatoes. *Flower Grower* **34**: 435.
Rice, E.L. 1984. *Allelopathy*. Second Edition. Academic Press, Orlando.
Rice, T.R. 1954. Biotic influences affecting population growth of planktonic algae. *US Fish and Wildlife Service, Fisheries Bulletin* **54**: 227-245.
Robbins, W.W., D.S. Crafts and R.N. Raynor. 1942. *Weed Control*. New York.
Rodhe, W. 1948. Environmental requirements of fresh-water plankton algae. *Symbolae Botaicae Upsalienses* **10**: 1-149.
Rodriguez Rosillo, A. 1944. La alelopatia en las plantas. *Boletin de la Real Sociedad Española de Historia Natural* **42**: 225-262.
Roshchina, V.V. and Narwal, S.S. 1998. Pioneers of allelopathy. IX. Soviet allelpathy scientists. *Allelopathy Journal* **5**: 1-12.
Roux, E.R. 1953. The effect of antibiotics produced by *Trachypogon plumosus* on the germination of seeds of kakiebos (*Tagetes minuta*). *South African Journal of Science* **49**: 334.
Roux, E.R. 1954. The nitrogen sensitivity of *Eragrostis curvula* and *Trachypogon plumosus* in relation to grassland succession. *South African Journal of Science* **50**: 173-176.
Sakamura, T. 1922. Uber die Selbstvergiftung der Spirogyren im distillierten Wasser. *Botanical Magazine* **36**: 133-153.
Schander, H. 1956. *Die Bodenmüdigkeit bei Obstgehölzen*. Bayerischer Landwirtschaftsverlag, Bonn.
Schneiderhan, F.J. 1926. Apple disease studies in northern Virginia. *Virginia Agricultural Experiment Station Bulletin* **245**: 1-35.
Schneiderhan, F.J. 1927a. The black walnut (*Juglans nigra*) as a cause of the death of apple trees. *Phytopathology* **17**: 529-540.
Schneiderhan, F.J. 1927b. Recent developments in the control of fruit diseases. *Proceedings of the Virginia State Horticultural Society* **31**: 145-157.
Schönbeck, F. 1954. Die Bedeutung von Hemmstoffen in der Landwirtschaft. I. Mitteilung. Untersuchungen über wasserlösliche Hemmstoffe aus Getreidestroh und Getreideböden. *Madaus Jahresbericht* **7**: 81-91.
Schindler, H. 1936. Sammeln und Anbau von Arneipflanzen. *Jahrbuch Dr. Madaus* **1936**: 59-66.
Schreiner, E.J. 1949. Can black walnut poison pines? *Morris Arboretum Bulletin* **4**: 94-96.
Schreiner, E.J. 1950. Can black walnut poison pines? *Shade Tree* **23**(1): 2.
Schreiter, T. 1928. Untersuchungen über den Einfluss einer *Helodea*-wucherung auf das Netzplankton des Hirschberger Grossteiches in Böhnen in den Jahren 1921 bis 1925 incl. *Sbornik Vyzkumnykh Ustavu Zemedelskych* **1928**: 98
Schuphan, W. 1948. Ein Beitrag zur physiologischen Wirkung einer Pflanze auf die andere. *Botanica Oeconomica* **1**: 1-15.
Selander, S. 1950. Phytogeography of south-western Lule Lappmark. *Acta Phytographia Suecia* **27**: 117.
Sherendetskii, E.Y. 1952. 'The reciprocal toxicity of aqueous extracts from plants' [Russian]. *Agrobiologiya* **2**: 137.
Sideri, D.I. and V.P. Zolotun. 1952. 'Improvement of conditions for oak development in the eroded soils of Zaporozh'e' [Russian]. *Les i Step* **1952** (8):

Shields, L.M. 1956. Zonation of vegetation within Tularosa Basin, New Mexico. *Southwest Naturalist* **1**: 49-68.
Smith, J.R. 1941. One plant's meat is another plant's poison. *Northern Nut Growers Annual Report* **32**: 28-31.
Smith, R.M. 1942. Some effects of black locusts and black walnuts on southeastern Ohio pastures. *Soil Science* **53**: 385-398.
Spoehr, H.A. 1945. Chlorellin and similar antibiotic substances. *Carnegie Institute of Washington Publication* **44**: 66-71.
Stead, A. 1929. The deterioration of pineapple soils. *South African Journal of Science* **26**: 131-132.
Stevenson, I.L. 1954. Antibiotic production of actinomycetes in soil and their effect on root-rot of wheat (*Helminthosporium sativum*). In: *Huitième Congrès International de Botanique, Rapports et Communications, Sections 21-27*, pp. 69-71. Paris.
Stiven, G. 1952. Production of antibiotic substances by the roots of a grass (*Trachypogon plumosus* (H.B.K.) Nees) and of *Pentanisia* variabilis (E. May) Harv. (Rubiaceae). *Nature* **170**: 712-713.
Stolk, A. 1947. Plantaardige remstoffen. *Levende Natuur* **50**: 66.
Strong, M.C. 1944. Walnut wilt of tomato. *Michigan Quarterly Bulletin* **26**: 194-195.
Talling, J.F. 1957. The growth of two plankton diatoms in mixed culture. *Physiologia Plantarum* **10**: 215-223.
Theron, J.J. 1951. The influence of plants in the mineralization of nitrogen and the maintenance of organic matter in the soil. *Journal of Agricultural Science (Cambridge)* **41**: 289-296.
Theron, J.J. and Haylett, D.G. 1953. The regeneration of soil under a grass ley. *Empire Journal of Experimental Agriculture* **21**: 86-98.
Tokin, B.P. 1942. *Bakteritsidy Rastitl'nogo Proiskhozhdeniya (Fitontsidy)*. Medgiz, Moscow.
Tokin, B.P. 1956. *Phytonzide*. VEB Verlag Volk und Gesundheit, Berlin.
Tokin, B.P. 1974. *Tselebnye Yady Rastennii*. Lenizdat, Leningrad.
Tribunskaya, A.Y. 1953. 'Chemical factors in the interactions between plants' [Russian]. *Agrobiologiya* **3**: 165-166.
USDA. 1948. Test clears walnut's reputation. Clip Sheet 2375-48.
Vogel, F. 1929. Untersuchungen und Versuche über die Ernährung und die sogennante Bodenmüdigkeit von Gehölzen. *Mitteilungen der Deutschen Dendrologischen Gesellschaft* **1929**: 288-303.
Vogel, F. and E. Weber. 1931. Beitrage zur Frage der Bodenmüdigkeit in der Obstbaumschule. *Gartenbauwissenschaft* **5**: 508-524.
Waks, C. 1936. The influence of extract from *Robinia pseudacacia* on the growth of barley. *Publ. Fac. Sci. Univ. Charles* **150**: 84-85.
Waksman, S.A. 1936. *Humus: Origin, Chemical Composition, and Importance in Nature*. Ballière, Tindall & Cox, London.
Walker, F.T. and M. Smith. 1948. Seaweed culture. *Nature* **162**: 31-32.
Went, F.W. 1940. Soziologie der Epiphyten eines tropischen Urwaldes. *Annales du Jardin Botanique de Buitenzorg* **50**: 1-98.
Went, F. 1942. The dependence of certain annual plants on shrubs in southern California deserts. *Bulletin of the Torrey Botanical Club* **69**: 100-114.
Went, F.W. 1950. The role of environment in weed growth. *Proceedings of the North Central Weed Control Conference* **7**: 2-5.
Went, F.W. 1956. *Some Aspects of Plant Research in Australia*. CSIRO, Melbourne.
Went, F.W., G. Juhren, and M.C. Juhren. 1952. Fire and biotic factors affecting germination. *Ecology* **33**: 351-364.
Wildeman, E. de, 1946. Allélopathie ou télétoxie, en particulier dans le règne végétal. *Academie Royale de Belgique, Bulletin de la Classe des Sciences* **32**: 117-126.
Wilkinson, J.F. 1948. The toxic effect of walnut trees on apple trees and other plants. *Iowa State Horticultural Society Proceedings* **82**: 134-137.
Willis, R.J. 1997. Pioneers of allelopathy: Gerhard Grümmer. *Allelopathy Journal* **4**: 1-6.
Willis, R.J. 2000. *Juglans* spp., juglone and allelopathy. *Allelopathy Journal* **7**: 1-55.
Winter, [A.] G. Winter. 1952. Die Bodenmüdigkeit im Obstbau. *Zeitfragen der Baumschule* **7**: 26-34.
Winter, A.G. 1953. Die Bodenmüdigkeit im Obstbau. *Zeitfragen der Baumschule* **7**: 26-34.
Winter, A.G. 1953b. Untersuchungen über die flüchtigen Antibiotika aus der Kapuziner- (*Tropaelium maius*) und Gartenkresse (*Lepidium sativum*) und ihr Verhalten im menschlichen Körper bei Aufnahme von Kapuziner- bzw. Gartenkressensalat per os. *Madaus Jahresebericht* **6**: 43-92.

Winter, A.G. 1954. Untersuchungen über die Natur der antimikrobiellen Wirkstoffe aus der Kapuzinerkresse. *Madaus Jahresbericht* **7**: 7-20.
Winter, A.G. 1954a. Grundsätzliches zur Hemmstofforschung in der Land- und Forstwirtschaft. *Madaus Jahresbericht* **7**: 74-80.
Winter, A.G. and W. Bublitz. 1953a. Untersuchungen über antibakterielle Wirkungen im Bodenwasser der Fichtenstreu. *Naturwissenschaften* **40**: 345-346.
Winter, A.G. and W. Bublitz. 1953b. Über die keim- und entwicklungshemmende Wirkung der Buchenstreu. *Naturwissenschaften* **40**: 416.
Winter, A.G. and F. Schönbeck. 1953a. Untersuchungen über die Beeinflussung der Keimung und Entwicklung von Getreidesamen durch Kaltwasserauszüge aus Getreideböden. *Naturwissenschaften* **40**: 168-169.
Winter, A.G. and F. Schönbeck. 1953a. Untersuchungen über den Einfluss von Kaltwasserextrakten aus Getreidestroh und anderer Blattstreu auf Wurzelbildung und –wachstum. *Naturwissenschaften* **40**: 513-514.
Winter, A.G. and F. Schönbeck. 1954. Untersuchungen über wasserlösliche Hemmstoffe aus Getreideböden. *Naturwissenschaften* **41**: 145-146.
Winter, A.G. and E. Sievers. 1952. Untersuchungen über die Beeinflussung der Samenkeimung durch Kaltwasserextrakte aus der Blattstreu verschiedener Gramineen. *Naturwissenschaften* **39**: 191-192.
Winter, A.G. and L. Willeke. 1951. Über die Aufnahme von Antibioticis durch höhere Pflanzen und ihre Stabilität in natürlichen Böden. *Naturwissenschaften* **38**: 457-458.
Worthington, E.B. 1943. *Eleventh Annual Report of the Director, Year Ending March 31, 1943.*, pp. 17-18. Freshwater Biological Association of the British Empire, Ambleside, Westmorland.

INDEX

Common names of plants are cross-indexed to their Latin names.

A

Abu Hanifa, 49
Abu l-Jayr, 49
Acacia lebbek, 238
Acer campestre, 281
Acer pseudoplatanus, 84
Acer spp., 227
Achillea atrata, 182
Achillea moschata, 182
Aconitum spp., 84, 112
Adenostoma fasciculatum, 5, 276
Advisse-Deruisseaux, P., 238
Agrippa, 79
Agropyron repens, 269
Agrostemma githago, 86, 145, 270
Agrostis spp., 231, 261
Ailanthus altissima, 140
Ajuga spp., 84 n18
Albertus Magnus, 28 n20, 68, 69, 70, 71
alfalfa. See *Medicago sativa*
algae, 11, 60, 233, 273-275
allelochemics, 3
allelopathy, defined, 1-6, 254-255
allelopathy, methodology, 7-8, 228-229
allelopathy, terminology, 279, 283
Allium ampeloprasum, 87 n25, 89
Allium cepa, 30, 64, 84, 183, 231
Allium chamaemoly, 280
Allium sativum, 4, 56, 61, 73, 89, 90, 109, 110, 258
Allium schoenoprasum, 61
Allium ursinus, 112
almond. See *Prunus dulcis*
Alnus glutinosus, 258, 281
Alopecurus pratensis, 199, 201
Althaea officinalis, 258

Ambrosia dumosa, 276-277
amygdalin, 264
Anacardium occidentale, 238
Ananas comosus, 262
Anatractylis cancellata, 280
Anaxagoras, 17 n1
Andropogon pubescens, 280
Andropogon scoparius, 275
animals and plants compared, 10, 15-18, 20, 103, 105, 117, 119, 133, 141, 160, 175, 182, 217, 224
Anona spp., 205
Antennaria fallax, 275
Anethum sowa, 56
Anthriscus sylvestris, 112
Antiaris toxicaria, 9 n6, 64, 94-97, 95 Fig. 5.8, 95 Fig. 5.9, 111
antibiosis, 6, 182, 184, 251, 252, 259-261, 269, 273-275, 278, 279, 283, 288
antiherbivory, 10, 22, 61, 119, 254
antipathy and sympathy, 12, 15, 16, 23, 31, 32, 41-42, 45-48, 51, 68, 71-92, 108-110, 122, 168-170, 182, 183
apple. See *Malus* spp.
apricot. See *Prunus armeniaca*
Arctium spp., 89 n28
Aristotle, 16-19, 76
Armoracia rusticana, 33, 84
aromatherapy, 32
Artemisia absinthium, 257, 271
Artemisia tridentata, 5, 11 Fig. 1.6
artichoke. See *Cynara scolymus*
Artocarpus sp., 238
Arundina graminifolia, 62
Arundo arenarius, 132
Asclepias incarnata, 171

ash. See *Fraxinus* spp.
Asparagus, 5, 161
Aster macrophyllus, 275
Asterionella spp., 273, 275
astrology, 41, 77-78, 81-82, 85, 87
Athenaeus, 22-23, 82
Atractylis gummifera, 22 n9
Atriplex halimus, 21 n6
Atropa belladonna, 258
Austen, R., 87
Australia, 8, 113, 235-236, 255 n4, 270, 275, 281
autotoxicity, 252, 263, 280
Avena fatua, 112, 121
Avena spp., 67, 117, 119, 120, 127, 145, 160, 172, 186, 231, 234, 237, 238, 269
Avicenna. See Ibn Sina
Azadirachta, 55, 56
Azara, F. de, 113
Azteca, 254

B

Backer, G., 171
Bacon, F., 86-89, 90-91, 109
bacteria, nitrifying, 5, 260
bacteria, soil, 203, 259, 280, 289
bacteriotoxins, 236
Bailey, L.H., 221, 223, 234
bamboo, 57, 60
banana. See *Musa* spp.
Banks, J., 113
barberry. See *Berberis* spp.
barley. See *Hordeum* spp.
barnyardgrass. See *Echinochloa crus-galli*
basil. See *Ocimum basilicum*
Bässler, F., 258
bay (laurel). *See Laurus spp.*
Bayliss, J., 206
beans, 26, 29, 48, 54, 60, 61, 62, 81, 84, 139, 141, 146, 148, 173, 174, 177, 262, 266
beans, broad. See *Vicia faba*
beans, kidney. See *Phaseolus vulgaris*
beans, red, 60

Becker, Y., 279-280
Bedford, 11th Duke, 196-204, 197 Fig. 9.2
beech. See *Fagus* spp.
beet. See *Beta vulgaris*
Bellani, A., 169
Bellis annua, 280
Berberis spp., 111
Bertrand, G., 279
Berzelius, J., 172
Beta vulgaris, 48, 231, 266, 271, 276
Betula spp., 119, 258
Bible, 93
bindweed. See *Convolvulus arvensis*
bioassay, 216, 224, 226, 229, 233-234, 238, 269, 272, 274, 277
birch. See *Betula* spp.
Black, G.V., 188
black gram. See *Vigna mungo*
blackberry. See *Rubus* spp.
blackthorn. See *Prunus spinosa*
Boas, F., 259-260
Bock, H., 71
Bode, H., 271
Boehmeria nivea, 60
Boerhaave, H., 91, 105 Fig. 6.1, 106, 115, 127, 151
bogs, 153, 233-234, 239
Boletus, 260
Bonner, J., 271-273, 276-277, 281
Bonnet, C., 140
borage. See *Borago*
Borago sp., 258
Bottomley, W., 236-237
Bouchardat, A., 183
Boussingault, J.B., 169, 172
box. See *Buxus* spp.
Brachypodium pinnatum, 280
bracken. See *Pteridium aquilinum*
Braconnot, H., 130, 159, 169, 171-173, 175, 177
Brassica cretica, 20, 20 n5
Brassica napus, 5
Brassica nigra, 276
Brassica oleracea (cabbage), 20, 20 n4, 21 Fig. 2.4, 23, 25, 28, 30, 33, 34, 42, 43, 47, 48, 49, 51, 68, 69, 71, 73, 75, 76, 76 Fig. 5.4, 78, 81,

82, 83, 84, 85, 86, 88, 89 n30, 90, 91, 92, 109, 147, 173, 183, 231, 258, 262, 267, 268, 268 Fig. 11.4
Brassica oleraceus (cauliflower), 43, 48
Brassica oleraceus (kale), 48
Brassica rapa, 43, 47, 231, 236
Brauner, J., 112
Breazeale, J., 224
Brian, P., 261
Briggs, L., 210
Brisseau-Mirbel, C.F., 121, 131, 168
Bromus erectus, 280
Bromus sp., 269-270
Brooks, M., 268-269
Broussonetia papyrifera, 62
Browne, T., 88
Brugmans, S.J., 108, 115-121, 115 Fig. 6.3, 127, 129, 130, 132, 134, 135 n10, 137, 145, 149, 168, 171
Bublitz, W., 259
buckwheat. See *Fagopyrum esculentum*
Buel, J., 153
Burton, D., 113
Burton, R., 88-89
Butea monosperma, 56
butternut. See *Jglans cinerea*
Buxus spp., 184
Byzantium, 34-35

C

cabbage. See *Brassica oleraceus*
Caesalpinia crista, 56
Calceolaria sp., 237
Calluna vulgaris, 205
caltrops (caltrop). See *Tribulus terrestris*
Camelia sinensis, 62
Camelina sp., 238
Cameron, F.K., 210-218, 221, 226, 228, 234, 237
Cameron, V.L., 96-97
Candolle, Alph. de, 154, 170 n4
Candolle, Aug. P. de, 19, 32, 108, 125-154, 128 Fig. 7.1, 159, 161, 163, 164, 166, 167, 168, 169, 170, 171, 172, 173, 176, 177, 178, 179 n10, 182, 183, 184, 185, 188, 204, 206, 211, 216, 217, 219, 237, 240, 257, 273
Cannabis sativa, 60, 61, 62, 76, 83, 258
canola. See *Brassica napus*
Caragana arborescens, 281
Cardano, G., 71
Carex arenarius, 205
Carlowitz, H.C., 104, 108
Carpenter, W., 167, 179
Carpinus betulus, 127
Carum carvi, 271
Cassia spp., 60
Cassionos Bassos, 34
Castanea sativa, 72
Castanopsis sempervirens, 281
Cato, 25, 69, 70
cauliflower. See *Brassica oleraceus*
Cauvet, D., 183
Cecropia sp., 254
cedar, 43, 45, 47, 48
Celtis spp., 42 n10
Centaurea cyanus, 280
Centaurea diffusa, 8
Centaurea maculosa, 8, 10 Fig. 1.5
Ceratium spp., 273
Chakrapani, 55
Chamaecyparis obtusa, 281
chamise. See *Adenostoma fasciculatum*
Chatin, A., 183
Chelidonia majus, 260
Chernobrivenko, S., 12, 282, 284-286, 285 Fig. 11.9
cherry. See *Prunus cerasius*
chervil, wild. See *Anthriscus sylvestris*
chestnut. See *Castanea sativa*
chickpea. See *Cicer*
China, 12 n7, 57-63, 259
chives. See *Allium schoenoprasum*
Chlamydomonas reinhardii, 274
Chlorella kesleri, 274-275
Chlorella pyrenoidosa, 274
chlorellin, 273-275
choline, 283

Chondrilla muralis. See *Mycetis muralis*
Chouchak, D., 237
Chrysanthemum spp., 62
Cicer arietinum, 242
Cicer sp., 22, 28, 239
Cicero, 22, 28, 239
Cichorium spp., 161
Cinchona sp., 281
Cirsium arvense, 117, 119, 121, 145, 270
Cirsium spp., 161, 281
Citrullus vulgaris, 42
Citrus spp., 5, 262, 265
Citrus aurantifolia, 205
Citrus aurantium, 48, 63, 113
Citrus limon, 50
Citrus sinensis, 42, 43, 45, 49, 50, 64, 113, 265, 267
Clements, F., 120 n21, 239
Cleome gynandra, 56
Clitocybe spp., 206
clover. See *Trifolium* spp.
clover, alsike. See *Trifolium hybridum*
clover, red. See *Trifolium pratense*
colver, white. See *Trifolium repens*
coconut. See *Cocos nucifera*
Cocos nucifera, 55
Coffea arabica, 262
Cogan, T., 82
Columella, 23, 27-28, 31 n26, 33, 43, 69
companion planting, 21, 25, 30, 56, 60, 73 n8, 87 n25, 257-258
competition, 1, 5-7, 21, 25, 87, 90, 111, 119, 138, 180, 184, 270, 280, 281, 285
conifers, 202, 266
Conium maculatum, 23, 23 n12, 31, 73, 81, 84, 91 n32, 109, 110, 141, 258
Convallaria, 43
Convolvulus arvensis, 270
Convolvulus scammonia, 68
Cooper, T., 112
copepods, 6
corals, 3

corn (maize). See *Zea mais*
corncockle. See *Agrostemma githago*
Cornus mas, 258
Cornus sp., 227
Corylus spp., 28, 28 n20, 31 n29, 68, 69, 71, 84 n17, 90, 152
Cotta, H., 132
Cottam, G., 275
cotton. See *Gossypium hirsutum*
Coulon, J.V., 115-119, 116 Fig. 6.4, 129, 130
cowpea. See *Vigna unguiculata*
Cowles, H., 239
Cowley, A., 89
crabgrass. See *Digitaria sanguinalis*
Crataegus spp., 108
creosote bush. See *Larrea tridentata*
Crepis virens, 260
Crescenzi, Pietro de, 51, 70, 69, 71
crop residues, 5, 62, 84, 97, 144, 153, 234, 260, 263, 264
crop rotation, 25, 32, 54, 57, 60-61, 76, 129, 138, 139, 140, 144-150, 153, 154, 163, 166, 167, 169, 170, 172, 242
Cryptomeria japonica, 281
Cubbon, M., 205
cucumber. See *Cucumis sativa*
Cucumis melo, 43, 61, 62
Cucumis sativa, 43, 71, 83, 84, 90, 186, 262
Cucurbita spp., 42, 231, 232
cucurbits, 42, 84 n19
Culpeper, N., 82, 85-86, 89
Curtis, J. 275
Curzon, H., 92
Cuscuta spp., 84
custard-apple. See *Anona* spp.
Cuvier, G., 104 n7, 127
cyanogenesis, 264-267
Cyclamen spp., 30, 81, 82, 83, 89, 91, 92, 109, 183, 258
Cynara scolymus, 76
Cynosurus cristatus, 199
Cyperus sp., 238
Czapek, F., 188

D

Dactylis glomerata, 199, 201, 269
Dandeno, L., 232-233
Dara Shikah, 56
Darwin, C., 119, 178, 179, 287
Darwin, E., 94-96, 119
Datura sp., 271
Daubeny, C., 132, 159, 176-177, 177 Fig. 8.4, 184
Daucus carota, 117, 121, 145, 182, 231, 261
Davies, W., 111
Dean, W., 97
Delamétherie, J., 129
Deleuil, G., 279-280
Demokritos of Abdera, 16, 17 n1, 23 n11
Demokritos of Mendes, Bolos, 23, 31 n26, 73, 76
Demolon, A., 237, 279
density-dependence, 281
Deschampsia flexuosa, 206
Dickson, A., 110
Didymos, 34
Digitalis purpurea, 141
Digitaris sanguinalis, 278
dihydroxystearic acid, 203, 224, 225 Fig. 10.5b, 226, 228, 230, 232, 234, 239
Dioskorides, 24 Fig. 2.5, 25, 46 Fig. 3.3, 68, 81
Diospyros kaki, 63
doctrine of signatures, 77-78, 78 Fig. 5.5
Dodoens, Robert, 30 n25
dogwood. See *Cornus* sp.
Dolichos uniflorus, 56
Dombasle, M. de, 105
Dommergues, Y., 261
Draba verna, 280
Drakon, 18
Dryopteris filix-mas, 24
Duhamel du Monceau, H., 103, 106, 107, 120 n21, 132
Dureau de Lamalle, A., 131
durian. See *Durio zibethinus*

Durio zibethinus, 64
Dutrochet, R., 141, 159, 174

E

Echinochloa crus-galli, 11
Ecklonia bicyclis, 60
ectocrine, 273
effluvia, 18, 32, 88, 109, 112, 119, 121, 122, 185 n15
eggplant. See *Solanum melongena*
elder. See *Sambucus nigra*
elecampane. See *Inula helenium*
elements, four, 16-17, 17 Fig. 2.2, 42
elm. See *Ulmus* spp.
Elymus arenarius, 132
Elodea canadensis, 273
Embelia ribes, 55
Empedocles, 16, 17 n1, 18, 182
Encelia farinosa, 276, 277
epiphytes, 185
Erasmus, 73
Erica arborea, 279
Erica multiflora, 280
Erica spp., 161, 185
Erigeron acre, 117, 119, 120, 145
Erigeron canadensis, 278
Erigeron pulchellus, 275
Eriobotrya japonica, 205
Ervum gracile, 280
Estienne, C., 51, 74
Estioka, R., 238
ethnobotany, 64, 97, 98
ethylene, 12, 87, 255, 258
Eucalyptus dives, 281
Eucalyptus radiata, 281
Eucalyptus spp., 8, 113, 162, 258
Eucalyptus tereticornis, 8, 9 Fig. 1.4
Euphorbia cyparissias, 149
Euphorbia dulcis, 183
Euphorbia helioscopia, 183
Euphorbia lathyrus, 177
Euphorbia peplus, 120, 121, 171, 183
Euphorbia spp., 45, 48, 117, 119, 120, 121, 129, 145, 149, 161, 171, 177

Evelyn, J., 88
Évon, 163
excretion, 12, 16, 18, 19, 103-108, 111, 116, 118, 120
exosmosis, 141, 174
experimental stations, 187, 205, 209, 213, 214, 218, 220, 229-233, 264, 265, 266, 285. See also Rothamsted and Woburn.

F

Fagopyrum esculentum, 117, 118, 121, 171, 172, 231
Fagus spp., 119, 152, 167, 179, 184
Fagus sylvaticus, 184, 259
fairy-rings, 184, 206, 227, 275
fallowing, 25, 28, 57, 60, 108, 118-119, 144, 150, 160, 163
Féburier, 130
Feeny, P., 3
fennel. See *Foeniculum vulgare*
fenugreek. See *Trigonella foenum-graecum*
fern, 24, 25, 31, 81, 84, 85, 91, 92, 93, 97, 109, 269
fertilisers, 15, 129, 160, 175, 187, 210, 214, 216, 217, 219, 226, 232, 234, 236, 238, 265
Ferula asa-foetida, 55
Festuca duriuscula, 199
Festuca pratensis, 201-202
Festuca rubra, 269
Festus, 32
Ficus carica, 19, 22, 23, 42, 43, 45, 47, 48, 49, 50, 62, 73, 76, 80, 83, 90, 91, 109, 110, 184, 240, 265
fig. See *Ficus carica*
fig, strangler, 59
figwort. See *Scrophularia* spp.
flax. See *Linum usitatissimum*
Fletcher, F., 205, 238-239
Foeniculum vulgare, 56, 86, 271
Foersch, J.N., 94
forests, 23, 31, 63, 96, 108, 122, 139, 153, 167, 168, 178, 179, 188, 233, 237, 261

four elements, 16-17, 17 Fig. 2.2, 104
foxglove. See *Digitalis purpurea*
Fragaria spp., 87 n25, 108, 131, 161
Frank, A.B., 188
Fraxinus spp., 26, 49, 87, 110, 111, 132, 152, 179,
Fuchs, L., 71, 72, 72 Fig. 5.1, 74 n10, 77
Funke, G., 30 n25, 270-271

G

Gaillardia grandiflora, 8
Galega officinalis, 258
Galston, A., 272, 276, 277, 281
Gardner, F., 210, 216, 221
garlic. See *Allium sativum*
Gasparrini, G., 164, 165 Fig. 8.2
Gautier d'Agouty, 120
Geoponika, 23, 33-36, 35 Fig. 2.7, 48, 54, 67
Gerard, J., 84-85
germination inhibitor, 69, 120, 258, 259, 261, 267, 269-271, 276, 279-281, 284, 287
Gilbert, A., 265
Gilbert, J., 188, 204, 212 n4, 222
Gladiolus spp., 98
Globularia alypum, 280
Glycyrrhiza glabra, 60
Göppert, H., 141
Gossypium hirsutum, 55, 60, 224, 226, 238
Gottfried von Franken, 70
grafting, 49, 62, 63, 84, 120, 204, 255, 268
grass, 4, 29, 42, 59, 60, 93, 94, 96, 97, 130, 131, 139, 150, 152, 154, 179, 180, 186, 198, 199, 201-206, 260-262, 269, 270, 280, 281. See also specific grasses such as barnyardgrass, ryegrass, etc.
Greece, ancient, 12, 15-25, 35, 48, 57, 71
Green, T., 115
Greig-Smith, R., 235-236

Grümmer, G., 13, 253, 282-284, 282 Fig. 11.7, 287
guava. See *Psidium guajava*
guayule. See *Parthenium argenteum*
Guyot, L., 279-280
Gyde, A., 168-169, 173-174
Gyllenborg, G.A., 113
Gymnocladus canadensis, 227

H

Hadj of Granada, 47
Haematococcus pluvialis, 274
Hahnemann, S., 259
Hales, S., 106
Hall, A.D., 196, 204, 213, 234, 238, 239
Harper, J.L., 7
Hartwell, B., 231
hawthorn. See *Crataegus* spp.
hazelnut. See *Corylus* spp.
heat, effect on soil, 198, 203, 205, 222, 228, 232-238
Hedwig, J., 121, 171
Helianthemum lavandulaefolium, 280
Helianthus annuus, 5
Helianthus occidentalis, 275
Helianthus rigidus, 275
Helianthus spp., 262
Helianthus tuberosus, 275
Heliotropium sp., 237
Helleborus foetidus, 280
Helleborus spp., 68
hemlock. See *Conium maculatum*
hemp. See *Cannabis sativa*
herbicides, 4, 31, 253, 284
Heresbach, K., 28 n20, 71
Herrera, G.A. de, 51, 70-71
Hieracium murorum, 280
Hieracium pilosella, 260, 280
Hieracium umbellatum, 280
Hieracium vulgatum, 280
Hilgard, E.W., 212-217, 214 Fig. 10.2
Hill, T., 82
Hiltner, L., 188
Hippocrepis spp., 280

Hippomane mancinella, 96, 131, 140
Hoare, C., 152
Holcus mollis, 206
holly. See *Ilex* spp.
Home, F., 109, 113
Home, H., 111
homeopathy, 256, 259
Hopkins, C.G., 213, 214, 217-220, 219 Fig. 10.3
hops. See *Humulus lupulus*
Hordeum spp., 28, 54, 62, 83, 148, 177, 188, 234, 238, 269, 276, 280
hormesis, 2, 79, 223
hornbeam. See *Carpinus betulus*
horseradish. See *Armoracia rusticana*
horseweed. See *Erigeron canadensis*
Howard, A., 205
Huber, F., 120, 141
Humboldt, A. von, 118, 127, 130, 131
Humulus lupulus, 112
humus theory, 18, 104, 175
Hutchinsia petraea, 280
Hyacinthus spp., 107
Hyoseris scabra, 280
Hypericum perforatum, 79, 84 n18, 260
Hypochoeris radicata, 280

I

Ibn al-Awwam, 23, 36, 39, 40, 43-50, 45 Fig. 3.2, 67, 70
Ibn al-Bassal, 36, 44, 48, 49
Ibn al-Raqqam, 42-43
Ibn Luyun, 44, 49
Ibn Outayba, 44, 51
Ibn Rabban al-Tabari, 51
Ibn Sina (Avicenna), 70
Ibn Wafid, 44, 48
Ibn Wahshiya, 35, 36, 40-44, 67
Ilex spp., 184
India, 9, 35, 41, 53-57, 205, 238, 259
indigo. See *Indigofera tinctoria*
Indigofera tinctoria, 172
Ingenhousz, J., 118 n18, 140, 170

insects, 10, 22, 31, 56, 61, 79, 88, 117, 119, 163, 235, 289
intercropping, 57, 59, 60-62, 161, 266
International Allelopathy Society (IAS), 3
International Biological Program (IBP), 3
Inula helenium, 117, 120, 121, 127, 145
Inula spp., 281
Invertebrates, 3, 273
ivy, 21, 32, 81, 84, 89

J

Jacquin, N., 114, 131
Jaeger, G., 181-183, 182 Fig. 8.5, 255
Janzen, D., 254
Japan, 63-64, 264, 267
jazz, 9 n5
Johnson, C., 163
Johnson, S., 174
Johnson, T., 82
Johnston, J.F.W., 167
journals, 150, 255, 262
Juglans cinerea, 265-266
Juglans hindsii, 265
Juglans nigra, 4, 113-114, 179, 180, 188, 265-269, 268 Fig. 11.4, 278
Juglans regia, 4, 25-33, 42, 43, 45, 46 Fig. 3.3, 47-49, 63, 68, 69 Fig. 5.1, 70, 72-76, 84, 88, 90, 93, 108, 109, 131, 140, 179, 258, 265, 267
juglone, 42, 266-267, 269 Fig. 11.3, 278
jujube. See *Zizyphus vulgaris*
Juncus squarrosus, 275
Jung, J., 104
juniper, See *Junipeus* spp.
Juniperus spp., 26, 43, 48, 49, 94

K

kale. See *Brassica oleracea*
Kalmia angustifolia, 5
Kayapó, 98

Keever, C., 280-281
King, F.H., 212, 216-218, 223
Kircher, A., 82, 91-92
Kitab al-Filaha, 39, 40, 40 Fig. 3.1, 41 Fig. 3.2, 43-51, 67
knapweed, spotted. See *Centaurea maculosa*
Knapp, R., 261
Knautia arvensis, 117, 119, 121, 127, 145
Koch, A., 252
Konrad of Megenberg, 28 n20, 69, 71
Kooper, W., 270, 276
Kumazawa, Banzan, 63
Küster, E., 252

L

Lamarck, 103, 125, 126, 129, 130, 143, 287
Lambert, J., 152, 179
Lamiastrum galeobdolon, 206
Lapointe, L., 153
Larix spp., 150
Larrea tridentata, 276, 277
Lathrop, E., 224, 225, 227, 228
Lathyrus odoratus, 258
laurel (bay). See *Laurus nobilis*
Lauremberg, P., 74, 75, 75 Fig. 5.3
Laurus nobilis, 20, 22, 23, 28, 30, 43, 47, 72, 73, 84, 88, 90, 92, 93, 184
Lavandula spp., 84, 90
lavender. See *Lavandula* spp.
Lawes, J., 187, 212 n4, 220
leachate, leaf, 33, 59, 92, 111, 139, 140, 180, 200, 201, 271, 280
leachate, soil, 200, 301, 202 Fig. 9.6, 213, 238, 239, 261, 269, 272, 278
leek. See *Allium ampeloprasum*
Leersia hexandra, 238
Lefevre, M., 273, 275
legumes, 4, 5, 25, 28, 61, 62, 84, 131, 137-139, 144, 148, 173, 185, 251, 280, 286
lemon. See *Citrus limon*
lemon balm. See *Melissa officinalis*

Lens culinaris, 43, 89
lentil. See *Lens culinaris*
Lepidium spp., 188
Leuchs, E., 141
Levisticum officinale, 271
lianes, 185
licorice. See *Glycyrrhiza glabra*
Liebault, J., 74
Liebig, J. von, 27 n18, 159, 160, 164-167, 165 Fig. 8.3, 169, 176, 178, 182, 187, 254 n2
Liebig Co., 254
Lilium spp., 73, 183, 258
lily. See *Lilium* spp.
lily-of-the-valley. See *Convallaria*
lime. See *Citrus aurantifolia*
Lindley, J., 96, 140 n15, 151, 166, 176
Link, H.F., 121, 171
Linnaeus, C., 112
Linum glandulosum, 280
Linum strictum, 280
Linum usitatissimum, 26, 28, 43, 110, 114, 117, 119-121, 127, 145, 171, 177, 182, 186, 238, 262, 263, 270, 280
Liquidambar styriciflua 185 n16
Liriodendron tulipifera, 227
Litchi chinensis, 205
Lithospermum fruticosum, 280
litter, 29, 205, 237, 259, 276, 281
Livingston, B.F., 210 n2, 212, 216, 221, 222, 224, 227, 229, 233, 239, 240, 251
Lobelia longiflora, 131
Loehwing, W., 254
Lolium perenne, 199
Lolium spp., 112, 117, 171, 199
Lolium temulentum, 117, 131, 171
loosestrife. See *Lythrum salicaria*
loquat. See *Eriobotrya japonica*
Loroglossum hicinum, 280
Loudon, Jane, 162, 179 n11
Loudon, J.C., 131, 152, 179
Lucas, C.E., 206, 273
lucerne. See *Medicago* spp.
Lucretius, 27, 29 n21

Lumière, A., 237
lupine (lupin). See *Lupinus* spp.
Lupinus albus, 257
Lupinus argenteus, 8
Lupinus spp., 23, 26, 31, 33, 47, 84, 257
lusus naturae, 93
lychee. See *Litchi chinensis*
Lycopersicon esculentum, 201, 205, 260, 263-268, 272, 277
Lycoris spp., 64
Lyell, C., 178
Lysenko, T., 287
Lythrum salicaria, 127, 145

M

Macaire-Prinsep, J., 135-137, 141, 143, 144, 144 Fig. 7.5, 149, 150, 152, 153, 159-161, 164, 166, 167, 169-171, 173, 175, 182, 183, 185, 188
MacDaniels, L., 267-268
Madaus, G., 254, 256-259, 257 Fig. 11.2, 271
madder. See *Rubia tinctorum*
Magnolia spp., 62
Mago the Cathaginian, 33
Main, J., 150
Maksimovich-Ambodik, N., 121
Malacothrix spp., 276
Malpighi, M., 68, 77, 104, 120
Malus spp., 5, 22, 42, 43, 47, 69, 87, 114, 115, 179, 180, 199-201, 202 Fig. 9.5, 204, 262, 264-268
manchineel. See *Hippomane mancinella*
Mandeville, J., 92, 94
Mandragora officinarum, 24 Fig. 2.5, 84
mandrake. See *Mandragora officinarum*
Mangifera indica, 205, 238
mango. See *Mangifera indica*
maple. See *Acer* spp.
marasmin, 283
Marasmius spp., 206
Marcet, F., 141

Marcet, J., 129 n4, 132-135, 133 Fig. 7.3, 141, 151
Marchantia, 233
marjoram (oregano). See *Origanum* spp.
marshmallow. See *Althaea officinalis*
Martin, H., 6 n4, 12, 282, 288-289, 288 Fig. 11.12
Martin, J., 262, 265
Massey, A., 265, 266
Masters, M., 188
Matricaria maritima, 270
Maupin, 112
Means, T.H. 210
Mechnikov, I., 252
Medicago spp., 237, 262
Medicago arborea, 21, 21 n6
Medicago sativa, 5, 21, 21 n6, 237, 265-268, 280
medicinal plants, 257, 259, 271
Megenburg, Konrad von, 28 n20
Melampyrum arvense, 280
Melia azadirach, 63
Melinis minutiflora, 97 n39
Melissa officinalis, 45, 271
melon. See *Cucumis melo*
Melville, H., 96
Mentha piperita, 186, 258
Mentha spp., 42, 81, 186
Mercurialis annua, 147, 175
Mexico, 60, 97, 160, 161, 271
Meyen, F., 168
Meyer, E., 132
Micheli, P., 104
microorganisms, 3, 5, 7, 159, 219, 224-226, 236, 252, 255, 260, 261, 264, 277-279, 283, 286, 287, 289
Miller, P., 115
millet. See *Panicum* spp.
mint. See *Mentha spp.*
Miscanthus floridus, 5, 6 fig 1.3
mistletoe, 21, 284
Mitchell, J., 184, 185
Mohl, H., 168, 171
Moldenhawer, J., 159
Molinier, R., 279

Molisch, Hans, 1, 2, 2 Fig. 1.1, 188, 254-256
Molliard, M., 237
Monarda fistulosa, 275
Monochoria hastata, 238
monoculture, 150, 258, 261
Monotropa hypopitys, 164
Morus nigra, 43, 47, 49, 57, 60, 61-63, 183
Muenscher, W., 267, 268
mulberry. See *Morus nigra*
mulberry, paper. See *Broussonetia papyrifera*
Muller, C.H., 10, 276, 278
Muller, W.H., 278
Murray, J., 159
Musa spp., 45, 55, 61
mushrooms, 137, 185, 208
mustard. See *Sinapis alba*
Mycetis muralis, 146, 149, 171
mycorrhizae, 189, 262-263
myrtle. See *Myrtus communis*
Myrtus communis, 22, 23, 25, 42, 43, 47, 49, 79, 82, 83

N

Nabathean Agriculture, Book of, 36, 39-42, 46-48, 67
Narcissus spp., 43
Nardus stricta, 205
Nasturtium officinale, 258
Nasturtium spp., 43
Neckam, A., 68
neem. See *Azadirachta*
Neilson-Jones, W., 260
Nelumbo nucifera, 64
Nenuphar spp., 183
Nepeta catanaria, 271
Nerium oleander, 49, 162, 171
Nestor, 34
Neumann, J., 175
Nicholas of Damascus, 17
Nicol, W., 111
Nicotiana tabacum, 98, 177, 185, 201, 202 Fig. 9.6, 210 n1, 224, 276
Nieremberg, J.E., 81

nitrification, 5, 205, 236, 237, 260, 261
Nitzschia frustulum, 275
Nostoc punctiforme, 273
novel weapons hypothesis, 8
nutrition, 16, 17, 42, 63, 104, 105, 117, 118, 127, 140, 159, 160, 170, 172, 175-177
Nymphaea sp., 233

O

oak. See *Quercus* spp.
oats, wild. See *Avena fatua*
occult, 9, 31, 40, 41, 46, 51, 67, 71, 73, 76, 77, 79-82, 85, 90, 91, 112
Ocimum basilicum, 56, 84, 271
Odontites lutea, 280
Odoric, 94
Olea europaea, 25, 27, 30, 30 n23, 33, 42, 43, 46, 47, 48, 49, 62, 71, 73, 74, 80, 83, 84, 89, 90, 92, 108
oleander. See *Nerium oleander*
olive. See *Olea europaea*
onion. See *Allium cepa*
Onobrychis sativa, 131
Oosting, H., 278
opium, 64, 141, 146, 149, 167, 168
opium poppy. See *Papaver somniferum*
orange, bitter. See *Citrus aurantium*
orange (sweet). See *Citrus sinensis*
orchards, 130
orchid, bamboo. See *Arundina graminifolia*
oregano (marjoram). See *Origanum* spp.
Origanum spp., 30, 43, 48, 50, 73, 84, 90, 92, 258
Orobanche spp., 51, 81, 84, 104, 127, 145, 161
Oryza sativa, 4, 4 Fig. 1.2, 8, 11, 54, 55, 57, 63, 161, 238, 270
Osmanthus fragrans, 63
Osvald, H., 112, 269
Ovid, 26, 26 n17
Oxalis stricta, 270
oxidation, 186, 203, 223, 237

P

Palladius, 33, 47, 70
palm, 30, 43, 48, 49
Pandorina spp., 277-278
Panicum spp. 60, 61, 234
Papaver rhoeas, 86, 109, 110, 139, 149, 177
Papaver somniferum, 26, 149, 167, 171, 177, 185, 258, 282
Paracelsus, 2, 78, 79, 87
Parashara, 55
parasitic plants, 51, 113, 161, 255, 280, 284. See also *Orobanche*, *Cuscuta*
Parkinson, J., 85
parsley. See *Petroselinum sativum*.
parsnip. See *Pastinaca spp.*
Parthenium argentatum, 271, 272, 272 Fig. 11.5, 277
Pastinaca spp., 32, 43, 91
Payen, A., 172
pea. See *Pisum sativum*
peach. See *Prunus persica*
pear. See *Pyrus communis*
Penicillium, 261
Peralta, F. de, 238
Perilla frutescens, 59
Periturin, F., 238
persimmon. See *Diospyros kaki*
Peter, A., 220, 221
Petit, A., 237
Petroselinum sativum, 258
Pfeiffer, E., 256
Phacelia tanacetifolia, 276
Phaseolus vulgaris, 146, 148, 173, 262
Philips, J., 89
Phillips, H., 131
Philo of Alexandria, 23
Phleum pratense, 199, 231
phloridzin, 265
phytoncide, 32, 283, 283, 286, 288
phytopathology, 77, 111, 253
Picatrix, 51
Picea spp., 30, 161, 259
Pickering, S.U., 186, 195-206, 197 Fig. 9.1, 228, 235, 236, 240, 251, 252, 273, 278

picoline carboxylic acid, 203, 224-226, 225 Fig. 10.5a, 230
Piettre, M., 237
Pimelea pauciflora, 281
pine. See *Pinus* spp.
pine, Scotch. See *Pinus sylvestris*
pineapple. See *Ananas comosus*
Pinus densiflora, 281
Pinus montana, 265
Pinus palustris, 178
Pinus spp., 29, 29 n22, 30, 30 n24, 62, 63, 68, 72, 178, 179, 185, 205, 227, 260, 266
Pinus sylvestris, 150, 179, 258
Piobert, 153
Piper spp., 64
pistachio. See *Pistacia vera*
Pistacia terebinthus, 42, 84
Pistacia vera, 49, 57
Pisum sativum, 5, 48, 54, 99, 148, 150, 151, 173, 175, 182, 186, 237, 261, 262, 276
Pitt, W., 121
plane. See *Platanus* spp.
plant physiology, 8, 18, 106, 108, 118, 125, 129, 132, 134, 137, 141, 164, 167, 170, 176, 216, 222, 223, 251-253, 259, 271-273
plantain (banana). See *Musa* spp.
Plappart, J.F., 114, 114 Fig. 6.2, 179
Platanus spp., 30, 63
Plato, 16, 19, 76
Plechtrachne spp., 276
Plenk, J.J., 118-121, 127, 145
Pliny the Elder, 20 n5, 23, 27, 28-31, 29 Fig. 2.6, 34, 41 n9 n10, 63, 68, 70, 71, 73, 91
plum. See *Prunus* spp.
Plutarch, 18, 31, 31 n28, 74 n10
Poa annua, 164, 165 Fig. 8.2
Poa nemoralis, 199
Poa pratensis, 261, 268, 275
pollen, 253, 284
Polonsia viscosa, 270
Polstorff, 159, 174, 175
Polygonatum multiflorum, 84
Polygonum persicaria, 270
pomegranate. See *Punica granatum*

poplar. See *Populus* spp.
poppy. See *Papaver somniferum*
Populus spp., 26, 49, 73, 183
Porta, G. della, 77, 79, 80
potato. See *Solanum tuberosum*
Potentilla fruticosa, 265
Pouget, I., 237
Pratt, R., 265, 274, 275
Primula elatior, 182
Primula officinalis, 182
Proebsting, E., 262, 264
protozoa, soil, 203, 235, 287
Prunus armeniaca, 48, 264
Prunus cerasius, 30, 33, 48, 76, 84, 90, 114, 200, 227, 264
Prunus dulcis, 48, 57, 84 n17, 185
Prunus persica, 5, 43, 47, 48, 57, 63, 139, 185-186, 207, 266-267
Prunus spinosa, 108
Prunus spp. (plum) 30, 48, 62, 63, 84, 90, 120, 137, 184, 185, 205, 260, 262-265
Pryanishnikov, D., 238
Prymnesium parvum, 274
Psidium guajava, 56, 205
psychology, 9
Pteridium aquilinum, 31, 205, 206
pumpkin. See *Cucurbita* spp.
Punica granatum, 22, 23, 30, 35, 43, 47, 49, 50 Fig. 3.4, 56, 63, 72
Pyrus communis, 43, 47, 63, 69, 106, 200

Q

Quarterman, E., 278
Quercus ilex, 72
Quercus spp., 25, 27, 30, 30 n23 n24, 33, 62, 72-74, 84, 91, 92, 108, 109 n10, 112, 119, 130, 131, 137, 152, 161, 167, 172, 178, 179, 216, 227, 258, 279
Qutama, 41

R

Rabotnov, T., 8
radish. See *Raphanus sativa*

Rafinesquia neomexicana, 276
rainforest, 161, 185, 276
Randall, J., 109
Ranunculus spp., 183
Raphanus sativa, 20 n5, 22, 28, 33 n33, 43, 48, 73, 82, 84 n23
raspberry. See *Rubus idaeus*
Rattray, S., 83, 94
Rayner, M., 260
redtop. See *Agrostis* spp.
reed (cane), 24, 25, 60, 71, 81, 84, 85, 91, 92, 109
Reed, H., 206, 210 n2, 216, 222-227, 229, 240
Reid, F., 222, 226
Reinitzer, F. 182, 183
Rennie, J., 109 n12, 150, 154
replant problem, 4, 28, 30, 60, 130, 137, 138, 177, 184, 260, 263-265
Reveil, P., 183, 184 Fig. 8.6
Rhinanthus crista-galli, 139, 164
rhizobia, 188, 251
Rhododendron alpinum, 182
Rhododendron arboreum, 184
Rhododendron hirsutum, 182
Rhododendron spp., 184, 267
Rhus laurina, 276
Rhus sp. 43, 47, 216
Ribes roezli, 281
rice. See *Oryza sativa*
Rice, E.L., 3, 5, 261
Rice, T., 275
Richard, A., 122, 168
Robinia pseudoacacia, 281
Rodriguez Rosillo, A., 256
Roget, P., 152
Rome, ancient, 15, 25-34
Romeo, J., 7
root excretion (exudation), 12, 91, 103-122, 126-154, 159-177, 165 Fig. 8.2, 179, 182, 183, 185, 187-189, 216, 221, 227, 231, 232, 237-239, 252-254, 258, 260, 261, 263, 269, 270, 279-281, 283, 284, 287
root spongiole, 127, 142, 143, 151, 159, 167, 172, 174
Röper, J.C., 140
Rosa canina, 258

Rosa spp., 4, 18, 57, 83, 84, 87, 89, 90, 109, 110, 141, 183, 225, 258
rosemary. See *Rosmarinus officinalis*
roses. See *Rosa*
Rosmarinus officinalis, 90, 279, 280
Rothamsted Experimental Station, 187, 188, 195, 198, 203, 204, 212 n4, 220, 235, 238
Rubia tinctorum, 258
Rubus idaeus, 150, 152, 186
Rubus spp., 186, 268
rue. See *Ruta graveolens*
Ruel, J., 71
Rumex acetosella, 234
Rumphius, G., 64, 94
Russell, E., 196 n3 n4, 198, 203-206, 213, 232, 234, 235, 238, 239, 257
Russia, 187, 198, 252
Russian Revolution, 198, 204, 251
rust (fungus), 111
ruta-baga. See *Brassica campestris*
Ruta graveolens, 32, 43, 45, 48, 50, 72 Fig. 5.2, 75, 81, 83, 84, 87, 89-91, 109, 110, 183, 258
rye. See *Secale cereale*
ryegrass. See *Lolium* spp.

S

Saccharum officinale, 61, 163. 172, 239, 270
sage. See *Salvia officinalis*
sagebush. See *Artemisia tridentata*
saintfoin. See *Onobrychis sativa*
Saint Francis of Sales, 73
Saint-Hilaire, A. de, 97
Saint-Simon, M.H., 107, 108
Saint Thomas Aquinas, 40
Salisbury, E.J., 206
Salix spp., 48, 73, 84, 127, 145
Salvia leucophylla, 5, 10, 12 Fig. 1.6
Salvia mellifera, 276
Salvia officinalis, 89
Salvia sclarea, 2731
Sambucus nigra, 92, 161
Saraca asoca, 56
Satureja hortensis 30

Saussure, N. de, 140, 141
savory. See *Satureja*
Scabiosa arvensis. See *Knautia arvensis*
scammony. See *Convolvulus scammonia*
Scenedesmus quadricauda, 274, 275
Schleiden, J.M., 94, 161, 168
Schmalz, J., 169
Schneiderhan, F., 266, 267
Schrader, J., 175
Schreiner, E. 266
Schreiner, O., 203, 216, 218, 219, 221-235, 223 Fig. 10.4, 238-240, 251
Schübler, G.,141
Scilla maritima, 23, 43, 47
Scotland, 109, 135, 170, 173
Scrophularia spp., 183, 258
Scythian lamb, 93, 93 Fig. 5.7
Secale cereale, 5, 132, 148, 161, 164, 169, 172, 175, 205, 207, 231, 258, 270
secondary metabolites, 3, 4, 7, 10, 11, 21, 141, 221, 253
Senebier, J., 104, 108, 119, 120, 125, 126, 140, 141, 170, 175
Senecio spp., 175, 271
Sergios, 34
Seringe, N., 132, 136
Serratula arvensis. See *Cirsium arvense*
service tree. See *Sorbus domestica*
sesame. See *Sesamum indicum*
Sesamum indicum, 54, 56 n3, 59, 205, 224, 238, 239
Setaria italica, 59, 61
Setaria lutescens, 270
Setaria spp., 60, 231
Shakespeare, W., 90
Shen Ying, 59
Shirreff, P., 152
Shorey, E., 222, 224-227, 230, 233, 235
Sinapis alba, 61, 173, 201, 253, 276
Sinapis arvensis, 280
Siosteen, K., 112

skeleton weed. See *Chondrilla juncea*
Skinner, J., 222, 224-227, 230
Smith, J.E., 131
Smith, S., 212 n4
smother crop, 5, 270
soil enzymology, 225
soil infertility, 27, 108, 136, 210, 216, 221, 222, 224, 229, 235, 251, 261
soil repellency, 227
soil sickness (fatigue), 4, 25, 27, 30, 55, 60, 69, 108, 111, 129, 136, 137, 154, 161, 179, 181-183, 185, 222-224, 237, 252, 254, 258, 259, 261-264, 279, 284
soil organic compounds, 4, 187, 203, 221, 222, 224-226, 230 Table 10.1, 232, 234, 236-239, 272, 275 See also picoline carboxylic acid, dihydroxystearic acid
Solanum melongena, 43, 56
Solanum tuberosum, 105 n2, 132, 149, 165, 169, 172, 173, 185, 224, 231, 233, 266, 267
Solidago virgaurea, 280
Solomon's seal. See *Polygonatum multiflorum*
Solon, 18
Sonchus oleraceus, 147, 149
Sophora japonica, 62
Sorbus domestica, 48
Sorbus aucuparia, 281
Sorghum spp., 60, 232
Sorghum bicolor, 263
Sorghum vulgare, 205, 238, 239
souring (soil), 109, 113
species diversity, 11, 87, 111, 161, 163, 255, 273
Spergula arvensis, 117, 119, 121, 171
Spinacea oleracea, 43
spinach. See *Spinacea oleracea*
Spirogyra spp., 273
Spondias dulcis, 238
sponges, 3
spongioles, root, 151, 159, 167, 172, 174

Sprengel, C., 131, 160, 161, 164
Sprengel, K.P., 131 n6, 132
spruce. See *Picea* spp.
squill (sea onion). See *Scilla maritima*
Staehelina dubia, 280
Steiner, R., 256
Stellaria sp., 271
Stendahl, 131
Stigeoclonium, 233
St. Johnswort. See *Hypericum perforatum*
Stoesz, A., 275
strawberry. See *Fragaria* spp.
Strindberg, A., 183
succession (ecological), 139, 152, 161, 167, 177-179, 181, 206, 261, 278, 279
sugarcane. See *Saccharum officinale*
Sullivan, M., 222-225
sumac. See *Rhus*
sunflower. See *Helianthus annuus*
Surapala, 35, 53-56, 54 Fig. 4.1
Swedenborg, E., 88
sycamore. See *Acer pseudoplatanus*
sympathy. See antipathy

T

Tagetes minuta, 261
tannins, 137, 141, 149, 167, 170, 172, 216
Taraxacum officinale, 270
Taxus baccata, 29, 84, 109, 184
Taylor, J., 83
tea. See *Camelia sinensis*
télétoxie, 255 n4, 279
terebinth. See *Pistacia terebinthus*
Terminalia arjuna, 56
Thaer, A., 105
Thamnosma spp., 277
Theophrastus, 18 n2, 19-23, 19 Fig. 2.3, 29 n32, 34, 51 n11, 68, 81, 82, 104 n1, 117, 182
Thoreau, H., 177-179
thyme. See *Thymus vulgaris*
Thymus serphyllum, 43

Thymus vulgaris, 90
Tillandsia spp., 185
timothy grass. See *Phleum pratense*
tobacco. See *Nicotiana tabacum*
Tokin, B., 282, 283, 286-288, 287 Fig. 11.11
tomato. See *Lycopersicon esculentum*
Towers, G., 150-152
Trachypogon plumosus, 261
Tradescantia sp., 234
Tragus. See Bock, H.
transformation, element, 175
tree of heaven. See *Ailanthus altissima*
Treviranus, L.C., 116, 161, 168, 171
Tribulus terrestris, 22 n7
Trifolium hydridum, 231, 269
Trifolium pratense, 111, 231, 269
Trifolium repens, 269
Trifolium spp., 34, 129, 138, 139, 144, 160, 182, 183, 185, 187, 188, 201, 202 Fig. 9.6, 262, 269
Trigonella foenum-graecum, 43, 47
Trimmer, J., 162
Triodia spp., 276
Triticum spp., 5, 8, 33, 54, 59, 62, 67, 69, 83, 84, 86, 88, 107, 109, 111, 114, 120, 121, 127, 129, 131, 135, 138-140, 144, 145, 148, 150, 160, 161, 163, 169, 171-174, 212, 214, 221-224, 226-229, 228 Fig. 10.7, 234, 236-238, 258, 270
Tropaeolum majus, 259
Tsuga canadensis, 23 n12, 233
Tulipa spp., 162
Tull, J., 105
tung oil, 60
Turner, W., 74
turnip. See *Brassica rapa*

U

Ulmus spp., 26, 30, 47, 73, 79, 82, 88, 89, 107, 111, 130, 183
Unger, F., 168
upas. See *Antiaris toxicaria*
Urostigma spp., 59

USDA Bureau of Soils, 25, 187, 202, 203, 205, 209-240, 227 Fig. 10.6, 228 Fig. 10.7, 251
Uslar, J.J. von, 118, 119
Uslar, J.L. von, 97, 118 n19, 125, 159-162, 162 Fig. 8.1

V

Vaccinium corymbosum, 233
Vallemont, P., 90
Vanilla planifolia, 238
Varahamihara, 54
Varro, 23, 25, 34, 69, 70
Vedas, 53
Veronica beccabunga, 258
Veronica peregrina, 270
vetch, bitter. See *Vicia* spp.
Vicia faba, 43
Vicia spp., 22, 26, 29, 82, 139, 183
Vigna mungo, 54
Vigna unguiculata, 224
vine (grape). See *Vitis vinifera*
Vindonus Anatolius, 34
Viola spp., 43
Viola tricolor, 258, 260
violet. See *Viola* spp.
Virgil, 25, 26, 49, 89, 93
Vitis vinifera, 20, 22, 23, 25, 26, 28-31, 33, 34, 42, 43, 47-49, 51, 60, 62, 63, 68, 69, 71-73, 75, 78-89, 80 Fig. 5.6, 91, 92, 104, 109, 112, 152, 180, 183, 205, 252, 265, 267
Vrikshayurveda, 35, 54-56, 54 Fig. 4.1

W

Waller, G.R., 3
Wallerius, J., 112
walnut. See *Juglans regia*
walnut, black. See *Juglans nigra*.
Walser, E., 159, 169, 171
watercress. See *Nasturtium* spp.
water-lily, 60, 233, 238, 241
watermelon. See *Citrullus vulgaris*

weeds, 4, 5, 22, 54, 59-61, 76, 83, 87, 98, 111, 112, 121, 139, 144, 161, 164, 167, 171, 176, 179, 180, 199, 206, 217, 235, 261, 262, 269, 270, 280, 284. See also individual weeds, e.g. bindweed, horseweed, etc.
Went, F.W., 185 n17, 253, 270, 375-277, 277 Fig. 11.6, 281
Westerhoff, R., 184, 227
wheat. See *Triticum* spp.
Wheeler, H., 226, 230, 231
Whitney, M., 203, 209-221, 211 Fig. 10.1, 223, 228, 229, 231, 234, 235, 237, 238, 240
Whittaker, R., 3
Wiegmann, A., 159, 174, 175
Willdenow, C., 118, 121
Willis, J.J., 186
willow. See *Salix* spp.
Winter, A., 259, 261, 263
Woburn Experimental Fruit Farm, 196-204, 198 Fig. 9.3
World War I, 198, 210, 236, 251
World War II, 258, 267, 271, 274, 276, 278, 282
Worlidge, J. 59, 75, 76, 111
wormwood. See *Artemisia absinthium*
Wray, L. 163

X

Xanthium spp., 89 n28

Y

yew. See *Taxus baccata*
Yvart, V., 130, 136, 153

Z

Zea mais, 139, 161, 168, 188, 172, 231-233, 237, 239, 266, 270
Zeller, E., 141
Zizania, 68
Zizyphus vulgaris, 42 n11, 46, 62